Citizen Science and Geospatial Capacity Building

Citizen Science and Geospatial Capacity Building

Editors

Sultan Kocaman
Sameer Saran
Murat Durmaz
A. Senthil Kumar

MDPI • Basel • Beijing • Wuhan • Barcelona • Belgrade • Manchester • Tokyo • Cluj • Tianjin

Editors

Sultan Kocaman
Hacettepe University
Turkey

Sameer Saran
Indian Space Research Organisation
India

Murat Durmaz
Hacettepe University
Turkey

A. Senthil Kumar
Indian Institute of Remote
Sensing Campus
India

Editorial Office
MDPI
St. Alban-Anlage 66
4052 Basel, Switzerland

This is a reprint of articles from the Special Issue published online in the open access journal *ISPRS International Journal of Geo-Information* (ISSN 2220-9964) (available at: https://www.mdpi.com/journal/ijgi/special_issues/Citizen_Science_Geospatial_Capacity_Building).

For citation purposes, cite each article independently as indicated on the article page online and as indicated below:

LastName, A.A.; LastName, B.B.; LastName, C.C. Article Title. *Journal Name* **Year**, *Volume Number*, Page Range.

ISBN 978-3-0365-3713-9 (Hbk)
ISBN 978-3-0365-3714-6 (PDF)

Cover image courtesy of Guiming Zhang

Contents

About the Editors

Sultan Kocaman (PhD) holds a Ph.D. degree on photogrammetry and remote sensing, and has practical experience in the fields of cadastral surveying, GIS and spatial DBMS. She is currently Assoc. Prof. at Hacettepe University, Department of Geomatics Engineering, Ankara, Turkey and visiting scientist at ETH Zurich, Institute of Photogrammetry and Remote Sensing, Zurich, Switzerland. She is also Chair ISPRS WG V/3 on "Promotion of Regional Collaboration in Citizen Science and Geospatial Technology". Her research interests include the application of geospatial technologies and machine learning algorithms to various problem domains related to Earth Observation, natural hazards and land information management.

Sameer Saran (PhD) is working as Senior Scientist and Head, Geoinformatics Department, Indian Institute of Remote Sensing (ISRO), Dehradun, India. He has completed his PhD in the field of Geoinformatics in 2003. He is Course Director of IIRS-ITC Joint Education Programme (JEP) between IIRS and Faculty of Geoinformation Science and Earth Observation, University of Twente, The Netherlands. His research interest includes various areas of Geoinformatics like Geodata modelling, Spatial Databases, Distributed GIS, 3D CityModels, Citizen Science and Geohealth. He has published more than 70 papers in international and national peer reviewed journals. He is national coordinator of Indian Bioresource Information Network (IBIN)—a multi-institutional initiative with Department of Biotechnology. He is holding a position of Vice President, Indian Society of Remote Sensing, Deputy General Secretary, Asian Association of Remote Sensing (AARS) and Co-Chair ISPRS WG V/3. He is recipient of many National and International Awards to his credit.

Murat Durmaz (PhD) is currently an Assistant Prof. at Department of Geomatics Engineering in Hacettepe University. He has worked as a lead software architect for GIS product development and lead systems engineer in various projects delivering decision support, command and control and mechatronics solutions mostly in defense industry for more than 15 years. He received his PhD degree in the field of statistical learning on Ionosphere modeling at Middle East Technical University in 2013. His current research interest is in the development of real-time algorithms for precise navigation and tracking with its applications to autonomous systems, intelligent agents and augmented reality.

A. Senthil Kumar (PhD) earned his PhD from the Indian Institute of Science, Bangalore in the field of image processing in 1990. He joined ISRO in 1991 and was involved in Indian satellite programs in various capacities. His research includes sensor characterization, radiometric data processing, image restoration, data fusion, and soft computing. He served as director of UN-affiliated Centre for Space Science and Technology Education in Asia and the Pacific and as director of Indian Institute of Remote Sensing, Dehradun. He is the president of ISPRS Technical Commission V on Education and Outreach and past Chair and active member of CEOS Working Group on capacity building and data democracy. He has published 115 papers in international journals and conferences. He is a recipient of ISRO team awards for Chandrayaan-1 and Cartosat-1 missions, outstanding contribution award from the Asian Association of Remote Sensing and prestigious Bhaskara and Prof. Satish Dhawan awards conferred by the Indian Society of Remote Sensing.

International Journal of *Geo-Information*

MDPI

Editorial

Editorial on the Citizen Science and Geospatial Capacity Building

Sultan Kocaman [1,*], Sameer Saran [2], Murat Durmaz [1] and Senthil Kumar [3]

1 Department of Geomatics Engineering, Hacettepe University, Ankara 06800, Turkey; muratdurmaz@hacettepe.edu.tr
2 Indian Institute of Remote Sensing, Indian Space Research Organisation, 4 Kalidas Road, Dehradun 248001, India; sameer@iirs.gov.in
3 ISPRS TC-V President, Former Director Indian Institute of Remote Sensing (ISRO), Dehradun 248001, India; senthil.iirs@outlook.com
* Correspondence: sultankocaman@hacettepe.edu.tr

Abstract: This article introduces the Special Issue on "Citizen Science and Geospatial Capacity Building" and briefly evaluates the future trends in this field. This Special Issue was initiated for emphasizing the importance of citizen science (CitSci) and volunteered geographic information (VGI) in various stages of geodata collection, processing, analysis and visualization; and for demonstrating the capabilities and advantages of both approaches. The topic falls well within the main focus areas of ISPRS Commission V on Education and Outreach. The articles collected in the issue have shown the enormously wide application fields of geospatial technologies, and the need of CitSci and VGI support for efficient information extraction and synthesizing. They also pointed out various problems encountered during these processes. The needs and future research directions in this subject can broadly be categorized as; (a) data quality issues especially in the light of big data; (b) ontology studies for geospatial data suited for diverse user backgrounds, data integration, and sharing; (c) development of machine learning and artificial intelligence based online tools for pattern recognition and object identification using existing repositories of CitSci and VGI projects; and (d) open science and open data practices for increasing the efficiency, decreasing the redundancy, and acknowledgement of all stakeholders.

Keywords: geospatial capacity building; citizen science; volunteered geographic information; crowdsourcing; participatory GIS

Citation: Kocaman, S.; Saran, S.; Durmaz, M.; Kumar, S. Editorial on the Citizen Science and Geospatial Capacity Building. *ISPRS Int. J. Geo-Inf.* **2021**, *10*, 741. https://doi.org/10.3390/ijgi10110741

Received: 21 October 2021
Accepted: 23 October 2021
Published: 1 November 2021

1. Introduction

Citizen science and volunteered geographic information (VGI) are gaining importance with the ubiquitous use of mobile technologies. In this new era, ordinary citizens may contribute to scientific processes based on their interest and abilities. The activities they may contribute to range from biology to environmental monitoring to classification of galaxies, all of which have a spatiotemporal dimension. The increasing demands on this research agenda are encouraging scientists from diverse backgrounds to collaborate under the term of "Citizen Science (CitSci)". Geospatial tools and technologies enable many CitSci projects and also benefit from them. Geospatial capacity building, which is one of the main focus areas of ISPRS Commission V on Education and Outreach, also benefits from these developments.

The Special Issue on "Citizen Science and Geospatial Capacity Building" aimed at emphasizing the increasing importance of CitSci, open science and open data in the scientific world for capacity building. The research articles collected in the Special Issue on "Citizen Science and Geospatial Capacity Building" demonstrated the increased efforts on the capacity building by researchers and citizen scientists; and outlined the significance of data quality in various projects, the possibilities provided by web and mobile geographic information system (GIS) technologies, web-based sharing of resources, the role of social media and crowdsourcing data collection methods, and semantical issues. The published articles are briefly introduced in the next Section. The main contributions of the Special Issue are summarized in the final section, and the future directions derived therefrom are presented.

2. Contributions of the Special Issue

The study by Singh et al. [1] presented a novel approach for the prediction of future habitats for monitoring of a bird species, the Jacobin cuckoo (*Clamator jacobinus*), using a combination of spatial modeling and machine learning (ML) techniques, i.e., the maximum entropy (Maxent) algorithm. The predictions were carried out for the years 2030 and 2050 under various future global climate modeling approaches using the Coupled Model Intercomparison Project (CMIP5)'s climate data of the CNRM-ESM2-1 [2]. An extensive review on the use of CitSci for biodiversity research was provided in the study. The study demonstrated the usability and essentiality of CitSci collected geodata for this kind of global analysis, which cannot be carried out by using institutional observations solely. The species distribution data was obtained from the Global Biodiversity Information Facility (GBIF) repository [3].

In a similar context, Zhang [4] analyzed the spatial sampling pattern of eBird (www.ebird.org (accessed on 15 October 2021)) data for analyzing the geographical and temporal distributions of the participants using the Maxent algorithm. The study revealed various types of biases (i.e., spatial, temporal, contributor and observation) in the collected data. The outcomes of the study indicated that the volunteers' efforts are uncoordinated, concentrated in highly accessible areas, which yield to hotspots with poor geographical coverage and in specific time periods, such as weekends. The diversity in the backgrounds (e.g., expert birders are more interested in rare species) and the contribution frequency of the volunteers also cause biases in the eBird database. The study emphasized that such bias patterns should be taken into account when performing analyses and deriving conclusions from VGI data.

Samulowska et al. [5] addressed the data bias in VGI and developed a geospatial web platform with a robust quality assurance (QA) approach. The article includes an extensive literature review on the CitSci contribution for air pollution mapping. The authors proposed a validation method for the detection and removal of bias in CitSci collected data by using a logical workflow and rules specifically designed for air pollution assessment. They concluded that the citizens can provide reliable data together with scientists, and can contribute to the QA process as well. On the other hand, the motivating mechanisms for participation require further research.

Gulnerman et al. [6] also analyzed the anomaly and the bias in VGI obtained from a social network service (SNS), i.e., Twitter, by using the data from Istanbul, Turkey. The data was obtained from approximately 80 k users with a total amount of 4 million tweets. They focused on the spatiotemporal patterns of the data in a 1 km $\times$ 1 km grid. With their proposed statistical approach, it was possible to detect anomalies and biases and replace with the expected values. The study revealed several outcomes such as; highly active users produce the majority of the data; the anomaly patterns may vary and the normalization approach should be defined accordingly; the anomaly pattern is stronger in dense population areas; and the biases also exhibited different temporal patterns.

Gray et al. [7] developed an open source tool named Community Water Data Analysis Tool (CWDAT) for water quality monitoring. In the system design, they co-created with the citizen scientists and put special emphasis on the data quality assessment and user engagement. By developing and providing the tool as open source, they also contributed to geospatial capacity building. A major outcome of the study was that the engagement of volunteers in the design stage facilitates the participation and contributes to the citizen perception of data quality, which is a significant issue, as stated in many other CitSci projects. The authors also emphasized the importance of presentation (e.g., visualization) and availability of proper guidelines for an increased success.

Yalcin et al. [8] developed a mobile app ("I felt the quake") to aid emergency management after earthquakes and to support earthquake-related studies by providing timely and spatially accurate data. The study includes an extensive review on the use of CitSci in earthquake related hazards and the availability of similar CitSci apps. The proposed app utilizes a modified version of the Mercalli scale, a scientific standard to assess the earthquake intensity. The scale was modified by [8] for increasing its usability and understandability by non-professionals. The volunteers were also trained by giving specific guidelines. In the experimental application, it was proven that the CitSci data quality can be as high as those provided by professionals when appropriate tools and guidelines are provided.

Vahidnia and Vahidi [9] analyzed the spatial data infrastructure (SDI) aspect of geoportals with a focus on public participation, and proposed a model called "Open Community-Based Crowdsourcing Geoportal for Earth Observation Products" (OCCGEOP) particularly for Earth Observation (EO) data. The framework considered the well-known spatial data structures and methodological standards from Open Geospatial Consortium (OGC), quality control and engagement mechanisms, facilitating the communication between users for improved interaction and sharing, data search and discovery and relying on open source technologies. The authors emphasized the importance of ontology to resolve or to reduce the semantic heterogeneity and to contribute to semantic interoperability; and pointed out these issues as future work. Integration of existing SNS and overcoming the language barriers were also mentioned for the same purpose.

Perdana and Ostermann [10] analyzed the background of participatory toponym handling and addressed several issues on the use of CitSci for leveraging the knowledge in this field in collaboration with governments. In the study, a generic framework was proposed for toponym handling and modified for Indonesia by utilizing collaborative learning among multiple stakeholders. The study emphasized the importance of open science and showed the applicability of the generic framework locally for concrete cases. An important outcome of the study was the acceptance of CitSci by national government in the legal framework. The authors stated that the different user perspectives and backgrounds as well as the motivations remained among the key issues in participatory approaches.

3. Summary and Future Directions

The scientific contributions and lessons-learned from the articles published in the Special Issue can be summarized as following;

- The CitSci and VGI are highly valuable for geospatial capacity building and for facilitating scientific developments.
- Open science and open data are key for multi-stakeholder (e.g., citizens, local and national governments, multinational organizations) collaboration.
- The citizen scientists have diverse backgrounds, and co-creation activities help to overcome issues sourced from this diversity.
- Although the data quality remains among major challenges, several approaches such as statistical, logical, spatial analysis, machine learning (ML), and collaborative methods can be utilized for this purpose.
- Revealing the spatial, social and temporal patterns in the CitSci collected data also supports reliable knowledge extraction and increasing the data quality.
- The geospatial application fields of CitSci and VGI are diverse and can benefit from each other for efficient implementation and analysis of such projects.
- Ontological studies are essential for developing scalable frameworks and increasing the participation capability, interoperability, and building the necessary capacity at various levels (e.g., local, regional, national, etc.).
- Platform integration efforts, i.e., between SNS and specifically developed CitSci environments, can also improve the success of such studies.

Based on the lessons learned and the recent developments in geospatial technologies, the concrete proposals and future directions derived from the views of the Guest Editors can be listed as:

- Considering the advancements in communication infrastructures and increased accessibility to the mobile computing devices such as cellular phones, the amount of data collected by individuals in the form of digital photos have been boosted, in addition to digital sound and sensor data such as raw GNSS measurements and accelerometers. Efficient utilization of these sensors and their data in CitSci projects can be promoted further.
- Individuals most of the time share the data they collect through various social media environments such as Facebook, Twitter, Forums, Trip and sightseeing blogs. This unconsciously shared data may include highly valuable information for managing emergency situations, better decision making and scientific experimentation. However, the extraction of such information, especially location of geographic information, from such a heap requires development of advanced data crawlers that may consume available metadata, apply artificial intelligence (AI) tools such as natural language processing (NLP), knowledge engineering, social network analysis, classification and sorting. This may be considered as the next generation of search engines, which are intelligent enough to decide whether a piece of information fits a specific purpose. Even after a successful searching and sorting of such a dataset is achieved, extraction and transformation of usable information from the dataset requires a definition of quality metrics and also development of software tools.
- On the other hand, when guided by higher level objectives, volunteered collection of such information with predefined quality metrics, data model and user interface is possible. There are various success stories for the VGI such as OpenStreetmap, eBird and WikiMapia, just to name a few. Although there are important standardization efforts mostly by Open Geospatial Consortium (OGC) to increase the interoperability of geographic information, the applications of these standards for VGI collection, storage, search and retrieval are still important issues.
- In order to cultivate multi-disciplinary research on VGI, tools and techniques are needed to simplify the merge of observations collected by different volunteer groups. OGC Semantic Sensor Network Ontology and Sensor Web Enablement related standards for example may provide the necessary common language for crawling, collecting, storing, searching and integrating sensor data from different studies. In this context, advanced software tools such as mobile applications that collect location information, photos and sensor data from cellular phones, gateways for cheap Internet of Things (IoT) sensors, sensor databases and services for extraction, transformation, load and analysis of sensor data.
- There is a need to make use of existing repositories of CitSci and VGI projects for developing AI, ML and deep learning (DL) based innovative solutions for object identification with faster retrieval on near real time basis. This would enable the utilization of the voluminous data in a more meaningful and lucid manner.
- Research on re-usable/configurable software development frameworks that rely on the abovementioned standards may ease the establishment of SDIs for VGI. In addition, these frameworks may also overcome the common problems associated with data quality, privacy and abuse and also bi-directional feedback. With such advanced tooling and contribution from volunteers, we may get closer to better understanding of our local neighborhood as well as global events.

Author Contributions: Conceptualization, validation, formal analysis, writing—original draft preparation, writing—review and editing, Sultan Kocaman, Sameer Saran, Murat Durmaz, Senthil Kumar. All authors have read and agreed to the published version of the manuscript.

Conflicts of Interest: The authors declare no conflict of interest.

References

1. Singh, P.; Saran, S.; Kocaman, S. Role of Maximum Entropy and Citizen Science to Study Habitat Suitability of Jacobin Cuckoo in Different Climate Change Scenarios. *ISPRS Int. J. Geo-Inf.* **2021**, *10*, 463. [CrossRef]
2. Séférian, R.; Nabat, P.; Michou, M.; Saint-Martin, D.; Voldoire, A.; Colin, J.; Decharme, B.; Delire, C.; Berthet, S.; Chevallier, M.; et al. Evaluation of CNRM Earth System Model, CNRM-ESM2-1: Role of Earth System Processes in Present-Day and Future Climate. *J. Adv. Model. Earth Syst.* **2019**, *11*, 4182–4227. [CrossRef]
3. Wheeler, Q.D. What if GBIF? *Bioscience* **2004**, *54*, 718. [CrossRef]
4. Zhang, G. Spatial and Temporal Patterns in Volunteer Data Contribution Activities: A Case Study of eBird. *ISPRS Int. J. Geo-Inf.* **2020**, *9*, 597. [CrossRef]
5. Samulowska, M.; Chmielewski, S.; Raczko, E.; Lupa, M.; Myszkowska, D.; Zagajewski, B. Crowdsourcing without Data Bias: Building a Quality Assurance System for Air Pollution Symptom Mapping. *ISPRS Int. J. Geo-Inf.* **2021**, *10*, 46. [CrossRef]
6. Gulnerman, A.G.; Karaman, H.; Pekaslan, D.; Bilgi, S. Citizens' Spatial Footprint on Twitter—Anomaly, Trend and Bias Investigation in Istanbul. *ISPRS Int. J. Geo-Inf.* **2020**, *9*, 222. [CrossRef]
7. Gray, A.; Robertson, C.; Feick, R. CWDAT—An Open-Source Tool for the Visualization and Analysis of Community-Generated Water Quality Data. *ISPRS Int. J. Geo-Inf.* **2021**, *10*, 207. [CrossRef]
8. Yalcin, I.; Kocaman, S.; Gokceoglu, C. A CitSci Approach for Rapid Earthquake Intensity Mapping: A Case Study from Istanbul (Turkey). *ISPRS Int. J. Geo-Inf.* **2020**, *9*, 266. [CrossRef]
9. Vahidnia, M.H.; Vahidi, H. Open Community-Based Crowdsourcing Geoportal for Earth Observation Products: A Model Design and Prototype Implementation. *ISPRS Int. J. Geo-Inf.* **2021**, *10*, 24. [CrossRef]
10. Perdana, A.P.; Ostermann, F.O. Eliciting Knowledge on Technical and Legal Aspects of Participatory Toponym Handling. *ISPRS Int. J. Geo-Inf.* **2019**, *8*, 500. [CrossRef]

International Journal of
Geo-Information

MDPI

Article

Eliciting Knowledge on Technical and Legal Aspects of Participatory Toponym Handling

Aji Putra Perdana [1,2,*] **and Frank O. Ostermann** [1]

1 Faculty of Geo-Information Science and Earth Observation (ITC), University of Twente, 7500 AE Enschede, The Netherlands; f.o.ostermann@utwente.nl
2 Badan Informasi Geospasial (BIG)—Geospatial Information Agency of Indonesia, Jl. Raya Jakarta-Bogor Km. 46, Cibinong, Bogor 16911, Indonesia
* Correspondence: a.p.perdana@utwente.nl; Tel.: +62-813-1564-9101 or +31-614-182-830

Received: 26 September 2019; Accepted: 3 November 2019; Published: 5 November 2019

Abstract: There has been increased collaboration between citizens and scientists to achieve common goals in scientific or geographic data collection, analysis, and reporting. Geospatial technology is leveraging the power of citizens in such efforts. Governments have been exploring participatory approaches. This situation should be balanced by sharing knowledge and collaborative learning between stakeholders involved in the participatory activity. Training and education are enhanced by providing guidelines, sharing best practices, and developing toolkits. For toponym handling, a generic framework and capacity building are needed to increase public awareness and enable citizen toponymists. This paper addresses issues around citizen involvement in increasing toponymic knowledge through citizen science and geospatial capacity building. First, we examined the current practice of toponym handling and developed a generic framework. We then used stakeholder feedback and other resources to modify the framework for Indonesian use. Second, we conducted collaborative learning to share information and bridge the knowledge gaps among multiple stakeholders. Third, we applied insights and lessons learned from these activities to develop ideas, suggestions, and action plans to implement participatory toponym handling in Indonesia.

Keywords: participatory toponyms; knowledge sharing; public participation; citizen science; geospatial capacity building

1. Introduction and Research Objectives

Since 1959, the United Nations Group of Experts on Geographical Names (UNGEGN) has promoted the preservation of toponyms and encouraged each country to collect and maintain toponyms [1,2]. UNGEGN provides an umbrella for each country to develop its regulations by developing a legal framework for toponym collection and maintenance. A national names authority (NNA) or a group of experts on toponymy (toponymists) have the responsibility to share their knowledge and existing technical and legal framework with the stakeholders. Stakeholders may include groups of experts on toponyms from academic and research institutions, the government, non-governmental organizations (NGOs), and members of the public (individual volunteers or local communities). Toponym collection has usually been conducted by a government agency, such as a national mapping agency (NMA) as part of its topographic mapping activity, or any other institution like an NNA or coordinating body on toponyms [3,4].

Typically, the government agency and toponymists work together to provide standardized toponyms in the form of a geographical index or dictionary, known as a gazetteer. In many developing countries, limited resources impede the creation of a comprehensive national gazetteer. Another source of toponymic crowdsourced geographic information (CGI) are global gazetteers (e.g., Geonames, the Alexandria Digital Library Gazetteer, the Getty Thesaurus of Geographic Names, and DBPedia Places). Gazetteers must include minimum information (place names, types of features, and geographical coordinates) to be used or connected by linked data or ontology [5,6], but they have varying coverage and there is no common classification of features. Moreover, fitness for purpose and licensing may be issues. For example, Geonames data sources (http://www.geonames.org/datasources/) for Indonesia are harvested from the Open Data initiative, as provided by national agencies (Open Data Indonesia, Badan Informasi Geospasial (BIG), the Geospatial Information Agency, and Badan Pusat Statistik, the Central Agency on Statistics) and local governments (Bandung City and Sulawesi Tengah Province). In this situation, the government agency and toponymists cannot rely only on global gazetteers to populate their toponym files to build a national gazetteer. We should therefore consider using the power of multiple stakeholders working together—especially members of the public, that is, citizens—to contribute to such efforts.

Technological developments (e.g., Web 2.0, mobile internet access) have increased the involvement of citizens in various activities. This is often called crowdsourcing [7]. Toponymic geographic information harvested from CGI or volunteered geographic information (VGI) offers potential information to investigate place names changes of man-made features [8]. Yet toponymic practices have included only a few such efforts. These include crowdsourcing and GIS-based methods in Austria [9], and Kadaster Netherlands using crowdsourcing to get feedback from citizens regarding place names on topographic maps [3]. In addition, private sector engagement in crowdsourced toponym handling platforms is increasing. These include Google's Local Guide program and Facebook's community of Place Editors. Non-profit mapping communities (e.g., the Humanitarian OpenStreetMap Team and local participatory mapping organizations) also collect, maintain, and share their place-based information, including toponyms [10,11]. A different strategy would be to encourage citizens to become toponymists, using their attachment to the places and their ability to share knowledge on toponyms. Involving citizens in research-related activities, known as citizen science [12,13], is an established practice in many scientific disciplines. In this research, we contribute to the wider citizen science body of knowledge by investigating methods to foster collaboration among toponym stakeholders and to enable citizen toponymists through participatory toponym handling (PTH) [14].

The key research questions in this paper are:

- How can we streamline the working processes and link multiple stakeholders through PTH?
- How can PTH combine techniques in a flexible way to speed up data availability, completeness, and to meet user needs?
- What technical and legal elements of toponyms are needed for the collaboration of multiple stakeholders in handling toponyms?
- What lessons can be learned from collaborative learning and sharing knowledge among stakeholders?

Our approach includes collaborative workshops, focus group discussions (FGDs), interviews with the key actors, and qualitative fieldwork. The key contributions of this paper are:

- A participatory toponym handling framework that can be modified to accommodate multiple stakeholders.
- A strategy for collaborative learning of participatory toponym handling that involves multiple stakeholders.
- Outcomes and lessons learned from a case study to improve and adapt the generic framework for concrete implementation.

Our paper is structured as follows: In Section 2 we outline our research context, which consists of the framework of PTH and the concepts of collaborative learning and knowledge sharing. In Section 3, we outline our research workflow, including short explanations of the activities. Section 4 presents the results, Section 5 contains a discussion, and Section 6 offers conclusions and recommendations for future research.

2. Research Context and Related Work

2.1. Participatory Toponym Handling

Conventional toponym handling does not usually include active citizen involvement [1,15]. There have been several attempts to incorporate citizen participation by developing mobile and web-based applications. For example, a toponym collection "game" was created to clean up the toponyms database of The Instituto Geográfico Nacional (IGN), or National Geographic Institute [16] and a "web gazetteer" using the Ordnance Survey dataset was used to build a historical gazetteer of Great Britain [17]. These two examples illustrate the need for a standardized but adaptable framework to engage citizen toponymists in different contexts since current approaches vary and relate to particular contexts.

The uncertainty and ambiguity derived from citizen involvement should be considered prior to integration with the authoritative data. The ambiguity and uncertainty of toponymic CGI or VGI are being investigated increasingly [18,19]. These two problems would affect the evaluation process of toponymic files and gazetteer related to inconsistency and accuracy. The uncertainty of toponym collected by volunteers may be correlated to the missing or incorrect category tags [20]. While the ambiguity of place names might increase to a more detailed level, such as at the national, regional and local level, than across nations [20], it can be related to similarity place names information [21].

Citizen participation depends on the willingness of the government to consider the possible benefits such as supplementing the government's limited human resources and increasing the accessibility and extent of toponym information. To understand citizen participation in toponyms better, we have developed a framework to guide the handling of toponyms. In this framework, we consider a citizen science ontology [22] to connect participatory toponym handling as part of geographic citizen science. Handling toponyms usually consists of toponym collection, maintenance, and publication, as well as education and training, including how to utilize and enrich toponym databases. In this research, we use the term "participatory" to refer to a toponym handling processes that involves the collaboration of multiple stakeholders, including the active involvement of citizen toponymists.

Public participation can gather information and knowledge to enrich authoritative data. It requires members of the public to contribute or collaborate in specific activities to tackle some public or research problems. The United States Geological Survey (USGS) has a long history of using crowdsourcing data collection to support topographic mapping [23]. Also, national agencies develop VGI pilot projects for topographic data collection, for example, Map Gretel, developed by the National Land Survey of Finland (NLS) [24]. Indonesia's NMA, Badan Informasi Geospasial (BIG), developed PetaKita (https://petakita.big.go.id/, also available in the Google Play store) to support community involvement in participatory mapping. It was based on a one-map database designed NGOs in 2014, which was less successful in engaging users due to a lack of contributors. BIG continues to improve the PetaKita application user interface to make it simpler, more intuitive, and user-friendly. Recent studies and actual work on geographical data collection indicate that engaging the public remains a challenge.

2.2. Sharing Knowledge of Toponyms Through Collaborative Learning

In this research, collaborative learning is a constructive process in which multiple stakeholders learn together, are actively involved, and share their understanding of information and knowledge of toponyms based on their roles and capabilities. These stakeholders typically share knowledge during geospatial capacity building, using toponymic education and training. For example, an NNA can conduct toponymic workshops, hold meetings with stakeholders, and publish toponymic guidelines.

Knowledge is classified into two types: Tacit (or implicit) and explicit knowledge. Knowledge sharing (KS) in an organization or project is an essential process that transfers both tacit and explicit knowledge through people as social capital [25] and technology to speed up the process [26,27]. Tacit knowledge can be documented and shared by observation, coordination, and communication among stakeholders. Explicit knowledge is easy to transfer and share within an organization, such as experience and expertise [28,29].

Technical tacit knowledge can be transferred and communicated to multiple stakeholders on toponyms in many ways. For example, to know the process of toponymic investigation, we establish coordination and communication between multiple stakeholders, including scientists (the experts on toponyms) and citizens. The citizens can observe, learn, and conduct a systematic investigation of toponyms from the scientists. Technical stakeholders also include teachers and surveyors who can show how to create a basic form to collect toponyms using mobile apps.

Meetings were organized with multiple stakeholders on how to reveal the language, history, and meaning of toponyms. Local people can share explicit knowledge with scientists, and vice versa. It works if the interests and preferences of stakeholders on toponyms are shared based on their needs. Local citizens usually know how to pronounce toponyms and are often willing to share their local knowledge of places in their neighborhood.

3. Methods and Research Workflow

For this project, we used a research workflow to describe the development process of the PTH framework through collaborative learning (see Figure 1). The focus of this workflow is collaborative learning, which enables stakeholders to apply and share their knowledge about toponym handling.

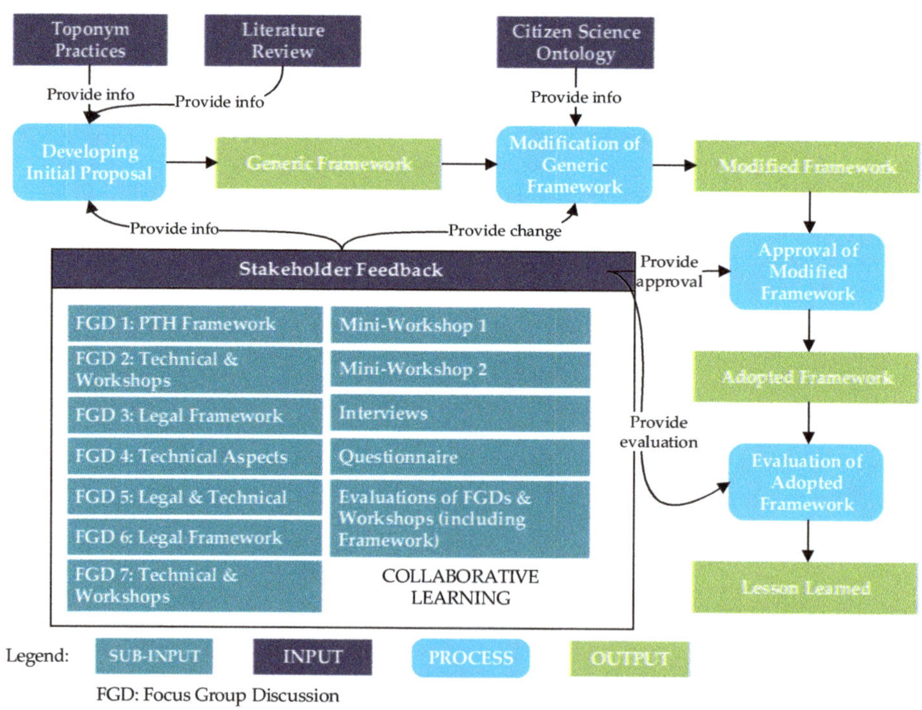

Figure 1. Research workflow.

We first investigated several relevant resources on toponymic guidelines, citizen science, VGI, and crowdsourcing. We combined parts of a citizen science ontology developed by a COST Action (https://cs-eu.net/wgs/wg5) [22] to modify the framework. Then, we improved the framework using existing knowledge and feedback from multiple stakeholders. Changes requested by stakeholders were made in an iterative process. When a new change request was made, the modification was approved or rejected after several discussions.

We employed gamification to stimulate participation and fuzzy cognitive mapping [30,31] to bridge the knowledge gaps between stakeholders. These methods were used to boost stakeholder interactivity in FGDs and workshops. We used two game elements (achievements and rewards) to encourage citizens to discuss or give feedback on specific issues and to suggest ideas for toponym handling projects. Each contribution to the discussion of a participatory toponym handling project earned a point, and badges were awarded. We provided rewards during the workshops and FGDs, and after the assessment was completed. The rewards included gifts (mugs, key chains, magnets), toponym books, and certificates. The strategy was to engage different stakeholders, build collaboration, and gain knowledge. The fuzzy cognitive mapping employed by stakeholders in the FGDs depicts the relationships between elements of toponym handling.

The scope of the FGDs ranged from technical to legal aspects of volunteer and scientific information on toponyms. The workshops enabled stakeholders to discuss and implement the modified framework. The workshops on technical aspects examined the exploration, adoption, and implementation of the plan for participatory toponym handling. A toponym handling system that used a mobile application for toponym collection also was discussed and introduced in the workshops. It is called SAKTI (http://sakti.big.go.id/sakti/webgis/) (Sistem Akusisi Data Toponim Indonesia/Indonesian Toponymic Data Acquisition System), and it was developed by BIG. The mini-workshops were conducted to manage and monitor progress and identify levels of engagement. In the workshops on legal drafting, we discussed and reviewed UNGEGN recommendations, best practices from other countries, and regulations in Indonesia related to toponyms. Recommended texts on the definition of toponyms and about technical issues was ere elaborated and promoted to be translated into a legal document.

In addition, we interviewed personnel involved in toponym practice, and we administered a questionnaire to relevant stakeholders and key actors. In the evaluation step, we identified the training conducted by the Indonesian NNA in 2018 as the baseline for the learning process and knowledge sharing mechanism. Then, we compared it with the feedback from the workshops and recommendations from interviews with participants.

4. Results

4.1. Fuzzy Cognitive Map and a Generic PTH Framework

During the FGDs, a fuzzy cognitive map was generated to illustrate the workflow and problems in toponym handling. The elements in this mind map and connections between elements were then incorporated into the PTH framework through discussions on essential toponymic content such as the meanings and interpretations of place names. To increase citizen involvement, *Organization* must be addressed as the focal point for problems in Indonesia's toponyms practice.

Figure 2 shows a fuzzy cognitive map of (participatory) toponym handling as one of the outcomes of the discussions about organizational challenges (represented in red lines) and opportunities in toponym handling (represented in blue line: support and black line: regular relationship). Most participants strongly recommended the integration of knowledge sharing and participatory approaches. This could be achieved through policy change and capacity building for multiple stakeholders. Such efforts increase collaboration between academics, local government, NGOs or the private sector, and members of the public. Challenges facing current toponymic practices in Indonesia cut across financing, human resource needs, standardization, and policy.

A framework for PTH was developed through several steps shown in the workflow (Figure 1). The participatory approach is relatively new in toponym handling. In this case, we start learning from the existing workflow of VGI and citizen science or crowdsourcing as the lowest level of citizen science (according to Haklay's typology of participation [14]. Currently, local governments are actively involved in toponym handling, and they sometimes involve university students to explore participation through crowdsourcing in toponymic surveys [32]. Based on these background situations, our initial proposal for a generic framework for PTH included three processes: Traditional survey, citizen participation, and toponymic CGI or VGI.

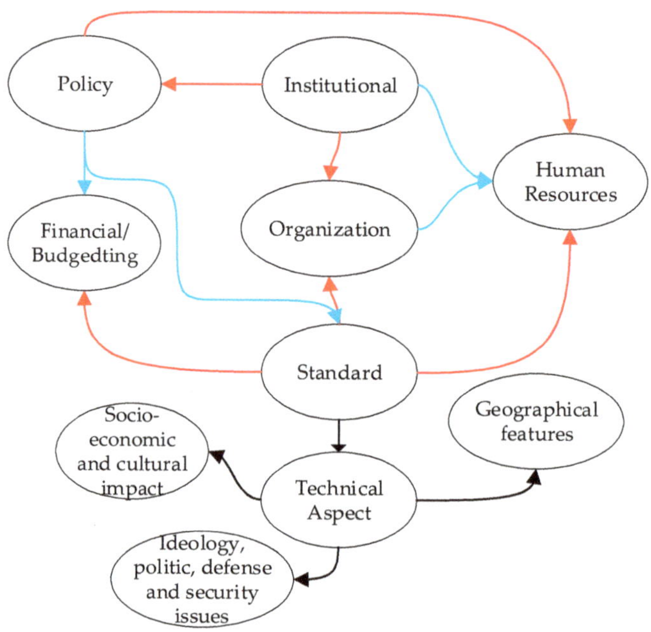

Figure 2. Fuzzy cognitive map of (participatory) toponyms handling in the Indonesian case study (blue line: support, red line: challenge, black line: regular relationship).

In the initial proposal, we focused on how toponym handling could help provide an enriched toponymic file and gazetteer. The initial generic framework on handling toponyms using a participatory approach was introduced and modified after several discussions. During these processes, stakeholders shared their knowledge and information on handling toponyms. In the modified framework (see Figure 3), we then changed the focus from data-oriented problems to toponym handling as a project-based approach. In the modified framework, we define eight classes: *Stakeholder*, *Actor*, *Role*, *Project*, *Task*, *Toolkit*, *Method*, and *Dataset*. The types of toponym handling in a proposed generic framework are covered in *Method*. Some categories in that framework were adopted from Citizen Science Ontology [22], for example, elements in *Role* class and *Project* class.

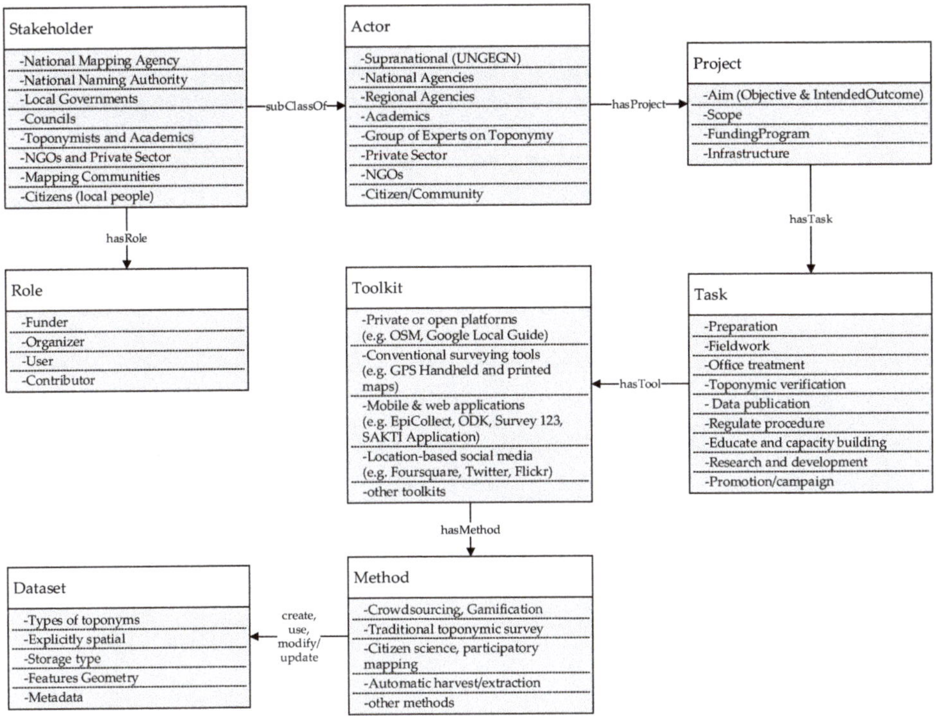

Figure 3. Modified participatory toponym handling (PTH) framework based on stakeholder feedback and citizen science ontology.

4.2. Testing Participatory Approach and Exploring Stakeholder Involvement.

There is a traditional concept of mutual assistance in Indonesia, known as *gotong-royong* [33]. A participatory approach is very similar to it. By testing or adopting the PTH framework, new information was generated through collaborative learning, which led to the adjustment of the generic framework into an actual condition in the Indonesian case study. We adapt the PTH framework in a concrete implementation on a regional case study in Yogyakarta through iteration of collaborative learning. Collaborative learning in the workshops and FGDs was achieved by citizen toponymists, researchers, and the government by cooperation in identifying, deciding and developing the toolkit for toponymic surveys. This includes both the development of mobile and web applications and the building of supportive project sustainability.

The questionnaires and interviews were conducted during June and July 2019, and we received feedback from 132 respondents and interviewees (a total of 150 participants in workshops and FGDs). The questionnaire was categorized into five groups. The first set of questions aimed at defining context and stakeholder groups. The second set explored the current organizational focus and involvement on toponyms. The third set examined the participant's preparation for the workshop and spatial literacy. The fourth group was intended to evaluate content and facilitation for the workshops. Finally, the fifth group explored basic knowledge on toponymy, personal appraisal, and viewpoint, including questions of voluntary participation in toponym handling. Some questions might not apply to all stakeholder groups.

The citizens who participated in the workshops opted for interviews instead of a questionnaire. In the interviews, we asked the same questions as in the questionnaires, but we allowed deviations to remain flexible in collecting information and establishing a connection with citizens. The interviews worked on exploring their fundamental knowledge about toponyms, their perspectives on current toponymic activities, and the prospect of inviting them to join toponyms data collection using the participatory toponymic approach. During the interview, two examples of questions arose: "In your words, what are the toponym collection problems we should be trying to solve? What would be your contribution to a participatory toponym handling project?"

There were four lists of questions on stakeholders' background, involvement, and willingness to share knowledge during PTH development and framework testing. Figure 4a shows the percentages of stakeholders interested in PTH and in participating. Some members of the public (8%) were willing to play an active role in participatory toponym handling projects. Figure 4b shows stakeholder participation in current toponyms practice. Most of the participants in the workshops and FGDs were from local committees at the district/city level and provincial level. However, thirty percent of respondents were not part of any toponym collection team or committee in Indonesia. Figure 4c indicates spatial familiarity. More than half of the participants in the workshops were not familiar with spatial data. This information allowed us to provide a better introduction to toponyms as part of geographic data collection using mobile and web applications.

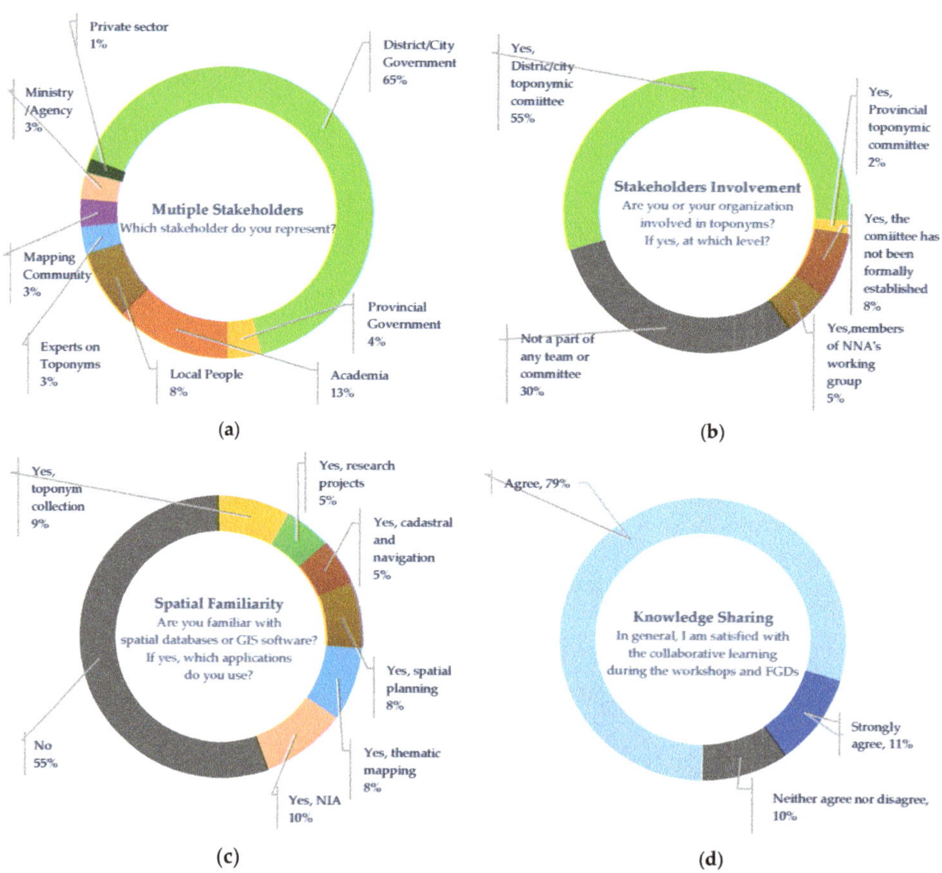

Figure 4. Multiple stakeholders' background and satisfaction on collaborative learning: (**a**) type of multiple stakeholders; (**b**) stakeholders involvement in toponyms; (**c**) stakeholder familiarity with spatial databases or GIS software; (**d**) knowledge sharing satisfaction.

In general, participants involved in the development and evaluation of the PTH framework were satisfied with knowledge sharing through collaborative learning activities (see Figure 4d). They realized and felt their considerations on the importance of place names in their daily lives and how elderly people gave names that had meanings linked to geographical phenomena. This was shown through their help collecting the toponyms for their neighborhood. The private sector was represented by a GIS consultant who works on capacity building of participatory GIS for villages. The workshops offered opportunities for future projects of collaboration. The private sector and one university in Yogyakarta are planning to set up a geospatial capacity building workshop to introduce geo-entrepreneurship and mobile apps for geographic data collection.

Understanding stakeholders and engaging with them can help to develop sustainable projects and promote organizational interoperability. Organizational interoperability focuses on how different stakeholders work together to reach common goals. Humanitarian OSM Indonesia shared its experience in building support for the ecosystem as part of the working environment and got the community to utilize their skills and knowledge.

4.3. Technical and Legal Elements of PTH for Consideration in Indonesia

It is important to consider the technical and legal aspects of any adaptation of a generic framework to a concrete case. As the responsible stakeholder on the national level, the recommendations of the FGDs are that BIG should consider and discuss developing comprehensive policies on how to link and cooperate on toponymic technical, organizational, and regulatory problems. In addition, BIG should continue outreach efforts to promote campaigns on toponymy and cooperate with the community such as Google Local Guide, Facebook Place Editor, OSM community, and local NGOs. Meanwhile, to motivate contributors, they might provide a certificate as a reward for the contributor's efforts in collecting toponyms.

BIG and other stakeholders in workshops on technical and legal issues are working together to address existing practices with Indonesian toponyms. Legal problems on toponyms in Indonesia began with the change from the previous organizational setting to BIG's responsibility as NNA. This resulted in a dualism of parallel regulations on toponym handling.

The legal aspect guides how to use crowdsourced geographic information (from global gazetteer, OSM, and many other sources, including local place names information from local participatory mapping activity) for completing and enriching the national gazetteer. NNAs should consider synchronization of existing toponyms with other potential sources (toponymic CGI, such as global gazetteer Geonames and OSM dataset). During the process of drafting legal documents, we realized that there is a need to evaluate the protocols and existing SOPs established by BIG or other agencies. Despite this, there is still no SOP or regulation on participatory toponym handling.

To deal with the verification and integration of multiple sources of toponymic files and gazetteers, we adopted criteria from three sources: the gazetteer quality criteria developed by Hill [34], the A4C4 quality requirements by Swisstopo [35], and the data quality requirements of Indonesian topographic base maps [36], especially for attribute accuracy for toponyms. Gazetteer quality criteria from Hill [34] were previously used to compare global gazetteers (Geonames and TGN) with gazetteers produced by mapping agencies from Ordnance Survey 50K and SwissNames 3D [37]. In addition, the stakeholders chose two quality criteria (Swisstopo and BIG) because they are fit to assess the mapping agency's gazetteer.

Table 1 shows the criteria for verification that stakeholders proposed during discussions in FGD 5 to FGD 7. The criteria and descriptions were modified from [34–36]. These criteria can be used by BIG to check and assess the toponymic data quality before the integration of crowdsourced toponym files. Appendix A shows the details in Table A1 complete with measuring units. Each criterion involves qualitative and quantitative considerations based on discussion sessions on implementing and testing PTH. These criteria must be reviewed again by multiple stakeholders before carrying out verification tests. Measurement results will be stored in metadata of toponymic files and gazetteer. Together, the toponymic dataset and other relevant information will be used to continue with the next verification steps with a group of experts on toponyms.

Table 1. Verification criteria for toponymic files and gazetteer.

Criterion	Description
Authority	Data source, production
Accuracy	Number of detectable errors in names, footprints, and feature types
Availability	The degree to which the toponymic files and gazetteers are freely available and not limited by restrictive conditions of use (data access)
Actuality	Date of data specific survey; the degree to which the toponymic files and gazetteer have incorporated changes
Completeness	Steered by toponymic surveying rules; the degree to which the scope of toponymic files and gazetteers are covered completely
Scope /Coverage	Small communal database, regional/national coverage, or worldwide coverage
Consistency	Data consistency and coherence
Richness of annotation	Amount and detail of descriptive information, beyond the basics of name, footprint, and feature type
Granularity	Whether data includes large, well-known features only or features of all sizes and those that are less well known
Balance	Uniform degree of detail, currency, accuracy, and granularity across scope of coverage

Source: Modified from [34–36] and measurement unit harvested based on FGDs (FGD 5 to FGD 7).

Another output from the workshops and FGDs on technical and legal aspects (FGD 3 to FGD 7) is a draft of a new government regulation. It was established through high-level discussions led by BIG and supported by the Faculty of Law, Universitas Gadjah Mada, Indonesia. The draft proposal for the new regulation on toponyms includes four recognized approaches to collecting toponyms (Figure 5).

A field survey is the traditional way that national agencies and local committees collect toponyms. Secondary data compilation is the collection of toponymic information from government or non-governmental organizations, published or unpublished. This method supports the field surveys and provides additional details and links to toponyms. There are two novel approaches in the draft of the new regulation on toponyms in Indonesia: participatory mapping and crowdsourcing. This regulation defines participatory mapping as activities to collect toponym data by involving community groups or organizations. In contrast, crowdsourcing is defined as toponym collection that involves citizens, especially individuals, where data are stored in the organizer's database. These two approaches allow a local government to encourage multiple stakeholders to be more active in toponym handling, specifically enabling citizen toponymists. A simple step to enable citizens to become toponymists is to ask people to help to preserve toponyms surrounding their neighborhood.

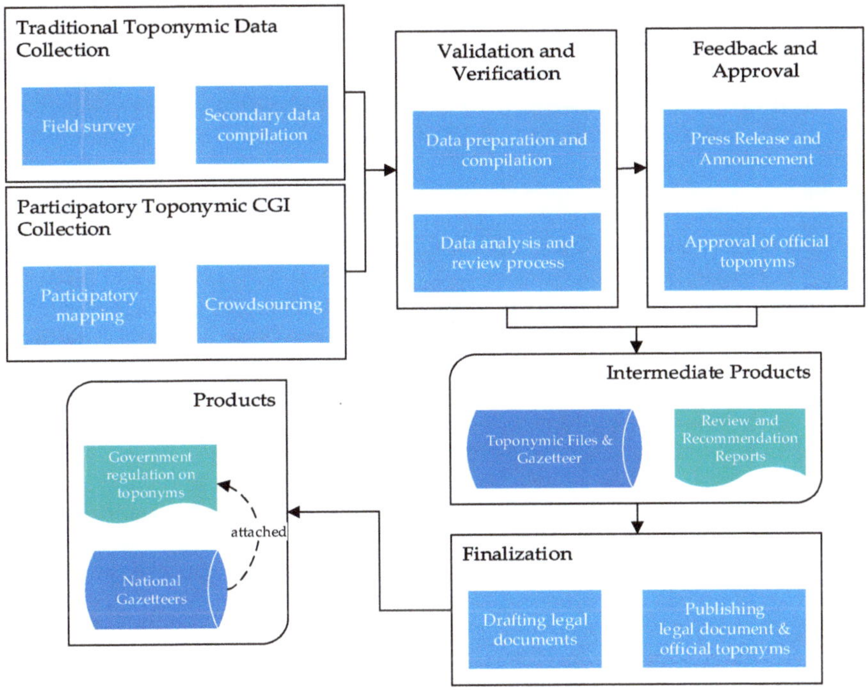

Figure 5. Proposal of toponym handling in a draft government regulation on toponyms in Indonesia (modified from BIG, 2019).

5. Discussion

Developing a generic framework and modifying it through collaborative learning must be a robust process to ensure that the generic framework can be adapted for different settings. Examples of citizen participation in geographic data collection (sometimes related to toponyms) have often been studied by researchers outside of the UNGEGN members. This includes toolkit development (gamified mobile apps for collecting points of interest, landmarks, or toponyms) in relation to specific themes. Discussion of citizen participation in the official toponymic workflow has been ongoing since a UNGEGN meeting in 2012. At that time, government officials and researchers (members of UNGEGN) shared their lessons learned. They tried to enable local governments or other related actors to collect toponyms voluntarily. This was a definite improvement, but it still lacked active involvement from citizens as toponymists. Any NNA or other associated organization needs a generic framework of PTH to be included in its current toponymic workflow.

The initial proposal for a generic framework focused on different approaches to enable citizen participation in official workflows of toponym collection. Participants in the discussions and modifications of the generic framework recommended having a participatory toponym handling as a project. Furthermore, a Citizen Science Ontology (developed by Working Group 5—Improve data standardization and interoperability of Citizen Science COST Action [22]) provided a foundation to establish or modify the initial generic framework into a more implementable workflow.

Streamlining the workflow of toponym handling needs common goals and agreements among stakeholders. Different perspectives, tasks, interests, and purposes also trigger new challenges in building and harmonizing participatory toponym handling. Several meetings are required to delineate a common understanding of the different roles played by multiple stakeholders before a mutually approved outcome can be achieved. This common understanding can help to address certain problems.

For example, where there are limited human resources for collecting toponyms, lecturers and students from an academic institution, who volunteer as part of their community service program, can support the local committee on toponyms in conducting a toponymic field survey. They may provide training to establish participatory toponym handling or work together with society. This kind of participatory approach is flexible enough to be implemented in Indonesia as long as it is supported with reliable information on handling toponyms. A pilot project showed that the participatory dataset could enrich toponymic information and include areas not covered by toponyms from the government.

Dealing with classical problems of toponym collection by the government (e.g., lack of human resources and budget limitation), academics and community members proposed a concrete action plan on how to involve citizens as toponymists—for example, student internships or fieldwork that connects academic study and community service practice. This concept remains a challenge because cooperation between academics, local authorities, and people is necessary. Moreover, it depends on every region's character and nature to encourage and involve the citizen. The modified framework can be adopted and modified according to the characteristics of each country or region. In addition, local conditions also help the implementation of this type of project. It would be useful to understand local characteristics before implementing this type of toponym handling project and to engage with multiple stakeholders.

On the other hand, the strategy of collaborative learning and knowledge sharing inspired BIG to develop their action plans. The involvement of multiple stakeholders on sessions concerning technical and legal concerns and the implementation of toponyms from multiple perspectives is going to be adopted in upcoming technical assistance to the local committee. An action plan to enable public participation in handling toponyms is being discussed by NNA in Indonesia. The adoption of participatory toponym handling is represented in the new Indonesian draft of a government regulation on toponyms. Regarding the two citizen-based approaches to collect toponyms, there is a need for research on the incentives and disincentives for toponym handling, and how organizers can actively engage citizens. Data collected or contributed by the citizens will be maintained as a citizen layer to be verified and integrated into official toponymic files and gazetteers. The future work is to develop toponymic guidelines or procedures by detailing technical steps of the toponymic workflow as proposed in the draft government regulation. Meanwhile, NNA is continuously working on high-level discussions with multiple ministries involved in reviewing the draft of toponymic regulation.

One suggestion from stakeholders is that training and education still need to leverage contributors for handling toponyms, either by the participatory or by the traditional approach. Each stakeholder recognizes their role in handling toponyms and offers their potential support to conduct or establish co-created or collaborative toponymic field survey projects. Local people learned directly from a group of experts on toponyms about toponymic diagnosing and the meaning of place names. The experts also realized their roles in promoting toponyms to the community by collaborating with local government. However, the intensity of meetings between multiple stakeholders in workshops or FGDs and the inclusion of citizen toponymists in toponymic collection projects should be considered to cover working areas, and the toolkit for toponym collection may help to improve the process of participatory toponyms. Nevertheless, the participatory approach demonstrated that citizen toponymists can provide toponymic data with voice recordings, photos, and other relevant information.

6. Conclusions

A long journey of sharing knowledge and collaborative learning on toponyms by multiple stakeholders provided meaningful feedback to improve and modify the framework. The results revitalized the relationship between them and should be maintained by the NNA in Indonesia. The next steps are to implement the framework through toponymic survey projects and promoting the roles of citizen toponymists in participatory toponym handling. Embedded knowledge on toponyms maintained in toponymic files and gazetteers and toponymy itself should be part of open science where a citizen could share and learn about toponyms in their neighborhood and other regions.

From this research, it can be concluded that establishing participatory toponym handling in Indonesia requires a collaborative approach and openness from multiple stakeholders. The case study of the participatory approach in Yogyakarta has shown that the generic framework can be implemented in concrete situations. The spirit of *gotong-royong* would be a valuable factor for enabling Indonesia's citizen toponymists and for conducting collaborative learning as part of geospatial capacity building on toponyms. One achievement is that the national government agreed to include citizen participation (participatory mapping and crowdsourcing) in the new draft of the toponymic legal framework. Concrete action plans still need to be established and initiated by NNA.

The limitation on funding and human resources becomes the primary concern in several regions due to the minimum understanding by multiple stakeholders at the regional, provincial, city/district, and sub-district levels. In this case, Indonesian NNA should provide the national program on toponyms as an umbrella for local governments and guidance to manage their budgeting and human resource allocation. Continued communication through toponymic training and education can leverage and sustain public participation in toponym preservation as part of their daily communication and activities. Toponymic guidelines should be published or provided through local committees until the lowest level of administrative areas are publicly available on the website or other platforms.

Different perspectives, backgrounds, characteristics, and motivations of citizens to participate as toponymists in participatory toponym handling remain one of the issues to be discussed. Knowledge sharing and collaborative learning through comprehensive information on technical and legal aspects encourage people to contribute or be involved in toponym collection and preservation. Finally, the NNA should consider how to handle toponymic files (including crowdsourced and volunteered geographic information) in their technical workflow and national program. Multiple stakeholders involved in toponym handling are ready to enable citizen-based approaches and optimize existing toolkits and methods.

Author Contributions: A.P.P. and F.O.O. discussed the idea; A.P.P. undertook the field work and analyses; A.P.P. and F.O.O. both contributed to the writing of this manuscript.

Funding: The authors would like to acknowledge financial support provided by Indonesia Endowment Fund for Education (LPDP), within code of LPDP: PRJ-2569/LPDP/2015 provided to Aji Putra Perdana, Faculty of Geo-Information Science and Earth Observation (ITC), University of Twente for his Ph.D. research funding.

Acknowledgments: The first author would like to thank the Ministry of Finance for the Republic of Indonesia's Indonesian Endowment Fund for Education (LPDP) for supporting his Ph.D. research. This work is supported by Badan Informasi Geospasial (BIG) as National Naming Authority of Indonesia. The authors would like to thank the anonymous reviewers for their valuable comments and suggestions.

Conflicts of Interest: The authors declare no conflict of interest. The founding sponsors had no role in the design of the study; in the collection, analyses, or interpretation of data; in the writing of the manuscript, and in the decision to publish the results.

Appendix A Verification Criteria for Toponymic Files and Gazetteer

Table A1 describes the criteria, descriptions, and measurement units for verifying toponymic files and gazetteer.

Table A1. Details of verification criteria for toponymic files and gazetteer.

Criterion	Description	Measurement Unit
Authority	Data source, production	Various sources, experts, field survey, mapping agency, NNA, or other authorities. Authorization status: 1: verified by authoritative agencies/experts 0: not verified yet
Accuracy	Number of detectable errors in names, footprints, and feature types	Calculated in percentage (%) by comparing detectable errors toponymic features to total data collection.
Availability	The degree to which the toponymic files and gazetteers are freely available and not limited by restrictive conditions of use (data access)	Information on details of data accessibility (e.g., types of data license) Degree of availability: 0: Restricted or limited access 1: Free 2: downloadable 3: Online access
Actuality	Date of data specific survey; the degree to which the toponymic files and gazetteer have incorporated changes	Date of survey (data collection timestamp) and date of changes (last update information)
Completeness	Steered by toponymic surveying rules; the degree to which the scope of toponymic files and gazetteers are covered completely	Cannot be calculated (?), calculated based on administrative area coverage (✓) or other boundaries (such as research boundaries) – calculated in percentage (%)
Scope /Coverage	Small communal database, regional/national coverage, or worldwide coverage	Worldwide, national (country name), regional (province name), local (district, sub-district), small communal (village or neighborhood), specific area (study/research area)
Consistency	Data consistency and coherence	Calculated in percentage (%) by comparing inconsistent data to total data collection. 1: Consistent - If all the database value is registered following the structure of the toponym files and toponymic writing rules. 0: Inconsistent – If there is any information not following the rules.
Richness of annotation	Amount and detail of descriptive information, beyond the basics of name, footprint, and feature type	Qualitative approach based on toponymic content richness. Complete information may include pictures and voice recording (High), place names and feature types or footprint information (Medium), limited to place names information only (Low).
Granularity	Whether data includes large, well-known features only or features of all sizes and those that are less well known	1: Very coarse—only specific large toponym with minimum information distributed 2: Coarse—large features only 3: Medium – well-known features only or feature of all sizes 4: Fine—covered all types of features and those that are less well known
Balance	Uniform degree of detail, currency, accuracy, and granularity across scope of coverage	0: Cannot be calculated (?) 1: Balance in across coverage area (Uniform)

Source: Modified from [34–36] and measurement unit harvested based on FGDs (FGD 5 to FGD 7).

References

1. Kerfoot, H.; Närhi, E. *Manual for the National Standardization of Geographical Names*; United Nations Publication: New York, NY, USA, 2006; ISBN 92-1-161490-2.
2. Zaccheddu, P. The UNGEGN Advanced Toponymy Manual. In Proceedings of the 11th United Nations Conference on the Standardization of Geographical Names (UNCSGN), New York, NY, USA, 8–17 August 2017.
3. Hogerwerf, J. Toponymic data and map production in the Netherlands: From field work to crowd sourcing. In Proceedings of the 11th United Nations Conference on the Standardization of Geographical Names (UNCSGN), New York, NY, USA, 8–17 August 2017.
4. Touya, G.; Antoniou, V.; Olteanu-Raimond, A.-M.; Van Damme, M.-D. Assessing Crowdsourced POI Quality: Combining Methods Based on Reference Data, History, and Spatial Relations. *ISPRS Int. J. Geo-Inf.* **2017**, *6*, 80. [CrossRef]
5. Zhu, R.; Hu, Y.; Janowicz, K.; McKenzie, G. Spatial signatures for geographic feature types: Examining gazetteer ontologies using spatial statistics. *Trans. GIS* **2016**, *20*, 333–355. [CrossRef]
6. Laurini, R. Gazetteers and Multilingualism. In *Geographic Knowledge Infrastructure*; Elsevier: New York, NY, USA, 2017; pp. 157–182. ISBN 9781785482434.
7. Howe, J. The Rise of Crowdsourcing. *Wired Mag.* **2006**, *14*, 1–4.
8. Ahmouda, A.; Hochmair, H.H. Using Volunteered Geographic Information to measure name changes of artificial geographical features as a result of political changes: A Libya case study. *GeoJournal* **2018**, *83*, 237–255. [CrossRef]
9. Rampl, G. Crowdsourcing and GIS-based Methods in a Field Name survey in Tyrol (Austria). In Proceedings of the Twenty-Eight Session on Geographical Names, New York, NY, USA, 28 April–2 May 2014.
10. Gercsák, G.; Mikesy, G. Does Google serve as a model for using place names? *Acta Geogr. Slov.* **2017**, *57*, 153–159. [CrossRef]
11. Moeller, M.S.; Furhmann, S. Mapping the World—A New Approach for Volunteered Geographic Information in the Cloud. *ISPRS Int. Arch. Photogramm. Remote Sens. Spat. Inf. Sci.* **2015**, *XL-6/W1*, 9–10. [CrossRef]
12. Bonney, R.; Cooper, C.B.; Dickinson, J.; Kelling, S.; Phillips, T.; Rosenberg, K.V.; Shirk, J. Citizen science: A developing tool for expanding science knowledge and scientific literacy. *BioScience* **2009**, *59*, 977–984. [CrossRef]
13. Serrano Sanz, F.; Holocher-Ertl, T.; Kieslinger, B.; Sanz Garcıa, F.; Silva, C. *White Paper on Citizen Science for Europe*; Socientize Consortium: European Commission: Brussels, Belgium, 2014.
14. Haklay, M. Citizen Science and Volunteered Geographic Information—Overview and typology of particpation. In *Crowdsourcing Geographic Knowledge: Volunteered Geographic Information (VGI) in Theory and Practice*; Springer: Dordrecht, The Netherlands, 2013; pp. 105–122. ISBN 978-94-007-4586-5.
15. UNGEGN. UNGEGN—Toponymic Guidelines for Map and Other Editors. Available online: https://unstats.un.org/UNSD/geoinfo/UNGEGN/toponymic.html (accessed on 17 May 2018).
16. Castellote, J.; Huerta Guijarro, J.; Pescador, J.; Brown, M. Towns Conquer: A Gamified application to collect geographical names (vernacular names/toponyms). In Proceedings of the The 16th AGILE International Conference on Geographic Information Science, Leuven, Belgium, 14–17 May 2013.
17. Southall, H.; Aucott, P.; Fleet, C.; Pert, T.; Stoner, M. GB1900: Engaging the Public in Very Large Scale Gazetteer Construction from the Ordnance Survey "County Series" 1:10,560 Mapping of Great Britain. *J. Map Geogr. Libr.* **2017**, *13*, 7–28. [CrossRef]
18. Jones, C.B.; Purves, R.S.; Clough, P.D.; Joho, H. Modelling Vague Regions with Knowledge From the Web. *Int. J. Geogr. Inf. Sci.* **2008**, *22*, 1045–1065. [CrossRef]
19. Vasardani, M.; Winter, S.; Richter, K.F. Locating place names from place descriptions. *Int. J. Geogr. Inf. Sci.* **2013**, *27*, 2509–2532. [CrossRef]
20. Camponovo, M.E.; Freundschuh, S.M. Assessing uncertainty in VGI for emergency response. *Cartogr. Geogr. Inf. Sci.* **2014**, *41*, 440–455. [CrossRef]
21. Kim, J.; Vasardani, M.; Winter, S. Similarity matching for integrating spatial information extracted from textual descriptions. *Int. J. Geogr. Inf. Sci.* **2017**, *31*, 56–80. [CrossRef]

22. Ceccaroni, L.; Schade, S.; Bastin, L.; Tsinaraki, C.; Lemmens, R.; Falquet, G.; Klan, F.; Piera, J.; Trojan, J.; Lendak, I.; et al. *WG5-Deliverable 1: Citizen-Science Ontology*; European Cooperation in Science Technology: Brussels, Belgium, 2018.
23. McCartney, E.A.; Craun, K.J.; Korris, E.; Brostuen, D.A.; Moore, L.R. Crowdsourcing the National Map. *Cartogr. Geogr. Inf. Sci.* **2015**, *42*, 54–57. [CrossRef]
24. Rönneberg, M.; Laakso, M.; Sarjakoski, T. Map Gretel: Social map service supporting a national mapping gency in data collection. *J. Geogr. Syst.* **2019**, *21*, 43–59. [CrossRef]
25. Zimmermann, A.; Oshri, I.; Lioliou, E.; Gerbasi, A. Sourcing in or out: Implications for social capital and knowledge sharing. *J. Strateg. Inf. Syst.* **2018**, *27*, 82–100. [CrossRef]
26. Alem, L.; Mclean, A.; Vercoustre, A. Knowledge sharing technologies to support community participation in natural resource management: A Research Agenda. In Proceedings of the Australian Conference for Knowledge Management & Intelligent Decision Support, Melbourne, Australia, 3 December 2003.
27. Teodoro, M.F.; Correia, A.; Nunes, P. Knowledge management in geospatial information context. A preliminary statistical approach—A case study. *WSEAS Trans. Bus. Econ.* **2017**, *14*, 74–80, ISSN 11099526.
28. Haradhan, M. Sharing of Tacit Knowledge in Organizations: A Review. *Am. J. Comput. Sci. Eng.* **2016**, *3*, 6–19.
29. Ngah, R.; Ibrahim, A.R. Tacit knowledge sharing and organizational performance: Malaysian SMEs perspective. **2007**, 276–281. [CrossRef]
30. Gray, S.; Mellor, D.; Jordan, R.; Crall, A.; Newman, G. Modeling with citizen scientists: Using community-based modeling tools to develop citizen science projects. In Proceedings of the 7th International Congress on Environmental Modelling and Software (iEMSs), San Diego, CA, USA, 15–19 June 2014.
31. Gray, S.A.; Gray, S.; de Kok, J.L.; Helfgott, A.E.R.; O'Dwyer, B.; Jordan, R.; Nyaki, A. Using fuzzy cognitive mapping as a participatory approach to analyze change, preferred states, and perceived resilience of social-ecological systems. *Ecol. Soc.* **2015**, *20*, 11. [CrossRef]
32. Perdana, A.; Ostermann, F. A Citizen Science Approach for Collecting Toponyms. *ISPRS Int. J. Geo-Inf.* **2018**, *7*, 222. [CrossRef]
33. Bowen, J.R. On the Political Construction of Tradition: Gotong Royong in Indonesia. *J. Asian Stud.* **1986**, *45*, 545-561. [CrossRef]
34. Hill, L.L. *Georeferencing: The Geographic Associations of Information*; MIT Press: Cambridge, MA, USA, 2006; ISBN 9780262083546.
35. Crowdsourcing in National Mapping 2017—An International Workshop. Available online: http://www.cs.nuim.ie/~{}pmooney/eurosdr2017/ (accessed on 17 May 2018).
36. Badan Informasi Geospasial. *Peraturan Kepala Badan Informasi Geospasial: No 15 Tahun 2014 Tentang Pedoman Teknis Ketelitian Peta Dasar*; Badan Informasi Geospasial: Cibinong, Indonesia, 2014. (In Bahasa Indonesia)
37. Acheson, E.; De Sabbata, S.; Purves, R.S. A quantitative analysis of global gazetteers: Patterns of coverage for common feature types. *Comput. Environ. Urban Syst.* **2017**, *64*, 309–320. [CrossRef]

International Journal of *Geo-Information*

MDPI

Article

Open Community-Based Crowdsourcing Geoportal for Earth Observation Products: A Model Design and Prototype Implementation

Mohammad H. Vahidnia [1,*] and Hossein Vahidi [2]

[1] Department of Remote Sensing and GIS, Faculty of Natural Resources and Environment, Science and Research Branch, Islamic Azad University, Tehran 14778-93855, Iran
[2] Graduate School of Media and Governance, Keio University, Fujisawa, Kanagawa 252-0882, Japan; vahidi@sfc.keio.ac.jp
* Correspondence: mhvahidnia@srbiau.ac.ir; Tel.: +98-21-4486-5154

Abstract: Over the past few decades, geoportals have been considered as the key technological solutions for easy access to Earth observation (EO) products, and the implementation of spatial data infrastructure (SDI). However, less attention has been paid to developing an efficient model for crowdsourcing EO products through geoportals. To this end, a new model called the "Open Community-Based Crowdsourcing Geoportal for Earth Observation Products" (OCCGEOP) was proposed in this study. The model was developed based on the concepts of volunteered geographic information (VGI) and community-based geoportals using the latest open technological solutions. The key contribution lies in the conceptualization of the frameworks for automated publishing of standard map services such as the Web Map Service (WMS) and the Web Coverage Service (WCS) from heterogeneous EO products prepared by volunteers as well as the communication portion to request voluntary publication of the map services and giving feedback for quality assessment and assurance. To evaluate the feasibility and performance of the proposed model, a prototype implementation was carried out by conducting a pilot study in Iran. The results showed that the OCCGEOP is compatible with the priorities of the new generations of geoportals, having some unique features and promising performance.

Keywords: community-based geoportal; citizen science; crowdsourced earth observation product; volunteered geographic information (VGI); remote sensing; spatial data infrastructure (SDI)

Citation: Vahidnia, M.H.; Vahidi, H. Open Community-Based Crowdsourcing Geoportal for Earth Observation Products: A Model Design and Prototype Implementation. *ISPRS Int. J. Geo-Inf.* **2021**, *10*, 24. https://doi.org/10.3390/ijgi10010024

Received: 27 November 2020
Accepted: 10 January 2021
Published: 12 January 2021

Publisher's Note: MDPI stays neutral with regard to jurisdictional claims in published maps and institutional affiliations.

1. Introduction

Efficient management, use, and sharing of geographic information is integral to the achievement of good governance and sustainable development objectives and brings significant economic, social, and environmental benefits to the countries. Over the past few decades, the concept of geoportals has emerged as one of the key technological solutions for improving the efficiency and effectiveness of geospatial activities. The geoportal allows the data consumers to access, search and discover geospatial data and enables data producers to publish and share geospatial data. Furthermore, this online infrastructure may provide other geographic information services such as data visualization, editing, and analysis to its various stakeholders [1,2]. The geospatial data are distributed and made available by the different data producers using a variety of technologies and formats. In this term, the geoportal provides effective solutions to the geospatial data interoperability; it facilitates the multi-source data integration and enables the stakeholders to access the geospatial information and maps in the standard formats [3]. The geoportals connect the geospatial data producers and consumers directly and improve collaboration and cooperation among the various stakeholders, leverage existing geospatial resources, and ease the finding of relevant geospatial products; hence, it plays a key role in preventing duplicated efforts, inconsistencies, delays, and wasted time and resources [4]. The geoportal is one of the key components that are needed for establishing spatial data infrastructure (SDI) [5]. It is considered as the most visible part of SDI and the entry point to it [6,7]. Due to the key functions and unique characteristics, and high demand for Earth observation (EO) products (raw and processed imagery), along with the general-purpose geoportals, the specialized geoportals have been designed and implemented exclusively for EO products [8–10].

Crowdsourcing [11] is an increasingly common means of obtaining of geospatial data in recent years as it can provide large volumes of low-cost, and up-to-date open geospatial data over large geographical extents in a short period [12–14]. This concept has been successfully used over the last 10 years in the new generation of online spatio-temporal mapping and monitoring projects in the various areas such as land use/land cover mapping projects (e.g., OpenStreetMap, Mapillary and Geo-Wiki), biodiversity mapping and monitoring projects (e.g., iNaturalist and eBird), and damage, hazard mapping and monitoring projects (e.g., Humanitarian OpenStreetMap and Did You Feel It?), and pollution mapping and monitoring projects (e.g., NoiseTube and Safecast) [15–18].

Most of the previous studies that aimed to blend the concept of crowdsourcing with airborne, and space-borne remote sensing (hereafter remote sensing) have mainly been focused on (1) providing crowdsourced ground truth samples to be used in training and validation steps of image classification, (2) using the crowdsourcing technique for geotagging and visual interpretation of remotely-sensed imagery, and (3) exploiting volunteers' power for manual modification and enhancement of the formal classification results [12,17,19,20]. A relatively small body of literature (e.g., [21–28]) has discussed the potential, application, and different dimensions of using the crowdsourcing technique for producing EO products.

By the rise of the citizen science [29] paradigm in the domain of remote sensing, the rapid increase in the availability of low-cost sensors and remote sensing platforms, and the growth of open data, free and open-source software solutions, and free and open training courses in recent years, a considerable volume of crowdsourced EO products have been produced by volunteers over recent years that have traditionally been produced by professionals. In this context, the deployment of inexpensive platforms, including unmanned aerial vehicles (UAV), balloons, and kites equipped with low-cost sensors for voluntary acquisition of EO data, has been increasingly prevalent [21]. Moreover, following a growth in the number of do-it-yourself (DIY) small satellite (e.g., DIY picosatellite) missions [30], the new citizen science applications for these relatively low-cost platforms, such as voluntary remote sensing, are gradually emerging. Similarly, over the past years, the light aircrafts equipped with cameras have been used as volunteer pilots for the voluntary acquisition of EO data [23]. The crowdsourced raw EO products that are collected through the voluntary remote sensing projects are partially openly shared through the few existing online platforms designed for hosting openly licensed remotely-sensed imagery (e.g., OpenAerialMap) [21,31]. Alongside the rise in production of the raw EO products, the increasing availability of open remotely-sensed data produced by volunteers and as well as professional EO data producers [32], free and open-source geospatial software, and free and open geospatial education, and the growth in the number and processing power of personal computing devices over the past few years have facilitated image processing tasks for the volunteers and have enabled them to produce various voluntary processed EO products according to their levels of expertise.

Some previous contributions have studied the different dimensions of the integration of volunteered geographical information (VGI) [33] or Web 2.0 [34] paradigm in SDI and geoportals to study the various advantages and features of them (e.g., [7,35–41]). However, until now, less attention has been paid to the development of geoportal models for hosting VGI—particularly the crowdsourced EO products. In this context, to the best of our knowledge, so far, no model has been proposed in the existing literature (especially those specifically developed for serving EO products) to provide the technological solutions for (1) supporting volunteer EO product providers to provide map services in accordance with the SDI interoperability standards, and (2) facilitating the communication between geoportal users (and facilitating the ordering of voluntary EO products, and control of their quality) simultaneously.

In this research, a schema for geoportals named "Open Community-Based Crowdsourcing Geoportal for Earth Observation Products" (OCCGEOP) was introduced. Furthermore, a prototype implementation of the proposed model was developed for crowdsourcing EO products, and then to test the prototype system, a pilot study was conducted in Iran. The proposed model was designed in compliance with open-source solutions and Open Geospatial Consortium (OGC) standard services. In the proposed model for our geoportal, the crowdsourcing concept plays a major role, meaning that the volunteers may share their EO products with others via standard structures and formats. The proposed model exploits the civic participation and integrates social community capabilities and local knowledge of volunteers into geoportal architecture to facilitate the user-to-user communication and directs and coordinates the production, sharing, and accessing of voluntary EO data in the geoportal. In this context, the main contributions of this study are (1) the conceptualization of an open SDI geoportal for voluntary earth observation products in a community-based setting, and (2) designing a model for the proposed concept and implementing a prototype for the developed model for the first time.

The remainder of this paper is organized as follows: Section 2 presents the related works to this research. Section 3 describes the features and properties of OCCGEOP. Section 4 proposes the architecture for OCCGEOP and presents the prototype implementation of OCCGEOP. Section 5 provides some results from the implemented OCCGEOP prototype system. Section 6 discusses the advantages, features and capabilities of OCCGEOP and evaluates the opinions and preferences of OCCGEOP's expert and practitioner users about them. Finally, the last section is reserved for the conclusion and provides some recommendations for future work.

2. Related Works

The first contributions on geoportals and explanations of its principles were carried out through the development of the United States national spatial data infrastructure (NSDI) in 1994 [42]. The development of the earliest major geoportal, the NSDI clearinghouse network, was organized by the United States Federal Geographic Data Committee (FGDC) [43]. NSDI clearinghouse network is now a distributed system of Internet-based agency servers containing field-level metadata of digital spatial data and searchable catalogs as well as available applications, and services. In 2003, the Geospatial One-Stop (GOS) geoportal was developed as part of the United States e-Government initiative [1]. GOS aimed to promote geospatial data collection and maintenance coordination and alignment across all levels of government [44]. One of the advantages of GOS over the NSDI clearinghouse network was that a web-based geoportal interface in GOS made it possible for users to be connected to data providers [45]. The GOS user may communicate with the system via a simple web browser (thin client) or a geographic information system (GIS) application (thick client). One of the most efficient examples of geoportals that extended the feature of sharing geographic information based on region or theme is the Infrastructure for Spatial Information in the European Community (INSPIRE) geoportal. INSPIRE was developed in 2007 to facilitate spatial or geographical information accessibility and interoperability for a wide range of sustainable development purposes in Europe [46]. Currently, many countries have taken fundamental steps in the development of geoportals at the national level [2]. Modern web-based geoportals such as NASA's Earth Observing System Data and Information System (EOSDIS) include direct access to raw data in different formats from various resources, such as satellites, aircrafts, field measurements, full metadata, and visual tools to interact with data on online maps [47]. In addition, the geoportals have been designed to be used in many other fields and applications such as agriculture, disaster management and early warning, land and water management, urban planning, air quality, and energy [2,48–54].

The main direction of studies on modern geoportals is now to provide effective ways to handle big data, develop web services shared with different parties, and application programming interfaces (APIs) for developers and the end-users [55,56]. De Longueville [7] discussed the possible strategies for the development of the new generation of geoportals including the facilitation of the user-to-user communication (for sharing of users' common interests and needs) and dataset and map sharing based on users requests and establishing a ranking mechanism to create "the most popular data" listings for geoportals.

The coordinated series of agreements on technology standards, institutional arrangements, and policies within an SDI provide an interactive connection of geospatial data, metadata, users, and resources, which can appear in a geoportal [57]. In this sense, the main advantage of this infrastructure is the sharing of spatial data produced by various public and private organizations in accordance with defined standards [58]. Currently, OGC and the International Organization for Standardization (ISO) have played a key role in standardizing web-based geospatial data and services to make them interoperable [59]. OGC provides the best open solutions and standards for achieving the geospatial data interoperability by providing a comprehensive framework of services and models [60]. Some OGC standard services such as Web Coverage Service (WCS), the Web Map Service (WMS), the Web Feature Service (WFS), and the Catalog Service for Web (CSW) have been frequently used in the design of geoportal architectures [61,62]. Service metadata can also be published based on standards such as ISO19115 and ISO19139 [63].

Among recent studies on geoportals, Granell et al. [64] presented a conceptual architecture and service-oriented implementation of a regional geoportal. Using their developed geoportal, they specifically focused on unified monitoring of rice crop at a regional scale. Iosifescu-Enescu et al. [65] proposed a cloud-based architecture for a Swiss academic geoportal so-called Geodata Versatile Information Transfer environment (GeoVITe). They discussed that the cloudification mechanism has a major impact on making the geoportals scalable on-demand. Furthermore, they discussed that the use of public clouds reduces the upfront costs of conventional computing infrastructures. Dareshiri et al. [66] have developed a recommender geoportal to enhance the functionalities of traditional geoportals. The proposed framework is able to evaluate the behaviors of users and suggest geospatial resources to the geoportal users according to their desires and preferences. Kadochnikov et al. [67] developed a real-time geoportal for air pollution and meteorological data monitoring. To create this system, they adopted mechanisms to provide real-time geospatial data as OGC web map service standards.

3. Features and Properties of OCCGEOP Model

The general workflow of the proposed OCCGEOP model for coordination, sharing, publishing, standardization, searching and discovery, and accessing of the voluntary EO products as well as facilitating users' communication, giving feedback, and rating published products, and management and maintenance of the proposed system have been depicted in Figure 1. In the proposed system, the volunteers (whose skills and competence were approved by the administrators of the system) are able to share their original EO products on the geoportal. All the users (data consumers) are able to use the system to search for volunteers as well as to search and discover map services using the generated data catalogs (published in standard CSW form in the geoportal) for the crowdsourced map services and access the generated standard crowdsourced map services (published in standard WMS and WCS forms in the geoportal). If an end-user needs a map service that has not been shared and published in the system, he/she can send his/her request to the volunteers for the production and sharing of his/her requested EO product. Furthermore, a user is able to give feedback on the contributions of volunteers and ratings of their shared EO products (more details about the workflow of the proposed model and its components will be provided in the following sections).

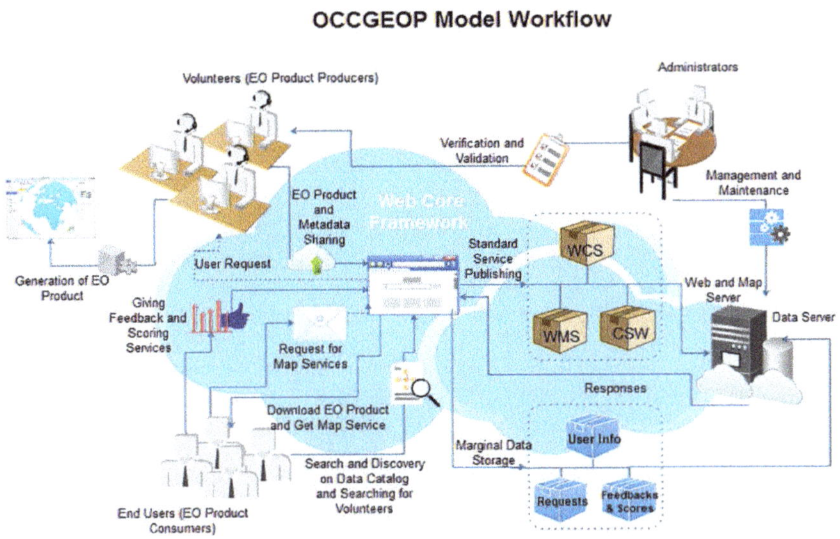

Figure 1. General workflow of the "Open Community-Based Crowdsourcing Geoportal for Earth Observation Products" (OCCGEOP) geoportal model.

The main features that have been taken into account in our proposal for the OCCGEOP model could be categorized into seven major areas and research lines (Figure 2). Some of these features are also available in some of the existing contemporary geoportals and do not show much difference at least in the concept; however, some features of OCCGEOP are novel and were designed consistent with the features and goals of new generations of geoportals. In Sections 3.1–3.7, the main features and properties of OCCGEOP will be introduced and discussed briefly.

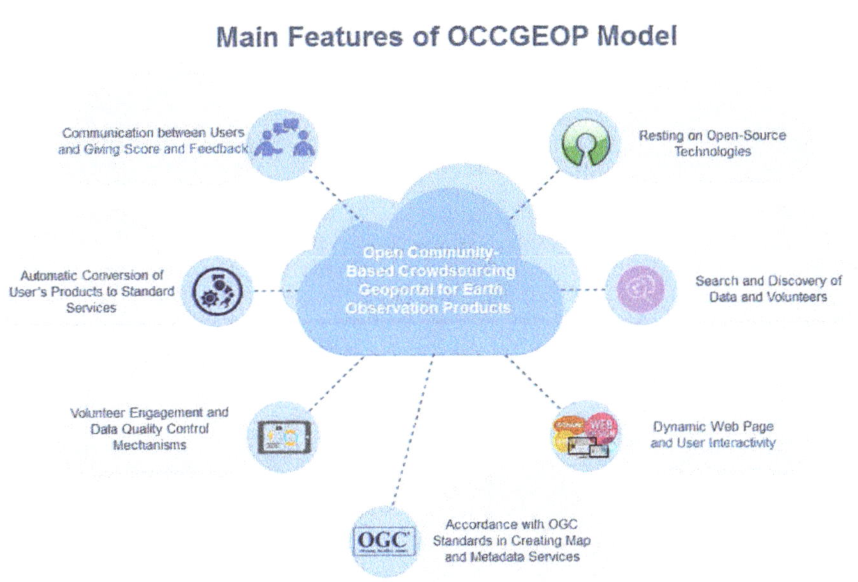

Figure 2. The main features of the OCCGEOP geoportal model.

3.1. Accordance with OGC Standards

OGC standards are developed to render discoverable, accessible, interoperable, and reusable location information and services. The OCCGEOP, as one of its goals, has considered in its plan homogenizing the heterogeneous crowdsourced EO products (in formats and themes) via interoperable and reusable standard OGC map services such as WMS and WCS as well as being discoverable through standard metadata services. This is consistent with the priorities of many modern geoportals where they concentrate on interoperability by implementing standards for the exploration and use of geographic data and services. As OCCGEOP is developed in accordance with OGC standards, GIS specialists and software developers can easily use the standard EO data of OCCGEOP with other open, interoperable, and reusable geospatial resources and integrate it in other standard interfaces and web-based GIS platforms.

3.2. Data Quality Control and Volunteer Engagement Mechanisms

The core of the OCCGEOP model is the crowdsourcing concept. Although VGI can potentially be used in different scientific research and practical projects, concerns over the quality of the crowdsourced data may remain as a barrier to its adoption by the data consumer [17,68,69]. Therefore, the assessment and assurance of VGI quality may reduce the concerns of the consumers of such data. The OCCGEOP users who do not want to share data in the system do not need to be authenticated. However, to reduce the aforementioned concerns, only the verified users can serve as the providers of crowdsourced EO products in OCCGEOP. In this term, upon the initial registration of a user who requested to take the data provider role in OCCGEOP, the administrators of OCCGEOP conduct a basic screening of the qualifications and experience of the user based on the information provided by the user in the online registration form. Then, if the qualifications and experience of the user meet the minimum requirements defined for data providers, the role of the user is promoted to the data provider role and a permit is granted to the user to access the tools for the generation of the new map service. In this basic approach for reducing the chance of sharing poor quality user-generated content [70] over the geoportal, it is assumed that a user with higher levels of self-declared skill and expertise in a particular area generally can produce a higher quality data in that area compared to the users with lower levels of skill and expertise [68,71]. This basic quality assurance approach has been used successfully in some other projects that deal with user-contributed information. The ratings and comments of other users on a crowdsourced geospatial product can serve as a proxy indicator for the quality of the product [68,72]. In this sense, the OCCGEOP model uses a star ranking mechanism for ranking a shared EO product in addition to the comments feature that enables the users to post their comments on a product. The indicators of data producers' provenance and reputation have been used in previous studies and projects to quantify the quality of the data [68,73]. The OCCGEOP model also uses a mechanism for ranking a provider of crowdsourced EO products using average star ratings (the feedbacks given by other users) and contribution history for his/her previous shared products. The computed score for a provider of a crowdsourced EO product through this mechanism can serve as an indicator for the trustworthiness level of the data provider and a proxy indicator for the quality of his/her shared products over the geoportal—including the products that have not been rated by other users yet.

Previous studies have emphasized the positive impact of recognition or reward (e.g., adding a score, token, or badge to users' online profiles based on the quality or quantity of their previous contributions) on sustaining the engagement of the volunteers in the participatory and citizen science projects [74]. In this context, the estimated score for the data providers in OCCGEOP can help to retain the engagement of the volunteers in the geoportal and enhance the popularity of the application of it.

3.3. Automatic Conversion of User's Products to Standard Services

Another goal of OCCGEOP is to provide embedded frameworks for automatically transforming heterogeneous user-generated EO products into standard services such as WMS and WCS. In most of the existing geoportals, there is no room for volunteers to present their geospatial products in the form of standard map services. Such activities are typically carried out by professional service providers and experienced mediators in geoportals. However, in the OCCGEOP design, the volunteers are able to share standard map services without engaging in a complicated process of publishing the services. In the OCCGEOP, the users can upload regular data formats such as GeoTIFFs or Shapefiles and use the automated mechanisms to transfer them into the standard map services on the GIS server and share them with other users. Such a functionality can lead to the realization of a crowdsourced geoportal.

3.4. Communication between Users

The organization of user communities is in line with the vision of the next generation of geoportals. In OCCGEOP, the users are enabled to request their desired product by exchanging messages within the system. The volunteer who receives the message can create the map service based on their expertise and then publish it. In OCCGEOP, a user's profile and related descriptions about specialties and capabilities of him/her, as well as the previously produced map services in the system, can be seen by other users. A user can interact with other users and their actions through giving feedback (comments) on the contributions of other users and rating of their products. Using the query features in OCCGEOP, individuals can also be aware of their community's geographic area of interest, subject and type of EO products, and the situation of constantly growing user-generated content.

3.5. Dynamic Web Page

A dynamic web page can display different content each time it is viewed in response to different contexts or conditions [75]. Using state-of-the-art technologies, dynamic and interactive web pages make a request to the server, interpret the data, and refresh the current screen in such a way that the user never knows that something had ever been sent to the server. As with many other geoportals, the interactive map component as well as the communication components for exchanging messages or scoring web services have been designated in OCCGEOP based on dynamic web page technologies. As the important Web 2.0 technologies, Asynchronous JavaScript and XML (AJAX) programming use JavaScript and the Document Object Model (DOM) to update selected regions of the page area without undergoing a full page reload. Using this method in OCCGEOP will result in a faster, more interactive, and more communicative geoportal.

3.6. Search and Discovery of Data and Volunteers

A common feature in all geoportals, as in OCCGEOP, is the search and discovery of geospatial information based on metadata such as products' bounding box, time limits, and other descriptions such as accuracy, spatial resolution of the products. In the OCCGEOP model, upon conducting a search and discovery, the service details (e.g., EO product thumbnail, an overview of the map, download link, and most importantly, the standard map service parameters) are provided for the user. Using the so-called standard map service parameters such as hostname, type of service, category name, and service name, etc., an EO product is easily reusable in another web-based GIS as an online geospatial layer. It is worth noting that the search and exploration capability in OCCGEOP is not limited to geospatial data but also applies to volunteers, i.e., finding their profile and related products.

3.7. Resting on Open-Source Technologies

Full reliance on open-source modules and components either for dealing with geospatial data or for other parts of the client and server is one of the most important points in OCCGEOP. This not only minimizes the expenses of the initial implementation of OCCGEOP but also makes the modification of the source code and software development easier. For instance, as will be explained in the next section, OCCGEOP will use GeoServer technology as a GIS engine in the background to publish standard WMS and WCS. Using such a strategy, no one has to worry about purchasing multiple licenses for internal components of OCCGEOP and installing these components several times.

4. Proposed Architecture and Prototype Implementation of OCCGEOP

4.1. Design and Technologies

Figure 3 presents the three-tier system architecture and adopted technologies for OCCGEOP. The system architecture is designed and implemented based on the state-of-the-art open-source technologies and components, the main components of which are discussed in the following. The open-source software technologies include Bootstrap (a free and open-source CSS framework for responsive web front-end development), JQuery (a JavaScript library designed to simplify traversal and manipulation of the HTML Document Object Model (DOM) tree), OpenLayers (a JavaScript open-source library for displaying map data in web browsers), Django (a Python-based free and open-source web framework that follows the model–template–views architectural pattern), GeoServer (a Java-written open-source server enabling users to publish, process and modify geospatial data), and PostgreSQL (a free and open-source relational database management system).

The presentation tier offers interfaces for the front-end framework from which user interactions such as communication with others, content generation, retrieving, and visualization of data are handled. Web pages, menus, icons, and widgets were designed using HTML, CSS, and Bootstrap libraries. To provide dynamic capabilities such as collecting the comments and scores about published geospatial services, participating in polls, filtering EO products and volunteers, the various Web 2.0 technologies such as AJAX, JavaScript, and JQuery as well as interfaces for using and presenting online map services such as OpenLayers API were used in the presentation tier.

The logic tier contains the server-side web core framework and application server along with some specific applications, including the request of EO products, the discovery of data and volunteers, scores and feedback, and adding map service and metadata. This layer offers a business logic for service and data connectivity and allows communication between end-users and remote data and services. The core logic tier of the proposed architecture is based on one of the most efficient web frameworks, Django. Django is a framework for developing high-level web applications in Python. Therefore, the main server-side programming language in this architecture is Python. Django follows the structure of model–view–controller pattern (MVC). In an MVC model, the code for working with the database (i.e., model) and the controller or business logic, which are the main modules of the system in Python, and the parts related to the rendering responses to the user interface (i.e., view) are separated. For example, the visual representation and template of the system do not contain any information such as the database and data storage parameters, the layer corresponding to respond to user requests, and the caching information for later use. Each information is related to a separate section and, if necessary, each section can exchange information or send a request to other sections. The appearance (i.e., HTML tags) or site template is stored in separate files. The control section is also created and stored as Python modules. In this case, the programmer will deal with the control modules and the designer with the HTML tags. If the purpose is to use a database, there is no need to write SQL statements, but this can be addressed through the internal mechanisms of Django with Python statements that enable the retrieval of the data, and deleting, updating and inserting a new record.

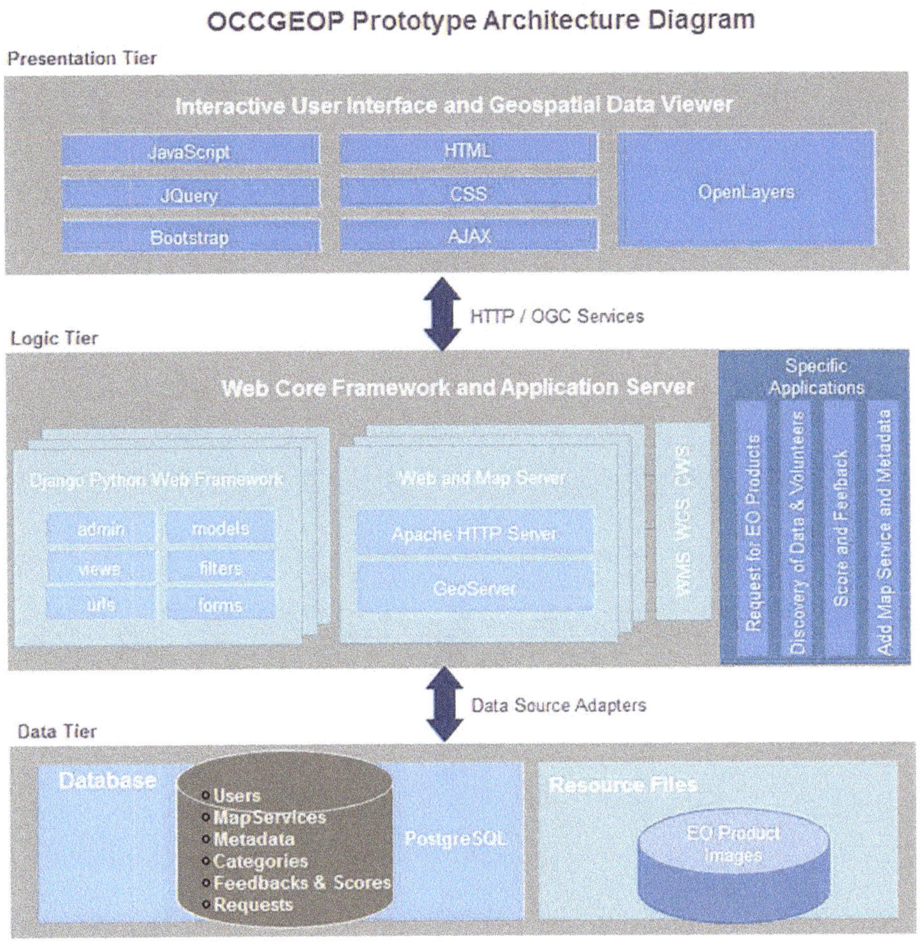

Figure 3. System architecture of OCCGEOP and the configuration of open-source technologies.

The most important modules as business logic can be divided into main items, including the admin module for the accessibility of admin pages, models for database design, filters for creating filters in user queries, forms for developing web forms for cases such as creating a new map service, URLs to structure links in the application, and views to process user requests and display responses on the web. In the logic tier, the get and post requests from clients are responded to via the open-source Apache HTTP server. In addition, GeoServer, which is an open-source GIS server technology for sharing and publishing map services and can publish data from any major spatial data source using OGC standards, was adopted in OCCGEOP. The logic tier enables the users to upload EO products and automatically transfer them to OGC standards such as WMS and WCS, and eventually share it with others. The OGC Web Map Tile Service (WMTS) [59] is also provided because the caching function is already enabled by default in GeoServer. Therefore, to increase the performance, at the time of displaying maps on OpenLayers, the tile-based map presentation is called.

Volunteers can log into the system and have access to map service generation tools after registering in the system and being verified by the admin. Any published EO product by volunteers can then be discovered and accessed by the search subsystem. To upload the

EO products by a volunteer, he/she is required to provide additional metadata information about the product such as additional description, time of acquisition of the base image, accuracy, sensor type, and geographical bounding rectangle of the product. The system supports popular geospatial data formats such as GeoTIFF, ArcGrid, and Shapefile for the input data. The system administrator can grant users access to upload geospatial data as well as delete or modify metadata. The administrator can also create a new category for EO products within the system.

The data tier focuses on databases, including user and volunteer data, map services, registered metadata, categories of EO products, reviews and ratings from users, and request messages. It supplies the logic layer and specific applications with data as well as information on data sources. The data tier also stores the source image files of volunteers' provided EO products. Django's default database, SQLite, was used in the programming phase of the data tier for this purpose, which will be replaced with PostgreSQL in the production phase. The proposed data model is linked to Django modules such as pycsw and GeoServer, where pycsw is an open-source server-side implementation of the CSW metadata standard (catalog service) written in Python. By using this technology, spatial metadata standards such as ISO 19115 and FGDC were provided in the proposed model.

4.2. Database Model

A data model developed for OCCGEOP has been demonstrated in Figure 4. The classes and details of this data model can be expanded as needed. The main classes in this data model are category, map service, user, user profile, request, and feedback.

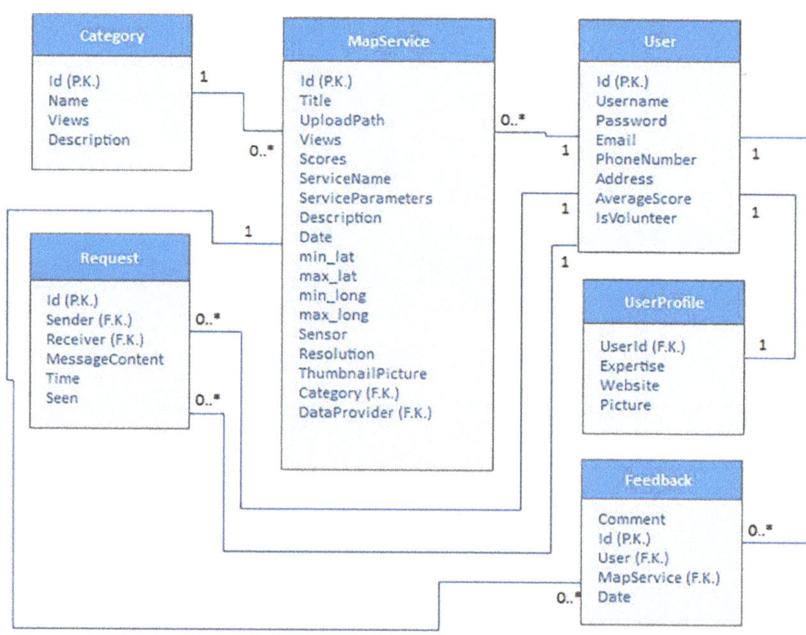

Figure 4. A data model for the implementation of OCCGEOP.

Category class contains the properties of a group of EO products such as a unique identifier, name, number of views, and description of that category. A category can include multiple generated map services. A map service stores the various metadata elements in the database, such as service name, description, the acquisition time of the base image, bounding rectangle (minimum latitude, maximum latitude, minimum longitude, and maximum longitude), type of satellite and sensor, and spatial resolution of the layer.

Metadata can also include other items such as the spatial reference system, the amount of cloud cover, and the accuracy measures for the data. In addition, information such as a unique identifier, number of views, average score, service parameters, and thumbnail picture for the service is provided with the EO product to the end-users. Each map service is essentially a subset of a category and is generated by one of the system's volunteer data providers, and the relations between a map service and a category or map service and a user have been established with the aid of foreign keys to keep the database integrated. Basic user information contains a unique identifier, username, password, email address, phone number, and postal address. As mentioned previously, a data provider is ranked based on the average scores of his/her published EO products in the OCCGEOP model. Other details, including user expertise, profile image, and personal website, are listed in the user profile as assets in a one-to-one relationship with the user class. A user may send a request to another volunteer regarding the production and sharing of a required EO product. In this sense, the records about the request's sender and recipient, as well as message content and associated time, are created based on the model's request class. Finally, the feedback class is in charge of establishing the link between a user and a map service to record and reflect relevant feedback and comments.

4.3. Use Cases and Activities

A representation of the main types of users and their interaction with the OCCGEOP and the different use cases in which the users are involved has been illustrated in Figure 5. The principal types of users in OCCGEOP include registered users, anonymous users, and administrators.

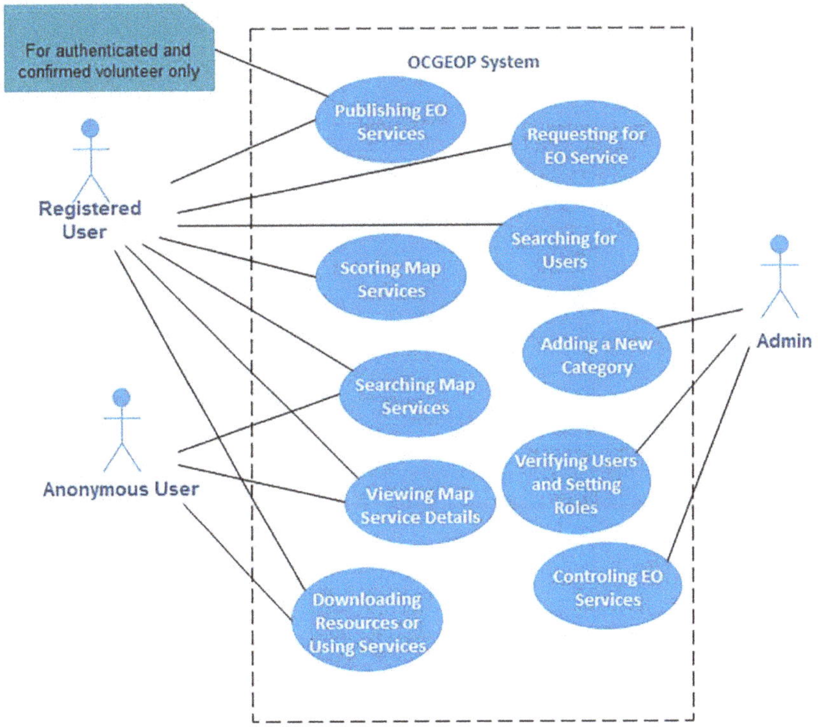

Figure 5. Important use cases and the interactions within the OCCGEOP.

All registered users are eligible to request EO services, search map services and other users, view map service details (e.g., preview the product on the map and view service metadata), score map service, and download resources or use services. As explained before, only the registered users whose competence has been approved can publish the EO product in the system. The anonymous or unregistered users can only search for map services, view map service details, and download resources or use services. The administrator is responsible for defining and adding new thematic categories, handling users and verifying them, controlling the resources, and other usual tasks such as monitoring the server status or creating the database backup.

Figure 6 presents the major activities in the OCCGEOP and the different decision paths that occur from a starting point to an ending point. For instance, a registered user can access one of several activities after logging into the system, namely advanced search, update profile, and/or search for users. If an eligible registered user aims to add a new map service, he/she must fill in the metadata elements, upload the EO product, and proceed to the automatic generation of a WMS or WCS service.

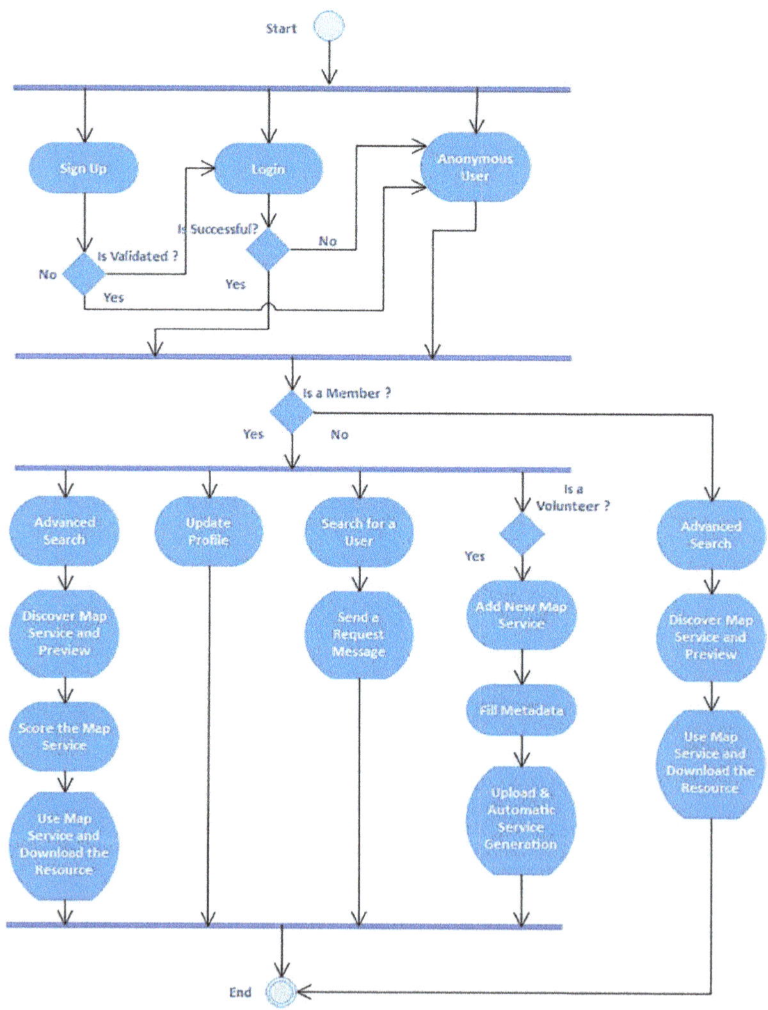

Figure 6. Activity diagram of the main decision paths that various types of users may confront.

4.4. Implementation of a Prototype

A prototype implementation of OCCGEOP was performed based on the proposed architecture, the technology, database design, and the required capabilities presented in the form of an activity diagram in the previous section.

The main components and features of the implemented system are presented in Figures 7–12. These figures demonstrate how the user interacts with the geoportal. Figure 7 shows the geoportal homepage, which displays the main menus before the user registers and logs into the system. Therefore, the menus only provide an advanced search for map services and a few general items. A public poll and a list of the most frequently visited map services are to the left of this webpage.

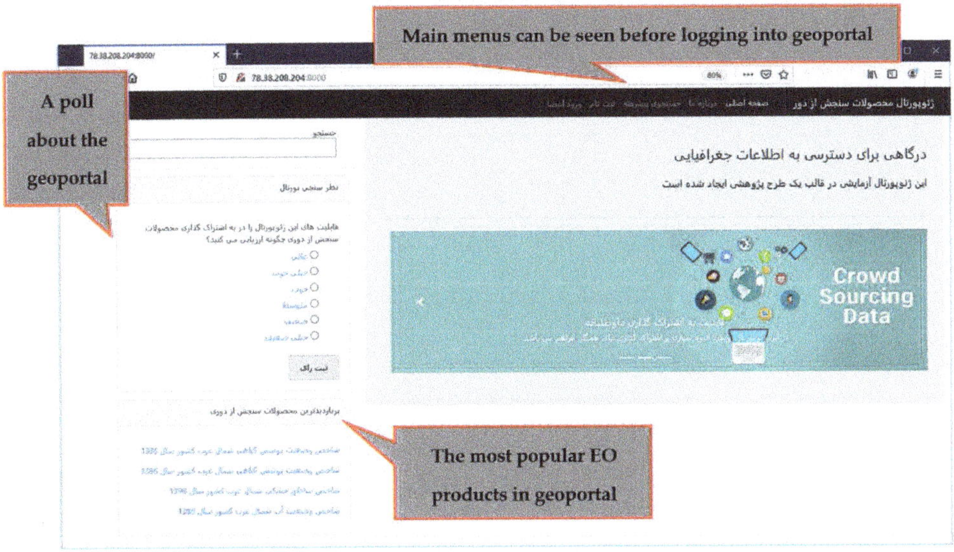

Figure 7. The home page of the OCCGEOP geoportal and menus.

Figure 8 illustrates the advanced search function of this system based on user-defined bounding boxes, as well as some metadata such as map category, service descriptions, time limits, etc. The search results include a list of names of map services, a snippet of the descriptions, the thumbnail of crowdsourced EO data products, and a link to details of the service. It should be noted that in the prototype implementation, the confirmed registered users voluntarily published their self-produced processed EO products (e.g., spectral indices images).

Figure 9 displays the detailed information of a map service. Previewing the remote sensing product overlaid on the OpenStreetMap base map, service rating (using rating stars), metadata, name, and profile of the volunteer who has prepared the product, WMS or WCS service parameters for calling the service, as well as the product download link in GeoTIFF format, are all among the features that the detail page provides to users, both members, and nonmembers.

Figure 10 is a screenshot after logging into the system where some new menus are accessible for the registered volunteer. In this figure, the user is adding a new map service. As described in the OCCGEOP model overview, after filling out a descriptive form of metadata elements and uploading the EO product, the user triggers the automated creation of WMS and WCS standards from data. In this model, automated processes for the development of standard map services and pre-designed metadata input web forms help to preserve data homogeneity and interoperability.

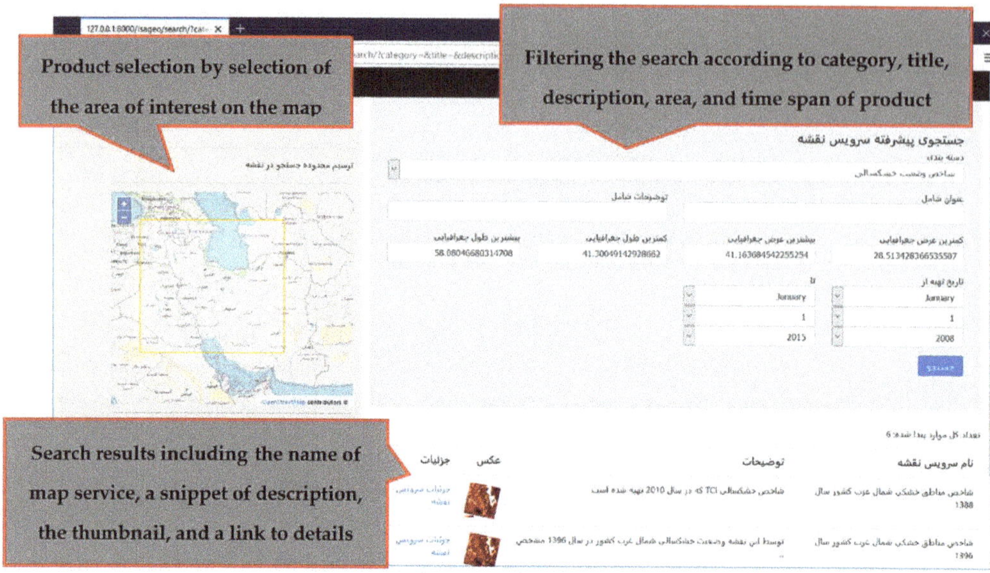

Figure 8. Advanced search for finding an Earth observation (EO) product service based on location boundary on the map and metadata such as category, title, a term in the description, and time span; the result includes the name of map service, a snippet of the description, the thumbnail, and a link to details.

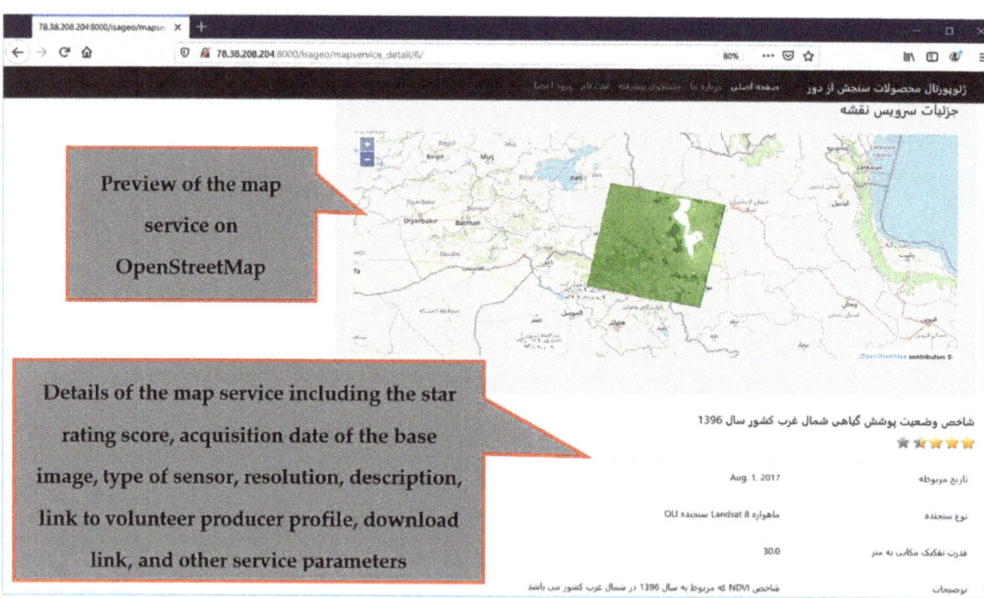

Figure 9. The details of map service including the preview, star score, metadata, download link, and service parameters.

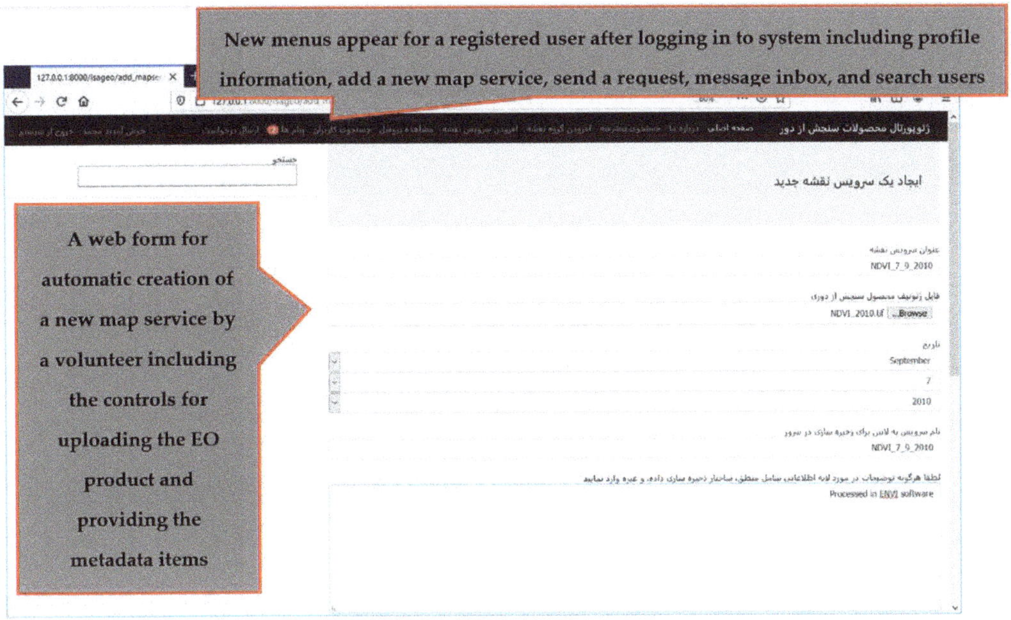

Figure 10. A web form for publishing a new EO product service by a volunteer.

Figure 11 shows the searching functionality for finding the members of the portal and the links to the members' profile. Within a profile, the area of expertise of the volunteer and the EO products that the volunteer has published are presented.

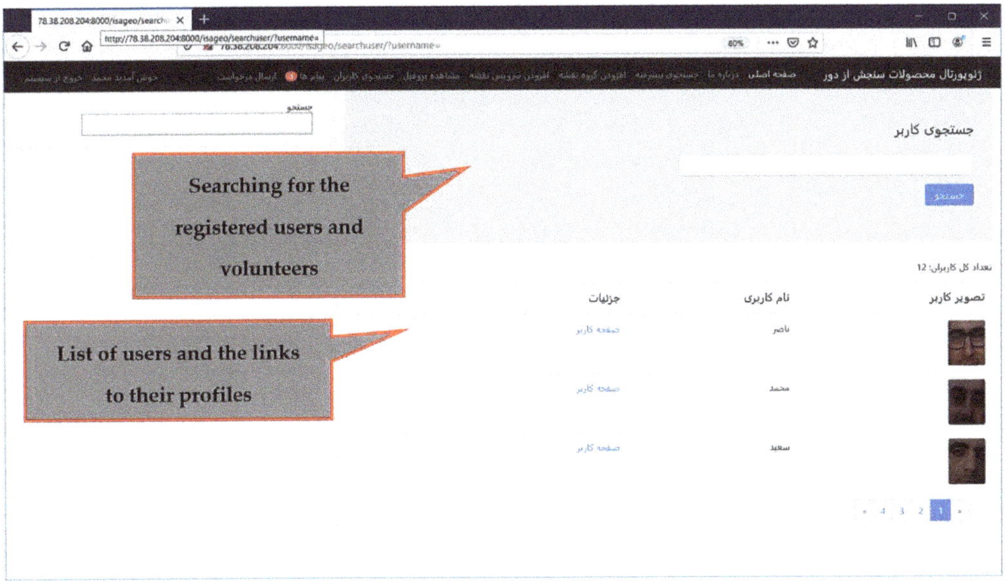

Figure 11. Searching for registered users and volunteers and visiting their profile.

Finally, Figure 12 shows the communication functionality of the OCCGEOP model, where a user can send request messages to other volunteers (e.g., for requesting EO products) or receive the request messages from other volunteers in the system.

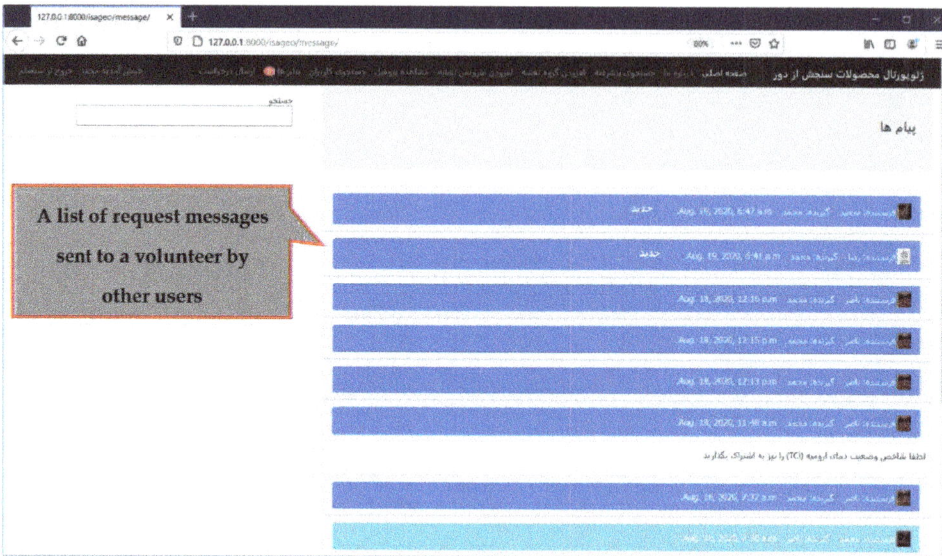

Figure 12. Sending or receiving request messages to publish an EO product.

The prototype implementation is now running experimentally on a Windows server temporarily available at http://78.38.208.204:8000. Python 3.4 and Django 1.10 were used for the development of this system. The Client URL (CURL) technology was used to convert the data to the standard formats on server-side and disseminate them through the GeoServer. The CURL commands, which run as a Python library, made it possible to make this data conversion possible through network protocols. The important note is that it takes a short time to convert data to standard map services, making it promising for providing a dynamic web portal. Restricting the data uploading and conversion mechanism in this way prevented heterogeneity in user-generated content.

5. Results

5.1. An Overview of Crowdsourced EO Products and Automatically Published Services via Prototype System

The proposed OCCGEOP model was generally developed for serving the crowdsourced raw and processed EO products in raster data format. The implemented prototype of this model was tested by conducting a pilot project for crowdsourcing EO products in Iran. In this sense, by using the OCCGEOP prototype system, a set of EO products was obtained through the crowdsourced approach from volunteers across the country. The crowdsourced data in the prototype system encompass both raw and processed EO products. The reviewing of the crowdsourced datasets has revealed that all of the shared raw EO products were acquired by the volunteers by employing the UAV-based optical sensors. Moreover, the crowdsourced processed EO products were voluntarily produced based on (1) the images acquired by the volunteers using UAV-based optical sensors and (2) the open images obtained by satellite-based optical and radar sensors. The crowdsourced processed EO products in the system were generated by volunteers using the different types of image processing techniques, including spectral indices calculation, supervised image classification, differential radar interferometry, and photogrammetry. The crowdsourced

EO products in the experimental geoportal cover different thematic EO areas, including the environment, natural hazards, and urban mapping and monitoring as well as base mapping.

Figure 13 presents examples of the created standard map services using the crowd-sourced EO products in the implemented prototype system. Figure 13a shows a shared very high resolution (VHR) image from a private garden in Mashhad, Iran, acquired by a UAV-based Canon EOS M3 sensor in 2020. Figure 13b presents a VHR digital terrain model (DTM) image for a rural area near Tangal-e Mazar village, Khorasan Province, Iran, produced based on the stereo images obtained by a UAV-based 1-inch 20MP CMOS sensor in 2020. According to the shared metadata for the product, the image was initially generated by the volunteer for a road construction project, and then he shared it voluntarily with the system.

Figure 13c,d illustrate a modified normalized difference water index (MNDWI) image and a temperature condition index (TCI) image for an area located in West Azerbaijan and West Azerbaijan Provinces, Iran. These EO products were produced using the images obtained by the Landsat 8 Operational Land Imager (OLI) sensor in 2017. After the sharing of the MNDWI image (Figure 13c) by the volunteer in the system, a user asked him if he could produce and provide the TCI image for the same study area. Upon this request, the volunteer shared a TCI image (Figure 13d) in the system. Figure 13e shows the classified image of a district in Tabriz, Iran. According to the metadata of the product, the image was produced by processing the data that were acquired through the Sentinel-2 Multi Spectral Instrument (MSI) sensor in 2019, using the support vector machine (SVM) classifier. The shared image contains seven classes (building, road, soil, tree, grass, crop, and water) with an overall accuracy of 81%. This processed EO product was shared by a volunteer via the system upon a request by a user of the system. Finally, Figure 13f presents an interferogram image for land surface displacement in an area in Kermanshah Province, Iran caused due to the 2017 Iran–Iraq earthquake. The product was generated by the processing of Sentinel-2 synthetic aperture radar (SAR) data (obtained in 2017) based on the differential radar interferometry technique.

As was mentioned in Section 4.4., the crowdsourced EO products can be downloaded directly from the implemented geoportal using the provided product's download link. In addition, the implemented prototype can visualize the published maps in the web browsers via WMS standard (Figure 13). Furthermore, for the sake of geospatial data interoperability, a URL for the WMS layer can be generated in the following general format by appending the required parameters for the GetMap operation that are provided by the prototype system: (http://Hostname:port/geoserver/CategoryName/wms?service=WMS&version=1.1.0&request=GetMap&layers=CategoryName:ServiceName&OtherParameters). Similarly, the client has access to the WCS service by appending parameters for the GetCoverage operation to the service's URL in the following format: (http://Hostname:port/geoserver/ows?service=WCS&version=2.0.0&request=GetCoverage&coverageId=CoverageId&OtherParameters). These capabilities are getting power from open technical solutions (including GeoServer technology and OGC data interoperability standards) and enable a user to simply and routinely import the EO product as a WMS (or a WCS) layer into their desktop GIS or Web GIS platforms (which support these capabilities) and view it using the provided URL. For example, Figure 14a demonstrates how a user (data consumer) add the WMS layer of a crowdsourced Normalized Difference Vegetation Index (NDVI) product (from an area located in West Azerbaijan and West Azerbaijan Provinces, Iran) that was published in OCCGEOP prototype to a GIS software (QGIS software). Basically, the metadata of an EO product are provided within the implemented prototype (Figure 9); however, the adopted OGC data interoperability standards in the implemented OCCGEOP also enable the user to access the service-level metadata via a web browser using different methods such as WMS GetCapabilities or WCS DescribeCoverage. For example, Figure 14b shows the service-level metadata for the WMS layer of a crowdsourced MNDWI product (for an area located in West Azerbaijan and West Azerbaijan Provinces, Iran) that was accessed through the WMS GetCapabilities method by a user. Similarly, the user also can invoke the WCS DescribeCoverage operation to request more information about the coverage of service, including the area occupied by the coverage, spatial reference system, information about its resolution, and available image bands.

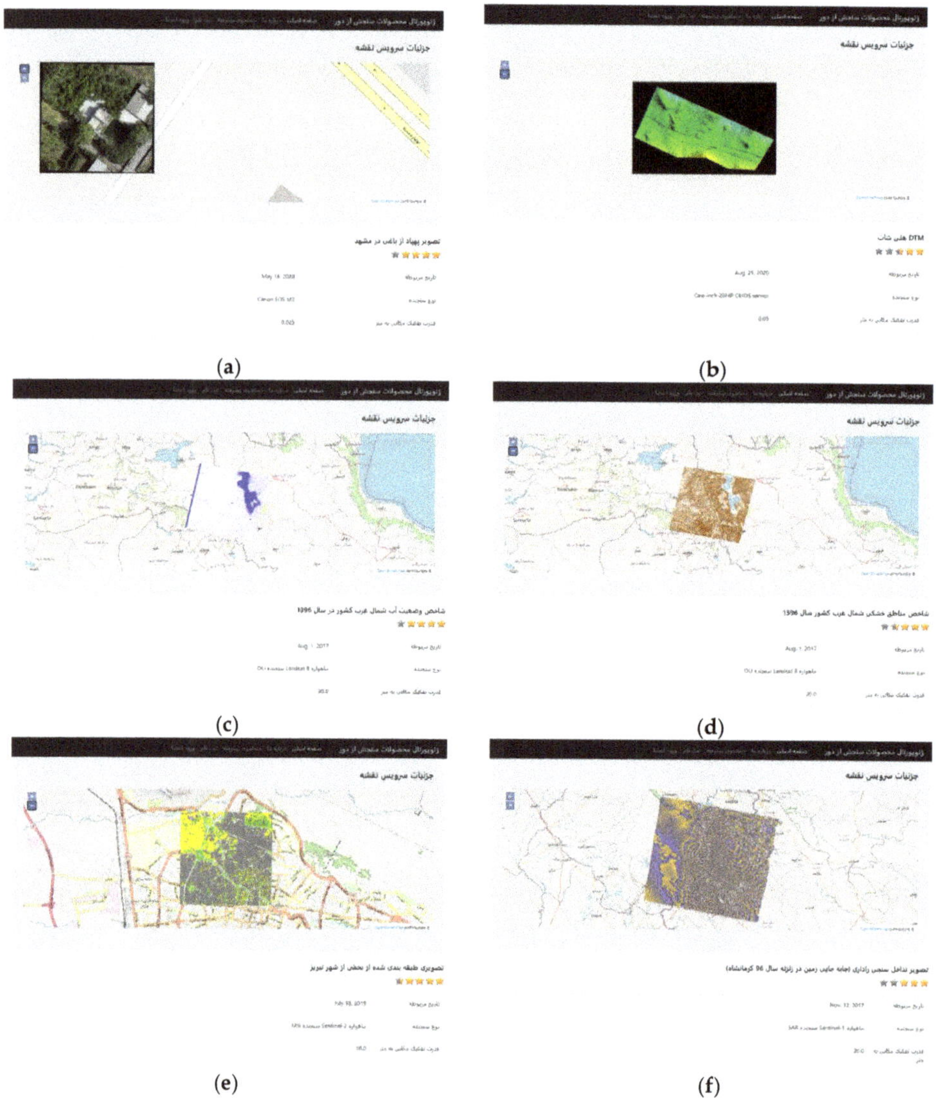

Figure 13. Examples of crowdsourced EO products in the implemented prototype system; (**a**) unmanned aerial vehicle (UAV)-based RGB image of a private garden in Mashhad, Iran, (**b**) UAV-based digital terrain model (DTM) image from a rural area near Tangal-e Mazar village, Khorasan Province, Iran, (**c**) modified normalized difference water index (MNDWI) image for an area located in West Azerbaijan and West Azerbaijan Provinces, Iran, (**d**) temperature condition index (TCI) image for an area located in West Azerbaijan and West Azerbaijan Provinces, Iran, (**e**) classified image of a district in Tabriz, Iran, (**f**) interferogram image for land surface displacement in an area in Kermanshah Province, Iran.

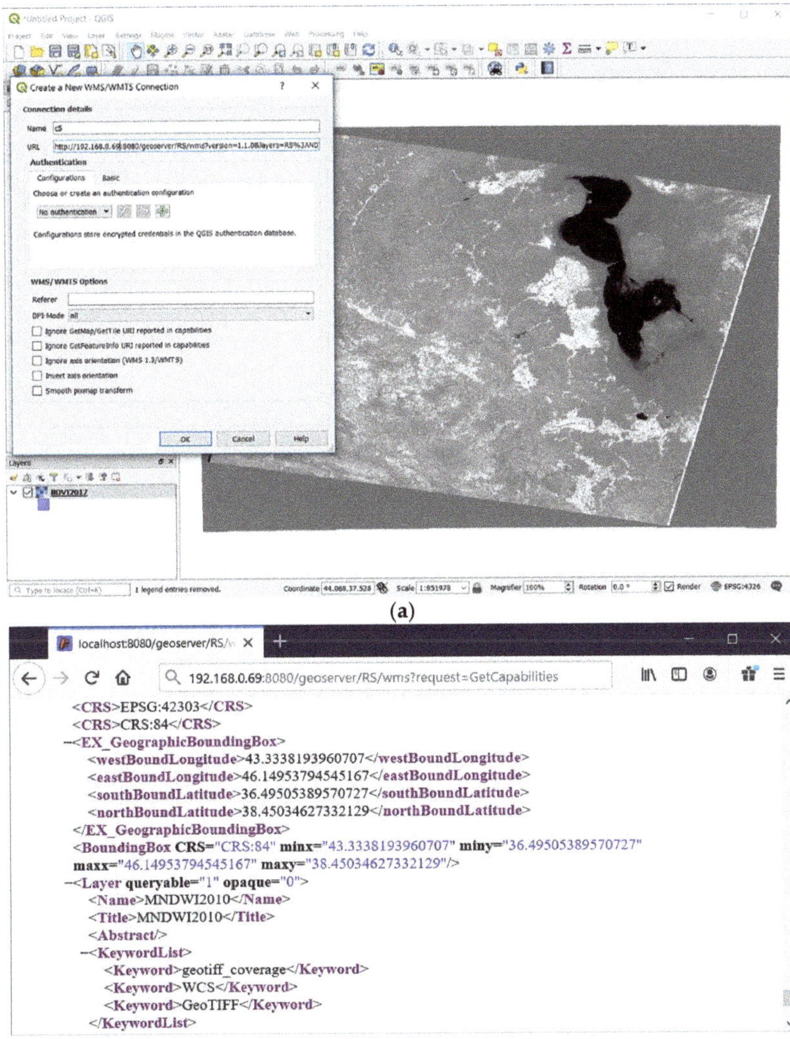

Figure 14. (**a**) Adding the Web Map Service (WMS) layer of a crowdsourced NDVI product (for an area located in West Azerbaijan and West Azerbaijan Provinces, Iran) into the QGIS software using the URL provided by the service; (**b**) accessing the service-level metadata for the WMS layer of crowdsourced MNDWI product (for an area located in West Azerbaijan and West Azerbaijan Provinces, Iran) using the WMS GetCapabilities method.

5.2. Performance Analysis and Optimization of the Prototype System

One of the most important analyses in creating prototype systems is performance testing. This can lead the developer to produce the final product with the desired quality to serve the end-users. In this sense, the GTmetrix (http://gtmetrix.com), a free tool to easily test the performance by crawling the web data, was used. This tool could analyze the performance of the implemented prototype system and recommend solutions for optimizing the system. Using the provided technical solutions by GTmetrix, we were able to improve the efficiency and optimize the OCCGEOP prototype system. As can be seen in Figure 15, using the GTmetrix tool an analysis was performed through GTmetrix test server located in London using a Chrome browser on 1 November 2020.

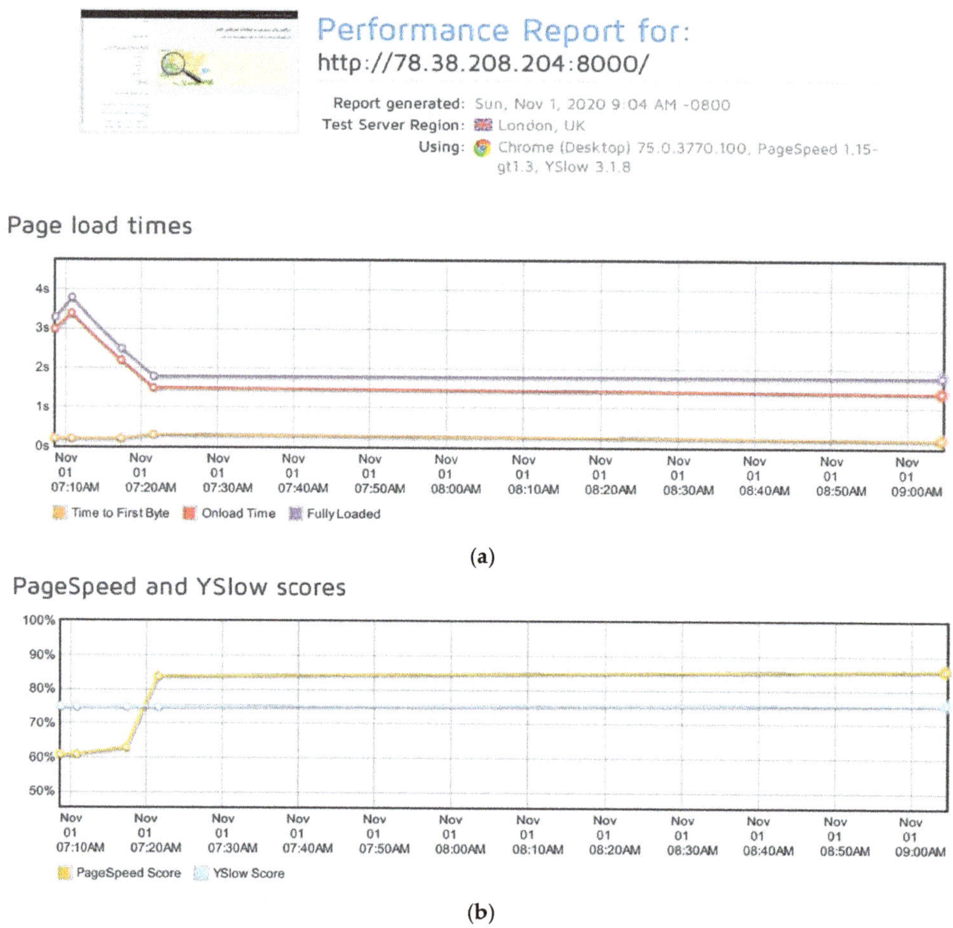

Figure 15. Analysis of the performance of the prototype system using GTmetrix tool based on (**a**) page load times and (**b**) speed scores.

Based on the results of this analysis, several solutions were adopted to improve the performance of the prototype system. For instance, using the provided recommendations, the size of the images was optimized in this study. In this sense, by loading optimum size images, we were enabled to reduce the load times of pages. Furthermore, a tile-based representation of map services was adopted as an effective approach. Another recommendation was to avoid using URL redirects. There are many reasons for redirecting the browser from one URL to another, such as indicating the new location of a resource that has moved or monitoring clicks and pages of reference logs. Regardless of the reason for this issue, redirects trigger an extra HTTP request–response loop and add latency for round-trip-time. Hence, the number of redirects provided by the web portal was minimized—particularly for the resources required to start the homepage. The analysis also recommended deferring the parsing of web scripts. Regularly, the browser must parse the contents of all JavaScript tags in order to load each web page, which adds additional time to the page load. Therefore, the initial load time of pages has been decreased by minimizing the amount of script required to render the page and preventing the parsing of the unneeded script until it needs to be performed. Setting a far-future expiration date for cached resources was another suggestion by GTmetrix. The resource expiration date specifies how long a file must be kept in the cache so that in future page views, the file does not have to be downloaded again. Using the far-future expiration strategy helped us to reduce the returning visitor load times. Finally, the removal of references to non-existent resources was another issue that was recommended by GTmetrix and consequently was addressed in this study. By optimization of the OCCGEOP implemented prototype according to the provided recommendations of GTmetrix tool, the two main indicators in measuring the system's performance, so-called YSlow and PageSpeed, were improved from 74% to 76% and from 61% to 86%, respectively. Furthermore, the pages' full load times were decreased on average from 3.7 s to 1.8 s. According to the recommendation of the GTmetrix tool, it is expected that the system's performance will be improved even more by employing other technical methods such as serving static files from a cloud service, avoid unnecessary cookie traffic, and deploying our content across multiple, geographically dispersed servers—the solutions that can be adopted in the complete implementation of OCCGEOP.

6. Discussion

The OCCGEOP integrated the social and participatory characteristics into the conventional attributes of geoportals. The synergy of this integration brings various benefits to an SDI and its stakeholders; several of them will be highlighted here. First, the proposed system builds the capacity for supplying both unused and used user-generated EO products. In this context, the OCCGEOP facilitates the crowdsourcing and sharing of the unavailable EO products and helps to integrate and publish the existing fragmented user-generated EO products on a voluntary basis. Second, the adopted solutions for publishing standard maps from the heterogeneous voluntarily shared EO products in the proposed system realize the interoperability of the shared products and their metadata. These consequently facilitate and accelerate the discoverability, accessibility, and utilizability (use or reuse of shared data [76] for the purpose defined by data consumers) of the shared data and reduce the cost and time of the analyses of these products. Third, the adopted community-based data quality assessment approach (using end-user-contributed scores and feedbacks) alongside the employed top-down approach for screening of the volunteer data producers may help to filter out the poor quality products and reduce the skepticisms in using or reusing such data. Fourth, the existing two-way communication mechanism between data producers and consumers may help to improve the quality of the products and expand the data coverage over time without a centralized management. Fifth, the crowdsourced EO products provided through data as a service (DaaS) [77] strategy to the end-users may benefit the research and applied projects that consume EO data by delivering the EO products on demand for free regardless of geographic locations and affiliations of data consumers. This advantage is more significant in developing countries such as Iran, where the lack of open EO products has always been an important obstacle to the projects. Sixth, the existence of volunteer data providers allows the system administrators and technical personnel to focus on geoportal maintenance and supervision tasks instead of data provision, data manipulation, and publication; thus, this allows for saving time and money. Seventh, the community-based and participatory nature of the proposed model connects the broader community with EO and EO products by increasing public participation and improving the citizens' engagement in EO, and disseminating open EO products among the public. Last but not least, similar to the citizen science projects, the OCCGEOP may provide learning opportunities for the system users, increase social interactions, and raise awareness of both crowdsourced EO data producers and consumers about the existing various challenges and opportunities on the Earth system spheres.

A comparative study of the main capabilities of OCCGEOP with the three worldwide well-known geoportals, including the INSPIRE, the NASA EOSDIS, and the Global Earth

Observation System of Systems (GEOSS) portal, has been performed in this study. INSPIRE is based on the infrastructures for spatial information established and operated by the member states of the European Union. NASA's EOSDIS has been designed as a vendor to provide key Earth observation data management capabilities from various sources (e.g., satellites, aircraft, field measurements, etc.). The GEOSS portal enables discovery and access to diverse data from independent Earth observation, information, and processing systems [78]. Jiang, van Genderen, Mazzetti, Koo and Chen [2] discussed the capabilities of these geoportals in detail.

Obviously, the functionalities of the implemented prototype and capabilities of OCCGEOP in its current experimental form are still far from the strong design and comprehensive capabilities of the well-established aforementioned geoportals. However, it is possible to compare the essence of OCCGEOP vision and its main capabilities with the vision and main capabilities of the aforementioned three geoportals, especially from the perspective of crowdsourcing. Table 1 presents the comparative analysis of OCCGEOP with INSPIRE, NASA EOSDIS, and the GEOSS Portal according to 15 key items.

Table 1. Comparative analysis of OCCGEOP vision and capabilities with the three target geoportals.

Item	INSPIRE Geoportal	GEOSS Geoportal	EOSDIS Geoportal	OCCGEOP Geoportal
Standard Services such as WMS and WCS	√	√	√	√
High Volume Data Coverage	√	√	√	×
Correspondence with Metadata Standard	√	√	√	√
Distributed Server	√	√	√	×
Data Preview	×	√	√	√
Visual Spatial Selection Search	√	√	√	√
Time Filter for Search	×	√	√	√
Providing Crowdsourced Data	×	×	×	√
Downloadable Resource	√	√	√	√
User Identification	×	√	√	√
Online Private Workplace	×	√	√	√
Online Translator	√	×	×	×
Automatic Conversion of User Data to Standard Map Services	×	×	×	√
Online Solution to Receive Feedbacks from Users	×	√	√	√
Interactive User Requests for Publishing Geospatial Web Services	×	×	×	√

In some aspects, such as providing standard map services (e.g., WMS and WCS), correspondence with metadata standards, visual and spatial selection search, and providing downloadable resources, OCCGEOP and the target geoportals all follow the same vision; however, some others are different. The three targeted geoportals adopt a top-down policy driven method to define processes of data entry, transfer, maintenance, and delivery. On the contrary, the OCEGEOP benefits from a bottom-up approach; hence, it is based on the crowdsourced data production paradigm. Although such an interactive communication is primarily designed for the bottom-up processes, it can serve as a coordinator in top-down processes too. Furthermore, compared with the three targeted geoportals, the OCCGEOP can provide unique capabilities such as the automatic conversion of crowdsourced geospatial data to standard map services and user-to-user interactive communication that facilitates the request for the provision of voluntary services. The OCCGEOP is equipped with functionalities that some of the target geoportals do not benefit from. Examples are data preview on the map, time filter for search, create and post metadata, user identification, online private workplace and profile for users, and online solution to receive feedback from users. While the three target geoportals have all been designated to address the big data challenges, OCCGEOP is still weak in this regard and needs to improve the efficiency of high volume and big data coverage. Nevertheless, this is more about using powerful hardware and distributed servers than about the portal's logical architecture. The three targeted geoportals have the capacity to access distributed servers at a large scale. Eventually, based on this comparative analysis and the 15 items evaluated, OCCGEOP has an acceptable and promising performance. As the present implementation of OCCGEOP is merely a prototype, to make the system more practical, the identified shortcomings can be improved on in future studies.

For further evaluation of the adopted approaches in the system and the capabilities of the proposed model, 40 volunteer experts and practitioners in the area of geoinformatics were asked to use the experimental implementation of the OCCGEOP and assess it by participating in a designated survey conducted in this study.

The participants of the survey were asked three fundamental questions about the visions behind the OCCGEOP (Table 2). The majority of the survey participants (1) agreed on the necessity of designing a new generation of geoportals for crowdsourced EO products, (2) expressed that a geoportal for crowdsourced EO products can supply some of their needs that cannot be addressed in other geoportals, and (3) believed that the visions behind the OCCGEOP will be pervasive in the new generation of geoportals in future.

Table 2. Feedbacks of survey participants on the OCCGEOP vision.

Questions	Answers
Do you agree with the necessity for designing community-based geoportals for crowdsourced EO products?	No: 12.50% Yes: 87.50%
Do you currently need the Earth observation products that cannot be obtained from the authoritative geoportals?	No: 40.00% Yes: 60.00%
Do you agree with the ideas used in OCCGEOP being pervasive in the new generation of geoportals?	No: 15.00% Yes: 85.00%
	■ No ■ Yes

Table 3 shows survey participants' feedback on five main features of the implemented prototype of OCCGEOP. These five main features that were evaluated by the survey participants include the spatial selection and searching of data, user-friendliness clarity of menus, uploading data and automatic standard map service generation, searching users and sending request message, and providing service detail and map preview.

The survey participants evaluated these features qualitatively by rating them using one of four categories: very strong, strong, moderate, and weak. According to the results, on average, 86% of survey participants were satisfied or very satisfied with each of the features and capabilities of the prototype system. These promising results of the survey on the adopted features in the prototype implementation of OCCGEOP showed that the implementation of the robust Django framework and Web 2.0 technologies in OCCGEOP could successfully create user-to-user communication, dynamic, and interactive environments in the geoportal. Generally, VGI is considered as data that are multi-source, unstructured, heterogeneous, uncertain, improperly documented, and loosely coupled with metadata; therefore, interoperability and standardization of VGI have always been considered as challenging issues in the integration of such data in the authoritative data sources and GISs [79–81]. In this term, integration of VGI, which is generated in a bottom-up process with the conventional geoportals and SDIs that are designed with the top-down models for the handling of authoritative data, is inherently a challenging issue [79]. However, in this study, the implemented prototype could generally address the aforementioned issues by creating simple web forms, automating conversion and standardization processes, and simple data quality control processes. The OCCGEOP can connect the professionals with amateurs tightly and interactively, direct, facilitate and accelerate the production and sharing of crowdsourced EO products. Moreover, the standard map services in OCCGEOP created through the bottom-up process could be integrated, at least structurally, with other standard maps created by the top-down strategies as well as standard platforms to create robust offline or online and distributed GIS.

Table 3. Feedback of survey participants on the main capabilities of the implemented prototype of OCCGEOP.

Criteria	Scores
Spatial Selection and Searching Data	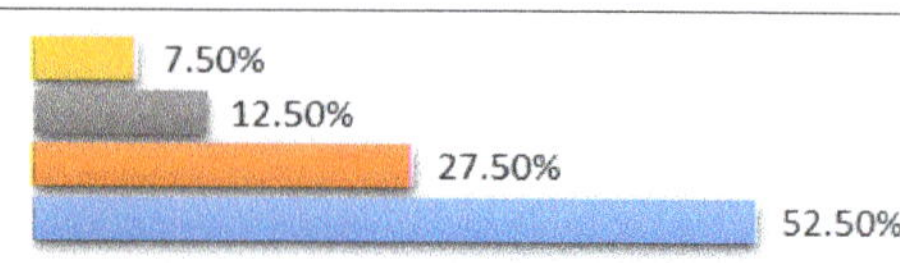
User Friendliness and Clarity of Menus	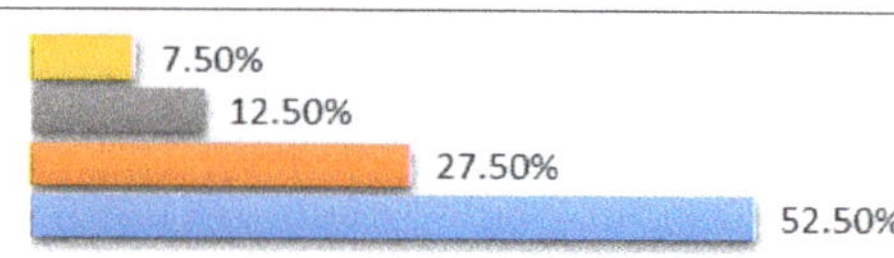
Data Uploading and Generation of Automatic Standard Map Service	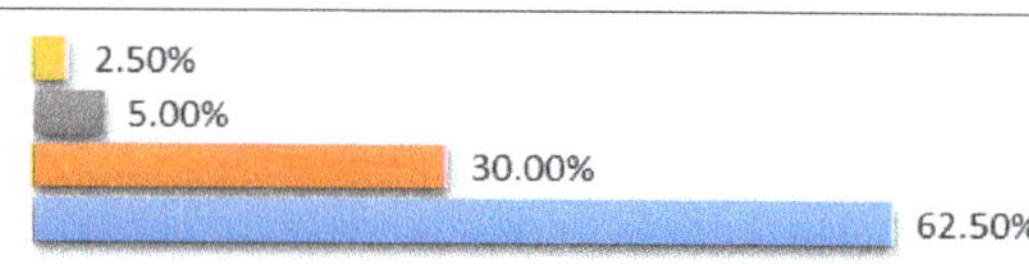
Searching Users and Sending Request Message	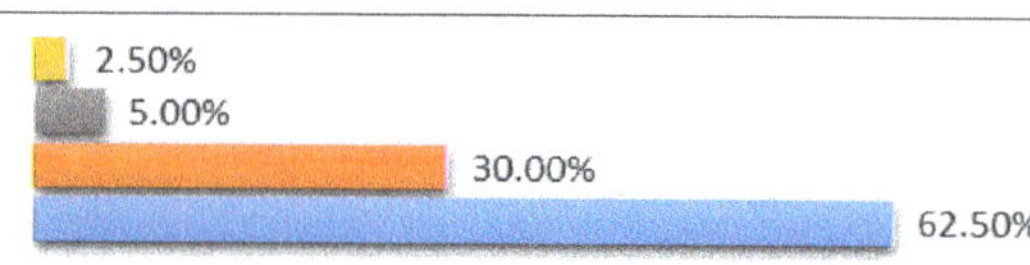
Providing Service Details and Map Preview	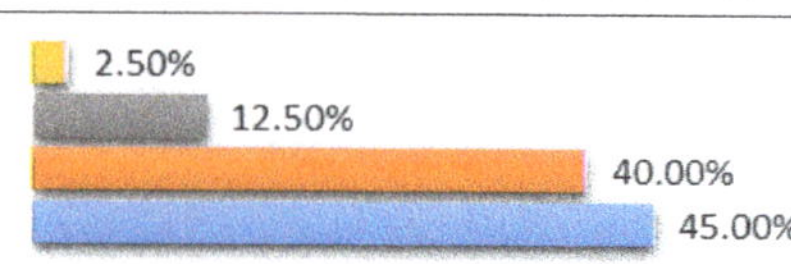

■ Weak ■ Moderate ■ Strong ■ Very Strong

7. Conclusions and Future Works

In the new generations of geoportals, taking the advantages of the VGI and developing a community-based environment for facilitating user-to-user communication are considered as two main priorities. In this context, this research introduced a new model for geoportals named "Open Community-Based Crowdsourcing Geoportal for Earth Observation Products" (OCCGEOP) based on the concepts of VGI and community-based geoportals and conducted a prototype implementation for the proposed model for environmental and climate change-related crowdsourced EO products. The proposed model enables user-to-user communication in the geoportal, eases the coordination of the production of crowdsourced EO data, as well as facilitating the administration, standardization and quality assurance, discovery, publishing, accessing, and sharing of the voluntary EO products. The heterogeneity of VGI is one of the main challenges in the integration of VGI in the geoportals. The automated mechanisms for transforming the heterogeneous data structure of crowdsourced EO products in OCCGEOP allow all voluntary maps to be generated in accordance with SDI standards. The conducted comparison of the different features and capabilities of the proposed model with the features and capabilities of three existing well-established geoportals in this study revealed that (1) the proposed OCCGEOP model is compatible with the priorities of the new generations of geoportals and (2) the proposed model has some unique features and capabilities for integration of the crowdsourcing paradigm into the geoportal that the other studied geoportals are missing. Furthermore, our survey about the system users' beliefs and preferences showed that the majority of the participants agreed with visions of the proposed model and on average, 86% of the participants in the survey are satisfied or very satisfied with each of the features and capabilities of the implemented prototype for the proposed model. The promising performance of the implemented prototype of OCCGEOP made it possible to consider the full implementation of OCCGEOP as a workaround geoportal that enables the handling of increasingly growing crowdsourced EO products.

Given that the selected names or descriptions in the voluntary map services can be expressed in different ways, one of the future directions of this research is to use ontology to resolve or reduce the semantic heterogeneity and contribute to semantic interoperability in OCCGEOP. The OCCGEOP model considered the approaches for assurance of crowdsourced EO data quality. However, in future works, the feasibility of using more robust approaches for the assessment of the credibility and trustworthiness of crowdsourced EO products in OCCGEOP should be investigated. The sharing of events related to crowdsourced data generated within OCCGEOP on social networks is another functionality that can be developed in future studies. In this sense, when a map service is produced in OCCGEOP, a user would be able to share it as an event (including a photo of the map, a general description, and the time of production with a link to the geoportal service details page) on social networks. The idea of sharing EO production events can contribute to the more direct and rapid diffusion of EO-derived information among the general public as well as attracting more viewers and volunteer contributors to the geoportal. In the current research, a survey on the beliefs and preferences of a group of Iranian geoinformatics experts and practitioners was conducted for assessing the quality of the design of the system. In the future, further study will be needed to obtain the opinions of a larger and more diverse group of the local audience, including the users with less experience in geoinformatics. Furthermore, in this study, the prototype of OCCGEOP was implemented in the Persian language to be used in Iran. Therefore, another future direction of this research is to implement the English version of the system to be used by international users. This will make the audience of the developed geoportal more diverse and enable us to conduct a more comprehensive survey on the beliefs and preferences of OCCGEOP users for enhancing the design of the system accordingly. As the OCCGEOP model was developed in accordance with interoperability standards, the various dimensions of integration of the OCCGEOP as a node into a national SDI (NSDI) (e.g., Iranian NSDI) are interesting research lines for future works. Conducting a further investigation on adopting the distributed servers to handle high volume and big crowdsourced EO data at a large-scale is necessary and is a high priority for the development of OCCGEOP. Another direction is to use the OGC APIs in developing OCCGEOP. In this sense, by using the resource-centric API solution presented by OGC APIs, reaching more modern, effective, and rapid web development would be possible. While OGC services usually use the Representational State Transfer (REST) protocol for communication, using OGC API in developing OCCGEOP can enable us to use any style of communication and improve interoperability in the (Information Technology) IT industry. Similar to the major existing SDI geoportals developed for EO products, OCCGEOP mainly focused on publishing, finding, and accessing EO products. However, future research could examine the feasibility of integrating geoprocessing services in a standardized way through OGC's Web Processing Service (WPS) as a marginal service for the system. In OCCGEOP, data are provided and evaluated by users for the users. Therefore, the ultimate success of OCCGEOP is tied to the participation and engagement of the citizens in the system. In this sense, alongside the technical and technological aspects of OCCGEOP, future research should be conducted to determine the effective approaches for attracting citizens and sustaining their engagement in the system.

Author Contributions: Conceptualization, Mohammad H. Vahidnia; formal analysis, Mohammad H. Vahidnia and Hossein Vahidi; investigation, Mohammad H. Vahidnia and Hossein Vahidi; methodology, Mohammad H. Vahidnia; software, Mohammad H. Vahidnia; visualization, Mohammad H. Vahidnia; writing—original draft, Mohammad H. Vahidnia and Hossein Vahidi; writing—review and editing, Mohammad H. Vahidnia and Hossein Vahidi. All authors have read and agreed to the published version of the manuscript.

Funding: This research was funded by Iran National Science Foundation (INSF), grant number 98011396. The authors gratefully acknowledge the financial support of INSF.

Institutional Review Board Statement: Not applicable.

Informed Consent Statement: Not applicable.

Data Availability Statement: Data sharing is not applicable to this article.

Conflicts of Interest: The authors declare no conflict of interest.

References

1. Maguire, D.J.; Longley, P.A. The emergence of geoportals and their role in spatial data infrastructures. *Comput. Environ. Urban Syst.* **2005**, *29*, 3–14. [CrossRef]
2. Jiang, H.; van Genderen, J.; Mazzetti, P.; Koo, H.; Chen, M. Current status and future directions of geoportals. *Int. J. Digit. Earth* **2020**, *13*, 1093–1114. [CrossRef]
3. Bernard, L.; Kanellopoulos, I.; Annoni, A.; Smits, P. The European geoportal—One step towards the establishment of a European Spatial Data Infrastructure. *Comput. Environ. Urban Syst.* **2005**, *29*, 15–31. [CrossRef]
4. Innerebner, M.; Costa, A.; Chuprikova, E.; Monsorno, R.; Ventura, B. Organizing earth observation data inside a spatial data infrastructure. *Earth Sci. Inform.* **2017**, *10*, 55–68. [CrossRef]
5. Borzacchiello, M.T.; Craglia, M. Estimating benefits of Spatial Data Infrastructures: A case study on e-Cadastres. *Comput. Environ. Urban Syst.* **2013**, *41*, 276–288. [CrossRef]
6. Coetzee, S.; Ivánová, I.; Mitasova, H.; Brovelli, M.A. Open geospatial software and data: A review of the current state and a perspective into the future. *ISPRS Int. J. Geo-Inf.* **2020**, *9*, 90. [CrossRef]
7. De Longueville, B. Community-based geoportals: The next generation? Concepts and methods for the geospatial Web 2.0. *Comput. Environ. Urban Syst.* **2010**, *34*, 299–308. [CrossRef]
8. Lehmann, A.; Chaplin-Kramer, R.; Lacayo, M.; Giuliani, G.; Thau, D.; Koy, K.; Goldberg, G. Lifting the information barriers to address sustainability challenges with data from physical geography and earth observation. *Sustainability* **2017**, *9*, 858.
9. Sánchez-Gallegos, D.D.; Gonzalez-Compean, J.; Sosa-Sosa, V.J.; Marin-Castro, H.M.; Tuxpan-Vargas, J. An interoperable cloud-based geoportal for discovery and management of earth observation products. *Comput. Sci. Inf. Technol. (CS IT)* **2018**. [CrossRef]
10. Nativi, S.; Mazzetti, P.; Santoro, M.; Papeschi, F.; Craglia, M.; Ochiai, O. Big data challenges in building the global earth observation system of systems. *Environ. Model. Softw.* **2015**, *68*, 1–26. [CrossRef]
11. See, L.; Mooney, P.; Foody, G.; Bastin, L.; Comber, A.; Estima, J.; Fritz, S.; Kerle, N.; Jiang, B.; Laakso, M. Crowdsourcing, citizen science or volunteered geographic information? The current state of crowdsourced geographic information. *ISPRS Int. J. Geo-Inf.* **2016**, *5*, 55. [CrossRef]
12. Su, W.; Sui, D.; Zhang, X. Satellite image analysis using crowdsourcing data for collaborative mapping: Current and opportunities. *Int. J. Digit. Earth* **2020**, *13*, 645–660. [CrossRef]
13. Comber, A.; Schade, S.; See, L.; Mooney, P.; Foody, G. Semantic analysis of citizen sensing, crowdsourcing and VGI. In Proceedings of the AGILE'2014 International Conference on Geographic Information Science, Castellón, Spain, 3–6 June 2014.
14. Heipke, C. Crowdsourcing geospatial data. *ISPRS J. Photogramm. Remote Sens.* **2010**, *65*, 550–557. [CrossRef]
15. Goodchild, M.F.; Glennon, J.A. Crowdsourcing geographic information for disaster response: A research frontier. *Int. J. Digit. Earth* **2010**, *3*, 231–241. [CrossRef]
16. Vahidnia, M.H.; Hosseinali, F.; Shafiei, M. Crowdsource mapping of target buildings in hazard: The utilization of smartphone technologies and geographic services. *Appl. Geomat.* **2020**, *12*, 3–14. [CrossRef]
17. Vahidi, H.; Klinkenberg, B.; Johnson, B.; Moskal, L.; Yan, W. Mapping the Individual Trees in Urban Orchards by Incorporating Volunteered Geographic Information and Very High Resolution Optical Remotely Sensed Data: A Template Matching-Based Approach. *Remote Sens.* **2018**, *10*, 1134. [CrossRef]
18. Jokar Arsanjani, J.; Zipf, A.; Mooney, P.; Helbich, M. An Introduction to OpenStreetMap in Geographic Information Science: Experiences, Research, and Applications. In *OpenStreetMap in GIScience: Experiences, Research, and Applications*; Jokar Arsanjani, J., Zipf, A., Mooney, P., Helbich, M., Eds.; Springer International Publishing: Cham, Switzerland, 2015; pp. 1–15. [CrossRef]

19. Saralioglu, E.; Gungor, O. Crowdsourcing in Remote Sensing: A Review of Applications and Future Directions. *Ieee Geosci. Remote Sens. Mag.* **2020**, *8*, 89–110. [CrossRef]
20. Johnson, B.A.; Iizuka, K. Integrating OpenStreetMap crowdsourced data and Landsat time-series imagery for rapid land use/land cover (LULC) mapping: Case study of the Laguna de Bay area of the Philippines. *Appl. Geogr.* **2016**, *67*, 140–149. [CrossRef]
21. Johnson, P.; Ricker, B.; Harrison, S. Volunteered Drone Imagery: Challenges and Constraints to the Development of An Open Shared Image Repository. In Proceedings of the 50th Hawaii International Conference on System Sciences, HICSS 2017, Hilton Waikoloa Village, HI, USA, 4–7 January 2017.
22. Hochmair, H.H.; Zielstra, D. Analysing user contribution patterns of drone pictures to the dronestagram photo sharing portal. *J. Spat. Sci.* **2015**, *60*, 79–98. [CrossRef]
23. Shanley, L.; Burns, R.; Bastian, Z.; Robson, E. Tweeting up a Storm: The Promise and Perils of Crisis Mapping. *Photogramm. Eng. Remote Sens.* **2013**, *79*, 865. [CrossRef]
24. See, L.; Estima, J.; Põdör, A.; Arsanjani, J.J.; Bayas, J.-C.L.; Vatseva, R. *Sources of VGI for Mapping*; Ubiquity Press Ltd.: London, UK, 2017.
25. Agapiou, A. Vegetation Extraction Using Visible-Bands from Openly Licensed Unmanned Aerial Vehicle Imagery. *Drones* **2020**, *4*, 27. [CrossRef]
26. Jorz, V. Open Aerial Map, Drones and Archaeology: The Implications of Using Drones to Contribute and Share Aerial Data on an Open Data Repository. Master's Thesis, University of Waterloo, Waterloo, ON, Canada, 2019.
27. Breen, J.; Dosemagen, S.; Warren, J.; Lippincott, M. Mapping Grassroots: Geodata and the structure of community-led open environmental science. *ACME Int. J. Crit. Geogr.* **2015**, *14*, 849–873.
28. Anderson, K.; Griffiths, D.; DeBell, L.; Hancock, S.; Duffy, J.P.; Shutler, J.D.; Reinhardt, W.; Griffiths, A. A grassroots remote sensing toolkit using live coding, smartphones, kites and lightweight drones. *PLoS ONE* **2016**, *11*, e0151564. [CrossRef] [PubMed]
29. Connors, J.P.; Lei, S.; Kelly, M. Citizen science in the age of neogeography: Utilizing volunteered geographic information for environmental monitoring. *Ann. Assoc. Am. Geogr.* **2012**, *102*, 1267–1289. [CrossRef]
30. Singh, J. Do-it-yourself satellites: Applications for citizen space. *Cent. Space Policy Strategy* **2019**, *3*, 1–12.
31. Bertolotto, M.; McArdle, G.; Schoen-Phelan, B. Volunteered and crowdsourced geographic information: The OpenStreetMap project. *J. Spat. Inf. Sci.* **2020**, *2020*, 65–70. [CrossRef]
32. Zhu, Z.; Wulder, M.A.; Roy, D.P.; Woodcock, C.E.; Hansen, M.C.; Radeloff, V.C.; Healey, S.P.; Schaaf, C.; Hostert, P.; Strobl, P. Benefits of the free and open Landsat data policy. *Remote Sens. Environ.* **2019**, *224*, 382–385. [CrossRef]
33. Goodchild, M.F. Citizens as sensors: The world of volunteered geography. *GeoJournal* **2007**, *69*, 211–221. [CrossRef]
34. Hudson-Smith, A.; Batty, M.; Crooks, A.; Milton, R. Mapping for the masses: Accessing Web 2.0 through crowdsourcing. *Soc. Sci. Comput. Rev.* **2009**, *27*, 524–538. [CrossRef]
35. Demetriou, D.; Campagna, M.; Racetin, I.; Konecny, M. Integrating Spatial Data Infrastructures (SDIs) with Volunteered Geographic Information (VGI) creating a Global GIS platform. In *Mapping and the Citizen Sensor*; Foody, G., See, L., Fritz, S., Mooney, P., Raimond, A., Fonte, C.C., Antoniou, V., Eds.; Ubiquity Press: London, UK, 2017; pp. 273–297.
36. Mooney, P.; Corcoran, P. Can Volunteered Geographic Information be a participant in eEnvironment and SDI? In *Proceedings of International Symposium on Environmental Software Systems*; Springer: Berlin/Heidelberg, Germany, 2011; pp. 115–122.
37. McDougall, K. Volunteered geographic information for building SDI. In Proceedings of the 2009 Surveying and Spatial Sciences Institute Biennial International Conference (SSC 2009), Adelaide, SA, Australia, 28 September–2 October 2009; pp. 645–653.
38. Wiemann, S.; Bernard, L. Linking crowdsourced observations with INSPIRE. In Proceedings of the 7th AGILE Conference on Geographic Information Science (AGILE 2014), Castellón, Spain, 3–6 June 2014; pp. 1–5.
39. Goodchild, M.F. Citizens as voluntary sensors: Spatial data infrastructure in the world of Web 2.0. *Int. J. Spat. Data Infrastruct. Res.* **2007**, *2*, 24–32.
40. Sterlacchini, S.; Bordogna, G.; Cappellini, G.; Voltolina, D. SIRENE: A spatial data infrastructure to enhance communities' resilience to disaster-related emergency. *Int. J. Disaster Risk Sci.* **2018**, *9*, 129–142. [CrossRef]
41. Bordogna, G.; Kliment, T.; Frigerio, L.; Brivio, P.A.; Crema, A.; Stroppiana, D.; Boschetti, M.; Sterlacchini, S. A spatial data infrastructure integrating multisource heterogeneous geospatial data and time series: A study case in agriculture. *ISPRS Int. J. Geo-Inf.* **2016**, *5*, 73. [CrossRef]
42. Cobb, D.A.; Olivero, A. Online GIS service. *J. Acad. Librariansh.* **1997**, *23*, 484–497. [CrossRef]
43. Kok, B.; Van Loenen, B. How to assess the success of National Spatial Data Infrastructures? *Comput. Environ. Urban Syst.* **2005**, *29*, 699–717. [CrossRef]
44. Koontz, L.D. *Geographic Information Systems: Challenges to Effective Data Sharing*; General Accounting Office: Washington, DC, USA, 2003.
45. Goodchild, M.F.; Fu, P.; Rich, P. Sharing geographic information: An assessment of the Geospatial One-Stop. *Ann. Assoc. Am. Geogr.* **2007**, *97*, 250–266. [CrossRef]
46. Dawidowicz, A.; Kulawiak, M.; Zysk, E.; Kocur-Bera, K. System architecture of an INSPIRE-compliant green cadastre system for the EU Member State of Poland. *Remote Sens. Appl. Soc. Environ.* **2020**, *20*, 100362. [CrossRef]
47. EOSDIS. EOSDIS Glossary. Available online: https://earthdata.nasa.gov/learn/user-resources/glossary (accessed on 2 October 2020).
48. Chen, N.; Zhang, X.; Wang, C. Integrated open geospatial web service enabled cyber-physical information infrastructure for precision agriculture monitoring. *Comput. Electron. Agric.* **2015**, *111*, 78–91. [CrossRef]

49. Mazzetti, P.; Roncella, R.; Mihon, D.; Bacu, V.; Lacroix, P.; Guigoz, Y.; Ray, N.; Giuliani, G.; Gorgan, D.; Nativi, S. Integration of data and computing infrastructures for earth science: An image mosaicking use-case. *Earth Sci. Inform.* **2016**, *9*, 325–342. [CrossRef]
50. Bourova, E.; Maldonado, E.; Leroy, J.-B.; Alouani, R.; Eckert, N.; Bonnefoy-Demongeot, M.; Deschatres, M. A new web-based system to improve the monitoring of snow avalanche hazard in France. *Nat. Hazards Earth Syst. Sci.* **2016**, *16*, 1205–1216. [CrossRef]
51. Karantzalos, K.; Bliziotis, D.; Karmas, A. A scalable geospatial web service for near real-time, high-resolution land cover mapping. *IEEE J. Sel. Top. Appl. Earth Obs. Remote Sens.* **2015**, *8*, 4665–4674. [CrossRef]
52. Dahlhaus, P.; Murphy, A.; MacLeod, A.; Thompson, H.; McKenna, K.; Ollerenshaw, A. Making the invisible visible: The impact of federating groundwater data in Victoria, Australia. *J. Hydroinformatics* **2016**, *18*, 238–255. [CrossRef]
53. Wiemann, S.; Brauner, J.; Karrasch, P.; Henzen, D.; Bernard, L. Design and prototype of an interoperable online air quality information system. *Environ. Model. Softw.* **2016**, *79*, 354–366. [CrossRef]
54. Vosgerau, H.; Mathiesen, A.; Sparre Andersen, M.; Boldreel, L.O.; Hjuler, M.L.; Kamla, E.; Kristensen, L.; Brogaard Pedersen, C.; Pjetursson, B.; Nielsen, L.H. A WebGIS portal for exploration of deep geothermal energy based on geological and geophysical data. *Geus Bull.* **2016**, *35*, 23–26. [CrossRef]
55. Granell, C.; Díaz, L.; Gould, M. Service-oriented applications for environmental models: Reusable geospatial services. *Environ. Model. Softw.* **2010**, *25*, 182–198. [CrossRef]
56. Sun, Z.; Di, L.; Gaigalas, J. SUIS: Simplify the use of geospatial web services in environmental modelling. *Environ. Model. Softw.* **2019**, *119*, 228–241. [CrossRef]
57. Scott, G.; Rajabifard, A. Sustainable development and geospatial information: A strategic framework for integrating a global policy agenda into national geospatial capabilities. *Geo-Spat. Inf. Sci.* **2017**, *20*, 59–76. [CrossRef]
58. Steven, A.R. The US National Spatial Data Infrastructure: What is new? In Proceedings of the ISPRS Workshop on Service and Application of Spatial Data Infrastructure, Hangzhou, China, 14–16 October 2005.
59. Bermudez, L. New frontiers on open standards for geo-spatial science. *Geo-Spat. Inf. Sci.* **2017**, *20*, 126–133. [CrossRef]
60. Percivall, G. The application of open standards to enhance the interoperability of geoscience information. *Int. J. Digit. Earth* **2008**, *3*, 14–30. [CrossRef]
61. Carr, T.R.; Rich, P.M.; Bartley, J.D. The NATCARB geoportal: Linking distributed data from the Carbon Sequestration Regional Partnerships. *J. Map Geogr. Libr.* **2008**, *4*, 131–147. [CrossRef]
62. El-Gamily, H.; Al-Awadhi, N.; El-Magd, I.A.; Watkins, D. Kuwait Integrated Environmental Information Network (KIEIN-IV): A way of developing national environmental indicators for better environmental information dissemination. *J. Spat. Sci.* **2015**, *60*, 403–414. [CrossRef]
63. Brodeur, J.; Coetzee, S.; Danko, D.; Garcia, S.; Hjelmager, J. Geographic Information Metadata—An Outlook from the International Standardization Perspective. *ISPRS Int. J. Geo-Inf.* **2019**, *8*, 280. [CrossRef]
64. Granell, C.; Miralles, I.; Rodríguez-Pupo, L.E.; González-Pérez, A.; Casteleyn, S.; Busetto, L.; Pepe, M.; Boschetti, M.; Huerta, J. Conceptual architecture and service-oriented implementation of a regional geoportal for rice monitoring. *ISPRS Int. J. Geo-Inf.* **2017**, *6*, 191. [CrossRef]
65. Iosifescu-Enescu, I.; Matthys, C.; Gkonos, C.; Iosifescu-Enescu, C.M.; Hurni, L. Cloud-based architectures for auto-scalable web Geoportals towards the Cloudification of the GeoVITe Swiss academic Geoportal. *ISPRS Int. J. Geo-Inf.* **2017**, *6*, 192. [CrossRef]
66. Dareshiri, S.; Farnaghi, M.; Sahelgozin, M. A recommender geoportal for geospatial resource discovery and recommendation. *J. Spat. Sci.* **2019**, *64*, 49–71. [CrossRef]
67. Kadochnikov, A.; Tokarev, A.; Zavoruev, V.; Yakubailik, O. Prototype of city environmental monitoring system based on geoportal technologies. In *Proceedings of IOP Conference Series: Materials Science and Engineering*; IOP Publishing: Bristol, UK, 2019; p. 062052.
68. Vahidi, H.; Klinkenberg, B.; Yan, W. Trust as a proxy indicator for intrinsic quality of Volunteered Geographic Information in biodiversity monitoring programs. *Giscience Remote Sens.* **2018**, *55*, 502–538. [CrossRef]
69. Fonte, C.C.; Antoniou, V.; Bastin, L.; Estima, J.; Arsanjani, J.J.; Bayas, J.-C.L.; See, L.; Vatseva, R. Assessing VGI data quality. In *Mapping and the Citizen Sensor*; Ubiquity Press: London, UK, 2017; pp. 137–163.
70. Yap, L.F.; Bessho, M.; Koshizuka, N.; Sakamura, K. User-generated content for location-based services: A review. In *Virtual Communities, Social Networks and Collaboration*; Lazakidou, A., Ed.; Springer: New York, NY, USA, 2012; Volume 15, pp. 163–179.
71. Yan, Y.; Feng, C.-C.; Wang, Y.-C. Utilizing fuzzy set theory to assure the quality of volunteered geographic information. *GeoJournal* **2017**, *82*, 517–532. [CrossRef]
72. Flanagin, A.J.; Metzger, M.J. The credibility of volunteered geographic information. *GeoJournal* **2008**, *72*, 137–148. [CrossRef]
73. Bishr, M.; Mantelas, L. A trust and reputation model for filtering and classifying knowledge about urban growth. *GeoJournal* **2008**, *72*, 229–237. [CrossRef]
74. West, S.E.; Pateman, R.M. Recruiting and retaining participants in citizen science: What can be learned from the volunteering literature? *Citiz. Sci. Theory Pract.* **2016**, *1*, 15. [CrossRef]
75. Doyle, B.; Lopes, C.V. Survey of technologies for web application development. *arXiv* **2008**, arXiv:0801.2618v1.
76. Custers, B.; Uršič, H. Big data and data reuse: A taxonomy of data reuse for balancing big data benefits and personal data protection. *Int. Data Priv. Law* **2016**, *6*, 4–15. [CrossRef]
77. Yang, C.; Goodchild, M.; Huang, Q.; Nebert, D.; Raskin, R.; Xu, Y.; Bambacus, M.; Fay, D. Spatial cloud computing: How can the geospatial sciences use and help shape cloud computing? *Int. J. Digit. Earth* **2011**, *4*, 305–329. [CrossRef]

78. Anderson, K.; Ryan, B.; Sonntag, W.; Kavvada, A.; Friedl, L. Earth observation in service of the 2030 Agenda for Sustainable Development. *Geo-Spat. Inf. Sci.* **2017**, *20*, 77–96. [CrossRef]
79. Elwood, S. Volunteered geographic information: Future research directions motivated by critical, participatory, and feminist GIS. *GeoJournal* **2008**, *72*, 173–183. [CrossRef]
80. Koswatte, S.; McDougall, K.; Liu, X. Ontology driven VGI filtering to empower next generation SDIs for disaster management. In Proceedings of the Research @ Locate 2014, Canberra, Australia, 7–9 April 2014.
81. Bordogna, G.; Carrara, P.; Kliment, T.; Frigerio, L.; Sterlacchini, S. Spatial data infrastructures empowered by interoperable volunteered geographic information. *Plurimondi* **2017**, *8*, 107–113.

International Journal of
Geo-Information

MDPI

Article

Citizens' Spatial Footprint on Twitter—Anomaly, Trend and Bias Investigation in Istanbul

Ayse Giz Gulnerman [1,*], Himmet Karaman [1], Direnc Pekaslan [2] and Serdar Bilgi [1]

1 Geomatics Engineering Department, Istanbul Technical University, 34469 Istanbul, Turkey; karamanhi@itu.edu.tr (H.K.); bilgi@itu.edu.tr (S.B.)

2 School of Computer Science, University of Nottingham, Nottingham NG8 1BB, UK; Direnc.Pekaslan@nottingham.ac.uk

* Correspondence: gulnerman@itu.edu.tr

Received: 11 March 2020; Accepted: 4 April 2020; Published: 7 April 2020

Abstract: Social media (SM) can be an invaluable resource in terms of understanding and managing the effects of catastrophic disasters. In order to use SM platforms for public participatory (PP) mapping of emergency management activities, a bias investigation should be undertaken with regard to the data related to the study area (urban, regional or national, etc.) to determine the spatial data dynamics. Thus, such determinations can be made on how SM can be used and interpreted in terms of PP. In this study, the city of Istanbul was chosen for social media data research area, as it is one of the most crowded cities in the world and expecting a major earthquake. The methodology for the data investigation is: 1. Obtain data and engage sampling, 2. Identify the representation and temporal biases in the data and normalize it in response to representation bias, 3. Identify general anomalies and spatial anomalies, 4. Manipulate the trend of the dataset with the discretization of anomalies and 5. Examine the spatiotemporal bias. Using this bias investigation methodology, citizen footprint dynamics in the city were determined and reference maps (most likely regional anomaly maps, representation maps, time-space bias maps, etc.) were produced. The outcomes of the study can be summarized in four steps. First, highly active users generate the majority of the data and removing this data as a general approach within a pseudo-cleaning process means concealing a large amount of data. Second, data normalization in terms of activity levels, changes the anomaly outcome resulting from diverse representation levels of users. Third, spatiotemporally normalized data present strong spatial anomaly tendency in some parts of the central area. Fourth, trend data is dense in the central area and the spatiotemporal bias assessments show the data density varies in terms of the time of day, day of week and season of the year. The methodology proposed in this study can be used to extract the unbiased daily routines of the social media data of the regions for the normal days and this can be referred for the emergency or unexpected event cases to detect the change or impacts.

Keywords: volunteered geographic information; social media; public participation; spatiotemporal bias

1. Introduction

Over the past two decades, public participatory (PP) mapping has evolved rapidly from paper to digital mapping [1–3]. Social media (SM) platforms that provide huge volume of crowdsourced data to digital mapping are not regarded as an appropriate participatory mapping resource, even though SM can be used as a pioneering platform that increases individual participation rates for PP mapping with its features of large data production capacity and uninterrupted data collection platforms. Global Navigation Satellite Systems' (GNSS) antennas in smart devices and public use of these devices, enable and foster location-based crowdsourced applications. The data is generated by the users of these applications and is referred to as Volunteered Geographic Information (VGI) [4,5]. The users can be thought of as unconscious volunteers for social media (SM) VGI, as deliberate volunteers for peer production VGI and as public participators for in citizen science based VGI [6–8]. The way of producing these forms of VGI is referred to as neo-geography in that it adopts neo-geographers (i.e., volunteers) who contributes to mapping activity without being expert [9]. This inexperience with regards to data production is questioned in the context of data quality [10–12], demographic bias (such as, gender, socioeconomic and educational aspects) [13,14] sampling bias (referring to volunteer sampling) and its impact on the generated data [15,16].

In their very first form, citizen science projects in the very first forms were carried out with the use of paper maps [1]. However, with the technological developments in computer and web sciences, nowadays they are mostly carried out with the help of a range of online platforms [17–19] designed for collecting data for citizen science purposes. These platforms are designed to collect data within a limited time period for specified purposes. On the other hand, there are also dedicated webpages deployed for local citizen science projects such as DYFI (Did You Feel It) [20]. DYFI served by the USGS, collects data from volunteers with regards to how intense they feel an earthquake, in order to show the extent of damage and shaking intensities on a map. Although the project is designed and structured to collect and process data, the count in terms of volunteers' response responses to individual earthquakes (the participation rate) to each earthquake is pretty low [21]. Although the project has been replicated for different countries such as New Zealand, Italy and Turkey [20,22] there is still not any organized approach by the relevant authorities [23].

SM platforms, although not seen as the proper way to organize citizen-based projects, still have a high and continuous data serving capacity around the world involving 3 billion of users [24]. In fact, SM with its wide, continuous, active data collection and serving capacity can be used as a pioneering platform for carrying out citizen-based projects especially for monitoring out of the ordinary events such as multi-emergency circumstances in big cities. However, bias in SM data can be seen an obstacle with regard to such projects. In order to use SM data as a citizen-based monitoring system, the data georeferenced in a particular area (such as a city, region or country) should be pre-assessed. In this way, ways in which to use and interpret the SM data as a tool with regard to city monitoring can be inferred.

1.1. SMD Studies on Emergency Mapping

SMD has already in use for disaster management for more than a decade. Houston, et al. [25] present a comprehensive literature on the functionality of SM in terms of the disaster management phases. The very first example of event detection with SMD was conducted by Sakaki, et al. [26]. Social media was also considered for disaster relief efforts by Gao, et al. [27] and Muralidharan, et al. [28], for crisis communication by Acar and Muraki [29] and McClendon and Robinson [30] and for evacuation ontology by Ishino, et al. [31] and Iwanaga, et al. [32].

Most of the former and latter studies have focused text-based filtering at first for detecting an event [33–36]. The filtering techniques were mostly used for a limited number of keywords related to a disaster domain (such as; hurricane, flood and storm for meteorological disasters) [36]. In respect to that, this kind of studies has selection bias due to determined keywords [37]. This might not be a problem for coarse-grained spatial analyses due to an abundance of data however; this may lead cause detection problems for the local events. Yet the studies are mostly basing on an event type instead of being a comprehensive monitoring system to detect any disastrous event anomalies. In addition to that, the spatial grain of the detection analyses are mostly coarse as county or city level [35,36], even the studies are focusing the spatial consideration at first [37].

Historical data exploration plays an important role in comprehensively monitoring unusual events in a fine-grain spatial level within a city [37,38]. That is why SMD should be assessed in terms of anomalies, trends and bias. In this way, citizen footprints on social media can be interpreted in several ways as base maps for further local event detection investigations. The most operational use of the proposed method can be on emergency mapping because of rapid succession and the ability to compare the difference with the daily life trends. Since, phases of emergency management requires rapid real-time data on the region of interest to compare the ongoing situation with the preparedness plans.

1.2. Aim and Region of the Study

In this study, the aim is to propose a methodology for bias investigation in order to reveal citizens' footprints in a city and to produce reference maps (most likely regional anomaly maps, representation maps, spatiotemporal bias maps, etc.). The city of Istanbul was chosen as the case study area as it is one of the biggest cities in the world with 18 million inhabitants and one which expects a major earthquake that could possibly have a catastrophic impact on the city [39,40]. Twitter platform is used as the data source, since it is one of the most commonly-used social media network for spreading the information all over the world [24,41]. Such data is referred to as Social Media Data (SMD) in this paper. With regard to the investigation of the SMD, the methodology includes the following steps: data acquisition and data tidying, determination of the representation and temporal bias in the data, data normalization for removing user representation bias, detection of anomalies in non-spatial and spatial data, the discretization of anomalies and the production of a trend map and the investigation of spatiotemporal bias. The data investigation outcomes in this study are discussed from the perspective of citizen-based event mapping with the use of SM data. In this way, the assessment techniques with regard to SM data is presented in this study for the benefit of the citizen-based geospatial mapping capacity building.

1.3. Bias in SM-VGI

Nearly half of the world population are unrepresented in SM due to internet censorship or access unavailability [42]. Consequently, this study of bias in social media data (SMD) starts with a consideration of the inadequacy of the technological and political infrastructure. Moreover, the usage rate of smart devices and computers affects the representation rate of societies and individuals. In addition to the representation of societies may not be equal which is mostly explained in terms of demographic (age, education, social status) differences. For several reasons, some parts of a society might be over-represented while other parts may be under or not represented at all [15,43]. However, determining demographic bias is mostly not possible due to the lack of availability of volunteers' personal data in VGI [16,44]. Additionally, volunteers of the platform might not even be a person, in that they can be a bot, a staff team member (who embodies and promotes a company) and/or a troll (a fake account).

Basiri, et al. [44] suggest that there are more than 300 types of bias and that crowdsourced data might tend to have some of those. Since volunteers are contributing directly without being asked to participate, SMD do not include *"selection bias."* However, the volunteers diversely show *"representation bias"* due to their immense activity rate. Also, population density possibly creates *"systematic bias"* over space. Due to the reputation of places, volunteers tend to share popular locations more, to flaunt and to be visible at the same page with others. This is referred to as *"Bandwagon bias"* and as which affects the spatial distribution of VGI. *"Status-quo bias"* is also a reflection of the demographic background of volunteers over space and is seen as specific types sharing a point of interest [44]. While Bandwagon and Status-quo biases plead to spatial bias, the temporal dimension creates varying patterns with regard to changing activities or participation rate s corresponding to the time of day, day of the week and season of the year. This changing trend is referred to as *"spatiotemporal bias"* and entails misinterpretation in the case of a comparison of improper temporal slices.

There have been number of attempts to identify spatial patterns, trends and biases with regard to the SMD. Li, et al. [45] conducted a research on Twitter and Flickr data at the county level in order to understand users' behavior as a result of demographic characteristics. The study also offers some exploratory graphs about the number of tweets over time and presents tweet density maps that are normalized by the population density at the county level. Another study is about understanding the demographic characteristics of users who enable location services on Twitter [41]. The study offers strong evidence based on demographic effects on the tendency of enabling geo-services and geotagging. Lansley and Longley [46] searched for the dynamics of the city of London using topic modelling and quantified the correspondence of topics with the users' characteristics and location. Arthur and Williams [47] conducted a research to identify regional identity and inter-communication between cities. The researchers found that regional identity that is quantified by text similarity and sentiment analysis of posted tweets in terms of several UK cities. Malik, et al. [48] conducted research to find the relationship between the census population and the number of geotagged tweets using statistical tests. They found that there were no impacts of population on tweets density. However, they did identify several other impacts such as the income level of the population, being in a city center and the age of the population. Another study with regard to the user bias in terms of the tweeting frequency of users, proposed the removal of the top 5% of active users from the data to avoid such biases [49].

In this respect, the studies carried out mostly focused on the demographic background of the users, the relationship between population and tweets density, representation bias or topic variances over coarse-grained space. However, those were not searching for a year based fine spatial data pattern and for biases that can allow the monitoring a city with a better interpretation of spatiotemporal data. However, this study is designed to present spatiotemporal variances with regard to representation diversity and anomalies and trends in the data, without blocking or removing any users' data that is a commonly adopted way of previous studies for data cleansing.

2. Materials and Methods

The methodology of this study is presented in five subsections and the conceptual flow of methodology can be followed from Figure 1. In the first Section 2.1, details of data acquisition techniques and data tidying steps are explained. In the second Section 2.2, the data investigation methodology in terms of users' activity levels and temporal levels are introduced. In addition, the application of user-weighted normalization techniques is introduced to investigate the impact of users' activity levels on the temporal data variation. In the third Section 2.3, details of the investigation of data anomalies are presented in two stages anomaly detection over non-spatial data and anomaly detection over spatially-indexed data. In the fourth Section 2.4, the methodology involved in obtaining regular data is explained in order to produce spatially-indexed overall trend data and a map. In the fifth Section 2.5, bias assessment details are incorporated into the methodology flow. Bias investigation in terms of temporal levels is explained step-by-step in this last part.

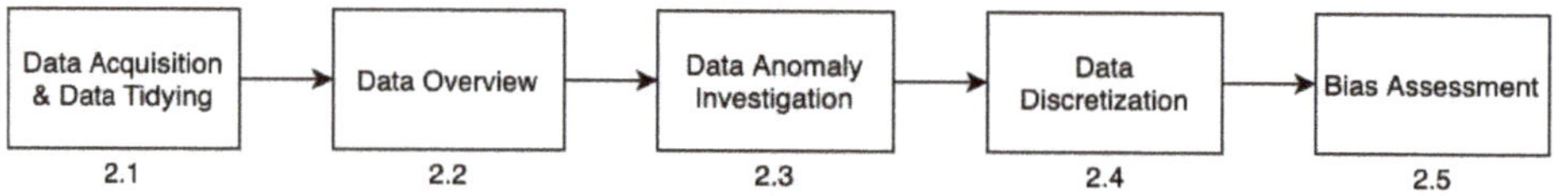

Figure 1. Conceptual Data Investigation Flow in the Subsections.

2.1. Data Acquisition and Data Tidying

The data flow of this part is composed of the following steps—data downloading, storing, sampling and tidying (Figure 2). Geo Tweets Downloader (GTD) [50] is chosen as the downloading software, since this study aims to monitor tweeting pattern within a spatial bounding box. GTD is a software that uses Twitter APIs to download georeferenced tweets and ingests this data into PostgreSQL in real time. GTD has acquired data during the year 2018 within the bounding box of Istanbul City. There were several interruptions such as electricity and internet cuts in the downloading server during this data acquisition process that lasted for one full year. Therefore, the acquired data has been sampled into weeks. Data continuity from Monday to Sunday inclusive was determined as the only sampling rule and starting from the first week of each month, each hour was checked as to whether or not there was a missing data. Based on this, the data was composed of 12 complete selected weeks belonging to each month of the 2018 year.

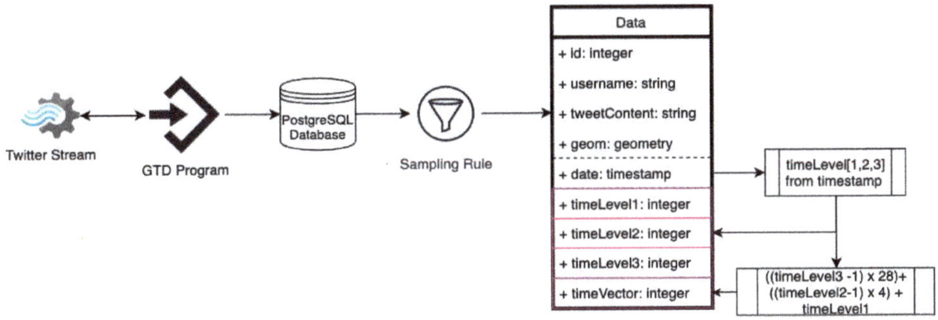

Figure 2. Data downloading and tidying flow.

Data was tidied with the addition of three temporal level columns for further investigation in the following sections. The first level (timeLevel1) shows four different time intervals of the day; night (00:00–06:00), before midday (6:00–12:00), after midday (12:00–18:00) and evening (18:00–00:00). The second level (timeLevel2) shows the day of the week from Monday to Sunday inclusive. The third level (timeLevel3) shows the month of the year from January to December inclusive. In the database time level values are represented by an integer value. On this basis, timeLevels 1, 2 and 3 have 4, 7 and 12 integer values, respectively. In addition to these time level columns, a time vector column was calculated with the formula given in Figure 2. According to this calculation, the timeVector has values from 1 (night, Monday, January) to 336 (evening, Sunday, December).

2.2. Data Overview

Data investigation was composed of a user representation level search, the tweet count variation in terms of time level and a normalized tweet count in terms of time level. The investigation started with data generators representation levels that may cause noisy weights over data. The activity level of each user was determined by using one of the most common k-means clustering technique on the overall users' activities. In terms of the activity level decision, 1. data was grouped by username, 2. a tweet count for each user was calculated, 3. the min, mean, standard deviation values of tweet counts were calculated (Figure 3). The histogram of the tweet count was not well represented since it was highly right skewed. However, a summary of tweet count is as follows:

Min.	1st Qu.	Median	Mean	3rd Qu.	Max.	Std.
1.00	1.00	4.00	53.91	17.00	4378.00	185.474

4. As the tweet number 1 is associated with many users in the overall dataset, it has been separated as the first cluster. Since the data is not normally distributed, the k-means clustering is applied on the remained dataset by separating it into 3 clusters. In the k-means clustering implementation, we followed the traditional approach as listed below.

1. 4 is chosen to be clusters number
2. Place the centroids c_1, c_2, c_3 and c_4 randomly
3. Repeat steps 4 and 5 until convergence or until the end of a fixed number of iterations
4. For each user's tweet number—find the nearest centroid (c_1, c_2, c_3 and c_4)—assign the user to that cluster
5. For each cluster j = 1..4 - new centroid = mean of all points assigned to that cluster
6. End

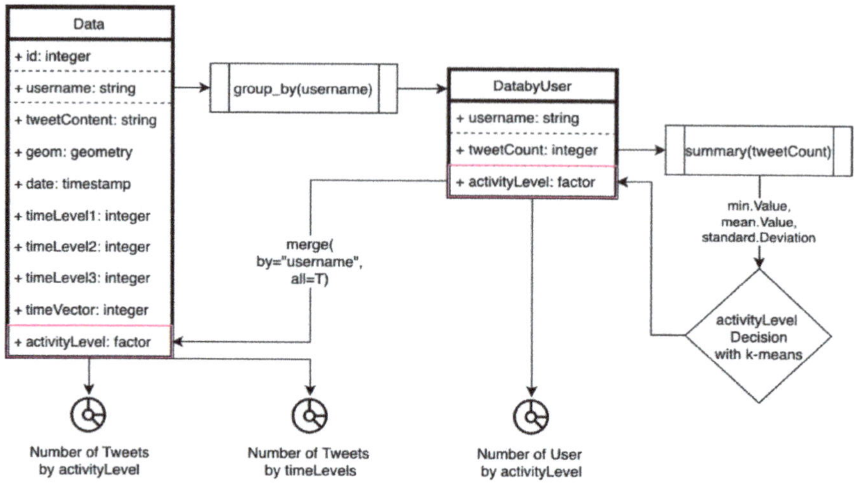

Figure 3. Data investigation methodology flow.

Each cluster representation can be seen in Table 1.

Table 1. User representation level clusters detail.

Cluster	Min	Max	Average	Std. Dev.
c_1	1	1	1	0
c_2	2	261	26.68	41.18
c_3	262	944	497.95	185.44
c_4	947	4378	1393.81	397

According to this, the min value 1 was specified as the single representation class of the activity level, the max value of cluster 2 (261) is bounding the upper part of the second activity class, the max value of cluster 3 (944) is bounding the upper part of the third activity class and lastly, the max the max value of cluster 4 (4378) is bounding the upper part of the fourth activity class. With the addition of the user activity levels to datasets as shown in Figure 3, the number of users and tweets in terms of activity level were investigated in terms of the plots in the "Results" section.

The number of tweets in terms of time levels was also investigated in terms of the circular bar plots in the "Results" section, to make general inferences with regard to any temporal bias. The variation in the number of tweets was firstly plotted over the raw data count without any weighting. In addition, the number of tweets is normalized with the technique described below. The normalized number of tweets in terms of time level is investigated in order to discuss the impacts of the representation level to temporal data variations.

Due to the nature of the SMD data, there is a noticeable variance in the number of tweets and user behaviors which lead to some data being "outlier," a term which is used to describe any unusual behaviors. However, it is often not possible to determine whether those outlier data are invalid or whether the data represent valid information for the overall dataset or task to achieve. Therefore, in this paper, those outlie tweet numbers/users' behaviors, which deviate noticeably from the overall data, are purposely not removed but rather all the users are assigned weights based on their activity levels. Through the use of these assigned weights, each user is represented at various levels in the used data latter. Thus, we followed two-step normalization procedures (entitled inter-normalization and intra-normalization) to capture and handle each particular user's behavior and the tweet information that exists.

In terms of intra-normalization, each users' weight is determined based on their tweeting numbers in comparison with the overall tweet dataset. In this paper, we follow the analogy that if user A tends to send large numbers of tweets (e.g., ~100) on a normal day, this number of tweets will increase–accordingly in the event of a disaster occurrence. On the other hand, user B, who tends to send a smaller number of tweets (e.g., 1 or 0) on a normal day, will increase the number of tweets in parallel with this normal daily tweet number. Therefore, user A is assigned with a smaller weight than is user B for each tweet. This approach can allow us to cope with any discrepancy in tweeting behavior between the users. The weight determination of each user is carried out on their overall tweet numbers in the overall dataset and each user's tweet number is divided by the maximum (c_{max}) number of tweets.

The user weighting is implemented as follows:

$$w_i = \left(1 - \frac{u_i}{c_{max}}\right), \tag{1}$$

where w_i is the determined weight of the i'th user, u_i is the total tweet numbers and c_{max} is the maximum number of tweets which were sent respectively.

After each user's weight is determined, the overall tweet number (n_i) of each user in the general pool is calculated by simply multiplying the user weight with the users' overall tweet number. Consequently, the inter-normalization is completed by allowing each user tweet's number to contribute in a 'compromising manner' to the general pool. In addition, by following this procedure, each user's contribution is utilized without being ignored or removed from the dataset. After gathering the weighted tweet numbers (n_i) from each user, in the intra-normalization, time-based tweet numbers are summed, and the commonly-used cube root transformation is implemented for each time-based tweet number (2) and the min-max normalization is applied on this gathered (c^t) tweet numbers dataset

$$c^t = \sqrt[3]{\sum_{i=1}^{N} n_i^t}, \tag{2}$$

where c^t is the transformed tweet number on a time t. N is the weighted user numbers which are tweeted at time t and n_i is the weighted tweet numbers from the intra-normalization. Consequently, that applying intra and inter-normalization procedures enable the representation of each user and each tweet in the final normalized dataset.

2.3. Anomaly in Data

The anomaly investigation was carried out in two stages that were applied to both non-spatial data and spatial data. The AnomalyDetection R package [51,52] was adopted for the applications. The package was created by Twitter for anomaly detection and for visualization where the input Twitter data is highly seasonal and also contains a trend. The package utilizes the Seasonal Hybrid Extreme Studentized Deviate test (S-H-ESD) which uses time series decomposition and robust statistical metrics along with the ordinary Extreme Studentized Deviate test (ESD). The S-H-ESD provides sensitive anomaly output specializing in Twitter data, with the ability to detect global anomalies as well as anomalies have a small magnitude and which are only visible locally. To compute the S-H-ESD test, Anomaly Detection package provides support for the time series method and for the vector of numerical values method where the time series method gets the timestamp values as inputs while the vector method requires an additional input variable "period" for serialization. Both methods require a maximum anomaly percentage, "max_anoms" (upper bound of ESD) and the "direction" of the anomaly (negative, positive or both) [51,52].

In this study, the vector method of the AnomalyDetection package is used. The period variable is set as 28, since 7 days of data is used and since each day is divided into 4 periods according to the hour of the given tweets. Hochenbaum, et al. [53] experimented with 0.05 and 0.001 as the maximum anomaly percentages in their experiments. They achieved better precision, recall and F-measure values with 0.001, though with very few differences between each other. Considering the experiment setup of Hochenbaum, Vallis and Kejariwal [53], the maximum anomaly percentage is chosen as 0.02 as the optimum value.

For the first anomaly application stage as presented in Figure 4, 1. data are grouped by timeVector and summarized as tweetCount, userCount and normalizedTweetCount, 2. the counts are ordered by timeVector, 3. anomaly detection is applied to the vectorized counts.

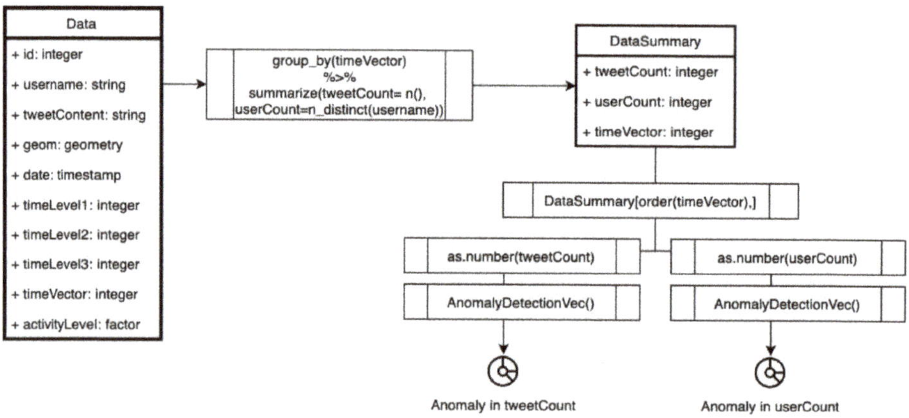

Figure 4. Anomaly Detection Stage 1.

For the second stage, spatiotemporal anomaly is assessed as presented in Figure 5. The steps are 1. spatial grids (1 × 1 km) within the Istanbul bounding box, which is spatially joined with the data, 2. data is grouped by timeVector and gridId and summarized as tweetCount (as the distinct count of the usernames) and grid geometry, 3. a figure and ground map is visualized to display unrepresented spatial grids, 4. anomaly detection is applied to the spatially normalized tweetCount for each grid, 5. anomaly assessment is done with the normalized anomaly rate in terms of time intervals, 6. an anomaly map is visualized and spatial pattern is tested with the Moran I.

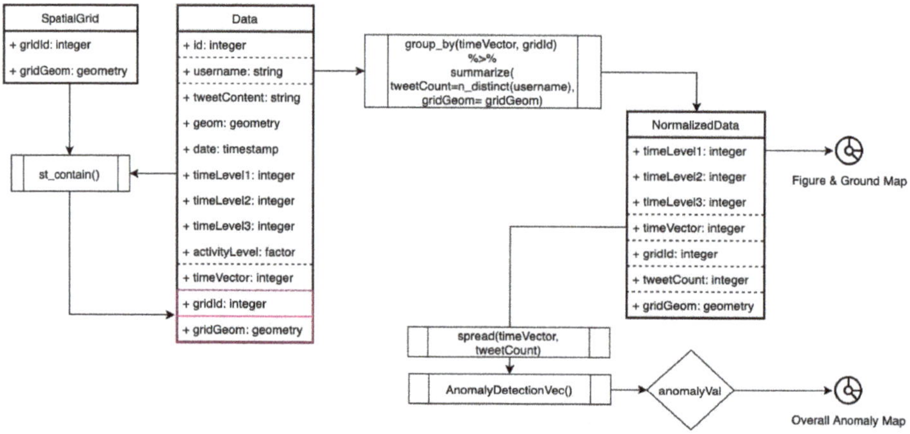

Figure 5. Anomaly Detection Stage 2.

In the second step, tweets from the same user within the same time interval and a 1 × 1 km grid are counted as 1 tweet. This is done to avoid over-representation of a user. In the fifth step, the normalized anomaly rate is formulated (3) for the anomaly assessment. The anomaly and expected values that are provided by the AnomalyDetectionVec part is used with the timeVector indexes (i) in this formula and the normalized anomaly rate is calculated for each grid and timeVector pair. The overall anomaly map is produced with the sum of the normalized anomaly rate for each grid. By this normalized anomaly rate values, the most anomalous spatial grids were plotted and tested with the Moran's *I* algorithm.

$$\text{normalized anomaly rate} = \frac{anomalyValue - expectedValue}{\sum_{i(timeVector)} \text{anomalyValue} - \text{expectedValue}} \tag{3}$$

Moran's *I* is the measure of global and local spatial autocorrelation. Global and local Moran's I are utilized for this part and for the following parts of the study, in order to determine the observed anomalies, trends and temporal differences, either clustered, dispersed or random in space. The "spdep" R package [54] was used to calculate global and local Moran's *I* which are formulated (4), (5) based on the feature's location and the values of the features [55].

$$I = \frac{n}{\sum_{i=1}^{n} \sum_{j=1}^{n} w_{ij}} \frac{\sum_{i=1}^{n} \sum_{j=1}^{n} w_{ij}(x_i - \overline{x})(x_j - \overline{x})}{\sum_{i=1}^{n} (x_i - \overline{x})^2} \tag{4}$$

$$I_i = \frac{(x_i - \overline{x})}{\sum_{k=1}^{n} (x_k - \overline{x})^2 / (n-1)} \sum_{j=1}^{n} w_{ij}(x_j - \overline{x}). \tag{5}$$

The variables are n = number of features indexed by i and j, x = spatial feature values, $\overline{x}$ = mean of x, w_{ij} = matrix of feature value weights. The values of Moran's *I* range from −1 (negative spatial autocorrelation) to 1 (positive spatial autocorrelation) and returns 0 value for a random distribution. The Spdep package provides moran.test() and localmoran() functions. In this study, both for global and local spatial autocorrelation calculations, moran.test() and localmoran() functions were used with the following arguments: x (the numeric vector of the feature attributes), listw (spatial weights for neighboring lists that is calculated by the nb2listw function in the spdep package), zero.policy (specified as TRUE to assign zero value for features with no neighbors). The functions' return values include Moran statistics (I, I_i) and the p-value of the statistics (p value, Pr()). A value of less than 0.05 for the p-value means that the hypothesis is accepted and is spatially correlated for Moran's *I* [54,55]. It is also taken into consideration for interpreting the results of all Moran's *I* tests in this study.

2.4. Data Discretization and Trends

In this part of the study, the trend dataset is manipulated as presented in Figure 6. Steps are as follows: 1. detected anomalies in the data were discretized, 2. the discretized anomalies were replaced with expected values in terms of gridId and timeVector and assigned as regular data, 3. data were grouped by gridId with the summary of the tweetCount mean value for the overall trend data, 4. an overall trend map was produced and tested with Moran I for the trend values' spatial pattern determination. The trend map so produced represented the general dynamics of the city and was also used as the reference in order to quantify the spatiotemporal bias in the next section.

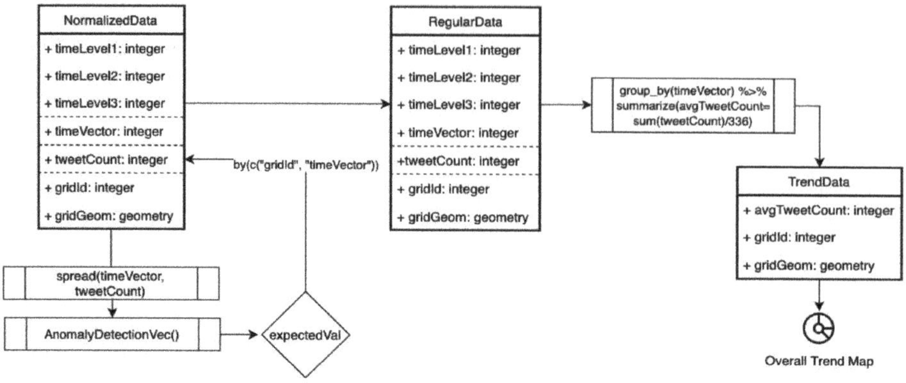

Figure 6. Data replacement with expected value.

2.5. Spatiotemporal Bias Assessment

The spatiotemporal bias was assessed in terms of hour, day and seasonal levels and the flow of the assessment was displayed in the Figure 7. The assessment steps are as follows: 1. data was divided into sub-datasets in terms of time levels as 4 sub-datasets (night, bmidday, amidday, evening) for timeLevel1, 7 sub datasets (from Monday to Sunday) for timeLevel2, 4 datasets (winter, spring, summer, autumn) for timeLevel3, 2. these sub-datasets were grouped by gridId and summarized as average tweetCount, 3. each grouped and summarized dataset was assessed with a comparison between the avgTweetCount of each grid in the sub-data and the trend data, 4. Maps of the bias assessment were visualized and tested with the Moran I for spatial pattern investigation.

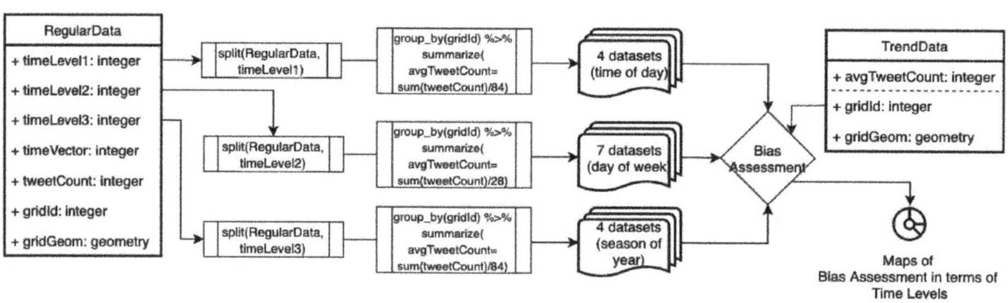

Figure 7. Bias assessment flow.

In the bias assessment part of the flow, the comparison was handled by taking the average tweet counts' difference in sub-datasets and trend data. This differences in value were plotted in five classes; two classes (low, less) for negative values, a trend class for 0 values, two classes (more, high) for positive values. These values were tested with the Moran I for quantifying the spatial correlation of the values.

3. Results

Data acquired and sampled for this study cover over 4 million tweets generated by nearly 76 thousand volunteers. The most active volunteer has 4378 tweets, while one third of all volunteers have just one tweet within all data. Average tweet count per user is 54 with 186 standard deviation. This reveals that the activity of volunteers is disunited and the most active group is highly overrepresented. This condition requires scrutinization to understand whether those groups are more active due to the unordinary situations or as their general behavior. As initial step to explore data, user's activity levels are classified by depending on the k-means clustering method instead using min, mean and standard deviation values of user's tweet numbers as explained in the methodology. In respect to this, users' representation levels (R.L.) are classified as a single (1), second level representation (2), third level (3) and fourth level (4) as the highest active class. The percentage of users per representation levels (a) and the percentage of tweet amounts corresponding to users' representation levels (b) are illustrated in Figure 8 as chord diagrams. The diagrams present, nearly 90% of users represent themselves one time or less than 262 times, just the opposite their total representation in data equals to less than 30%. This initial analysis uncovers the reality in diverse representation levels of users and this variation points to representation bias. Many studies in literature omit the data come from over-represented groups in order to decrease representation inequality but also cause a big chunk of data to conceal in this way.

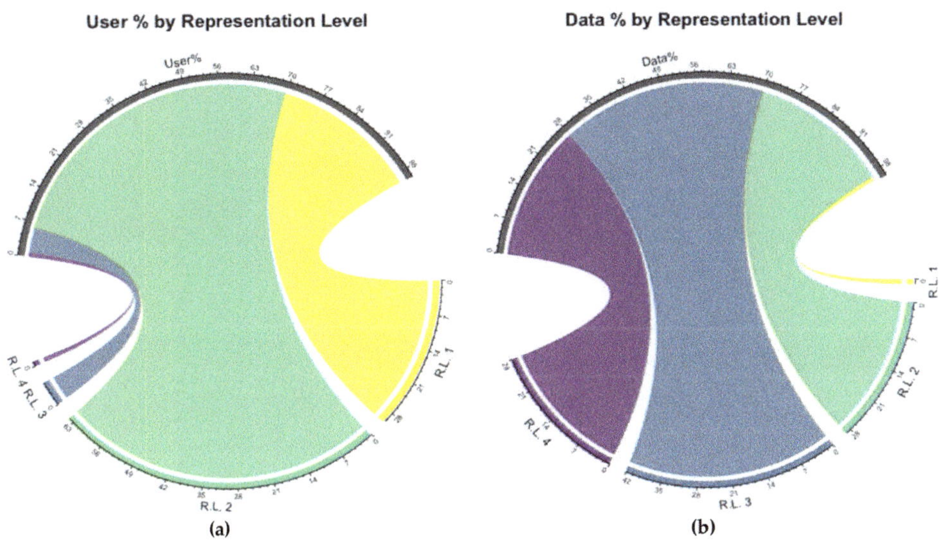

Figure 8. Representation level (R.L.) by (**a**) percentage of users; (**b**) percentage of data.

The total number of tweets belongs to each user helps to draw an overview of users' representation. However, this might be misinterpreted without consideration of temporal variation. The high representation of a user may indicate either over-representation regularly spread overtime or a specific situation that has greater importance for just one period of time. In other words, some of the users considered as over-represented themselves while they do this representation in a limited time interval but underrepresented for the rest of the time. In respect to that, representation varies among users, likewise, it varies for a user temporally due to several circumstances such as seasonal (summer or winter), emergencies (natural disasters, terror attacks), politics (election, referendum).

In order to explore a number of tweets to each temporal level in one hand, a circular bar plot is combined. This gives the general temporal variation view of tidied data with some delusions due to inequalities in user/bot representations. Each bar in the plot represents daily data size according to the stacked time slices of the day as it is explained in the methodology section. Data without any weights considering user representation and normalization is visualized in Figure 9a. With respect to that, data production during a nighttime interval is pretty low or almost none for some days, not because of system failure but because of temporal reasons. The biggest number of tweets is generated in the evening time for nearly every day though it is explicitly less than after middays for some days such as 20, 49, 50 (Figure 9a).

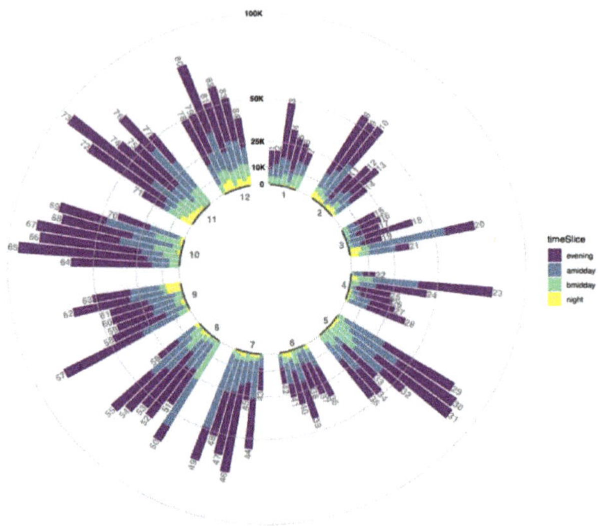

(a)

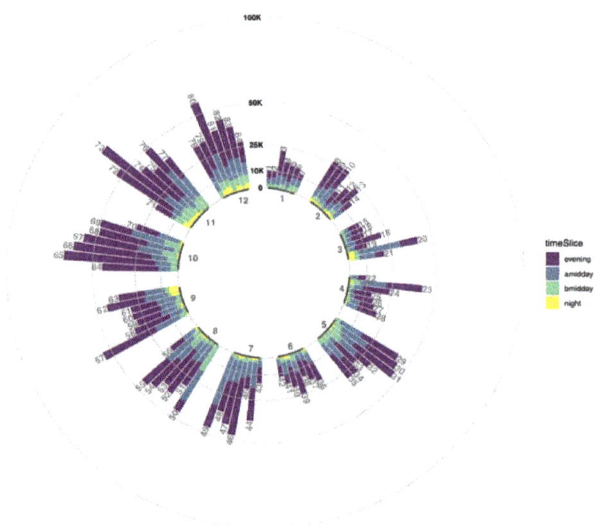

(b)

Figure 9. Circular stacked bar plots for the 2018-year data (**a**) number of tweets in each temporal level; (**b**) normalized number of tweets in each temporal level.

There might be several misinterpretations while assessing the number of tweets in regard to time slices due to diverse representation levels of users. In order to avoid this, number of tweets to temporal level 1 is normalized by weight assignment to each user. According to that normalization, tweet counts for each time slice are recalculated by taking the sum of each user tweet count multiplied with its user weight. Normalized data is displayed in Figure 9b as similar to Figure 9a. The number of tweets is obviously decreasing for all bars and a higher decreasing relative to previous numbers means higher overrepresentation level is normalized for time slices.

Diversity in representation level might create uncertainty on the number of data and requires to question whether data has any trend to do further inferences depending on that. In this general perspective without any location-based dimension, data is assessed with the anomaly detection algorithm in order to extract any trending activity. Three anomaly assessments are performed over tweet count (a), user count (b) and normalized tweet count (c) (Figure 10) respect to the time vector. While tweet count and user count have 6 anomaly slices which 4 of them are matched with each other, the normalized count has 3 anomaly slices that are all the common with both tweet and user count anomalies. According to these matches, diversity in representation creates three more anomalies than the normalized representation. However, the matching anomaly slices 195 between the tweet and user count is also notable to be considered even it is not in the normalized count anomaly. Anomaly values in overall data are detected at 1.79%, 1.79% and 0.89% respectively. This can be interpreted that data has strong trends for the assessment of 24 periodic time slices.

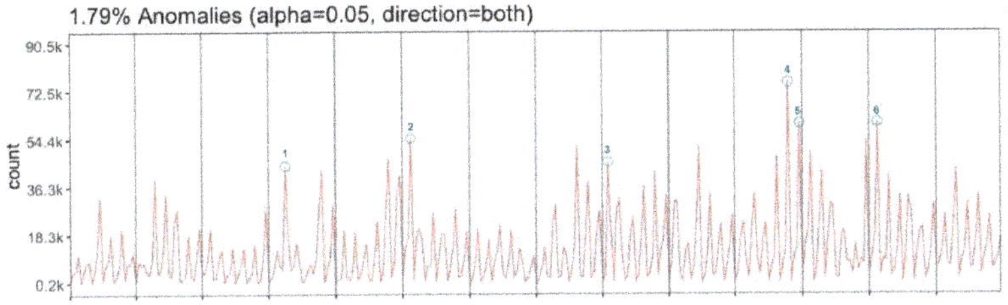

Legend Label	Time Index	Anom Value
1	79	442502
2	124	545293
3	195	455144
4	260	754735
5	264	599536
6	292	60343

(a)

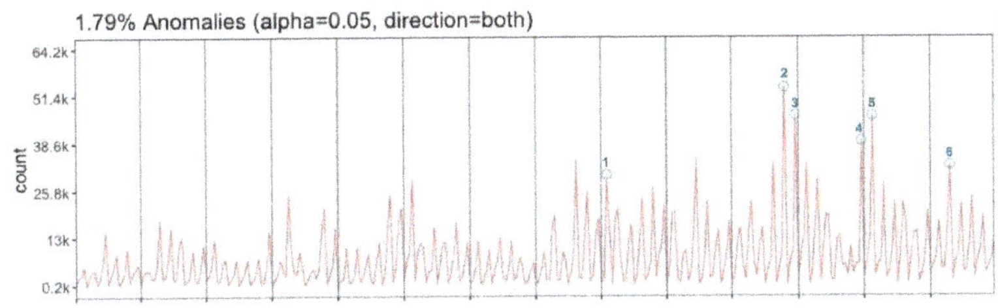

Legend Label	Time Index	Anom Value
1	195	299252
2	260	535513
3	264	459844
4	288	390525
5	292	45793
6	320	32429

(b)

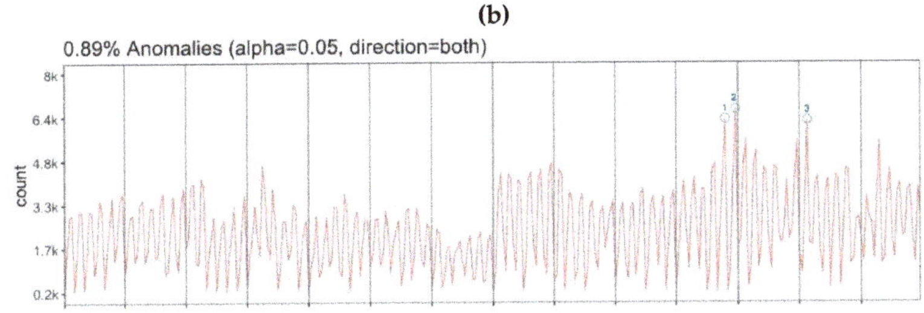

Legend Label	Time Index	Anom Value
1	260	63172
2	264	66643
3	292	6268

(c)

Figure 10. Anomaly in (**a**) number of tweets; (**b**) number of users; (**c**) normalized number of tweets.

Besides diverse representation levels of users, spatial representation is another aspect in order to understand data. Istanbul bounding box is divided into 100 m × 100 m grids to visualize this representation as represented and "unrepresented" grids as missing data. Figure and ground map (Figure 11a) of the missing data perfectly match the shape of the city (Figure 11b). This matching can be taken as Twitter is a living bionic tool for Istanbul that more or less represents the living area of the city.

(a)

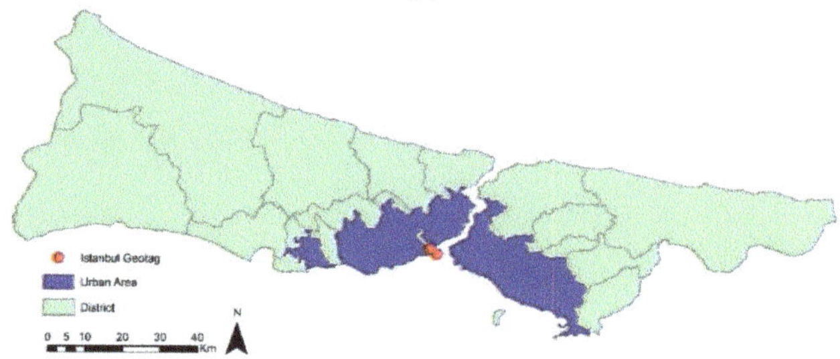

(b)

Figure 11. Istanbul city (**a**) figure and ground map for tweet representations; (**b**) urban area and social media geotags.

A lattice-based monitoring system was designed in order to understand the spatial footprints of users. Grid size is determined as 1 km × 1 km since it is well enough for fine-grained event detection. The number of tweets is spatiotemporally normalized corresponding to the grids. In order to explore the spatial dynamics of the Istanbul, anomaly analysis is performed for each grid over the spatially normalized tweet count values as time vector with 336 slices. Each detected anomaly values for a grid is normalized with the overall anomaly amount detected in its time interval. In this way, the detected anomaly magnitude for a grid is calculated regarding the overall anomaly. The normalized anomaly rate per grid was calculated by adding all anomaly magnitudes for a grid and visualized in Figure 12. This reveals the locations where most likely to have an anomaly in the city. In addition, this anomaly tendency map was tested with global and local Moran's I spatial correlation algorithm. The global I score and the p-value were found to be 0.24 and less than 0.0001, respectively, which means the values were slightly positively correlated and the significance of the test is pretty high. In order to assess the positive and negative spatial autocorrelation, the anomaly tendency map was also tested with local Moran's I. It appears from Figure 12a, there are high anomaly rates in the center part of Istanbul, the local Moran's test confirms that there is positive spatial autocorrelation with the high positive Ii (Figure 12b) and low p-value (Figure 12c) in this area.

In order to understand general dynamics of the city, data was discretized from its anomalies detected previously. The anomalous values were replaced with the expected values for the related part of the data. In Figure 13, the average value of this retrieved trend data was represented in 4 classes. The first class includes the two most active grids that have an average of 535 and 10,473 for 6-h. These spots are outliers and the closest value to the outliers is approximate to 50 tweets average in defined temporal level. There are two main reasons behind these outliers. The first, Istanbul geotags of social media platforms (Figure 11b) are located within these grids. The second, spots are located in the central area of Istanbul (Figure 11b) where the old city and touristic attractions are dense. The second and third classes grids by their activity have nearly the other 10% of the grids. The area covered with the second and third classes is matching with the urban area of Istanbul. These darker green grids show the central location bias of the data for general terms but also give the opportunity of monitoring them with the higher capacity of users' representation. The last class that covers nearly 90% of spatial grids include less than 1 tweet average within the 6-h interval. This lowest active class comprehends mostly residential areas, rural parts and seaside. While general activities within the darker areas presumed to be easily inferred from the tweet contents, the lowest active area could be easily spotted when there are extraordinary events.

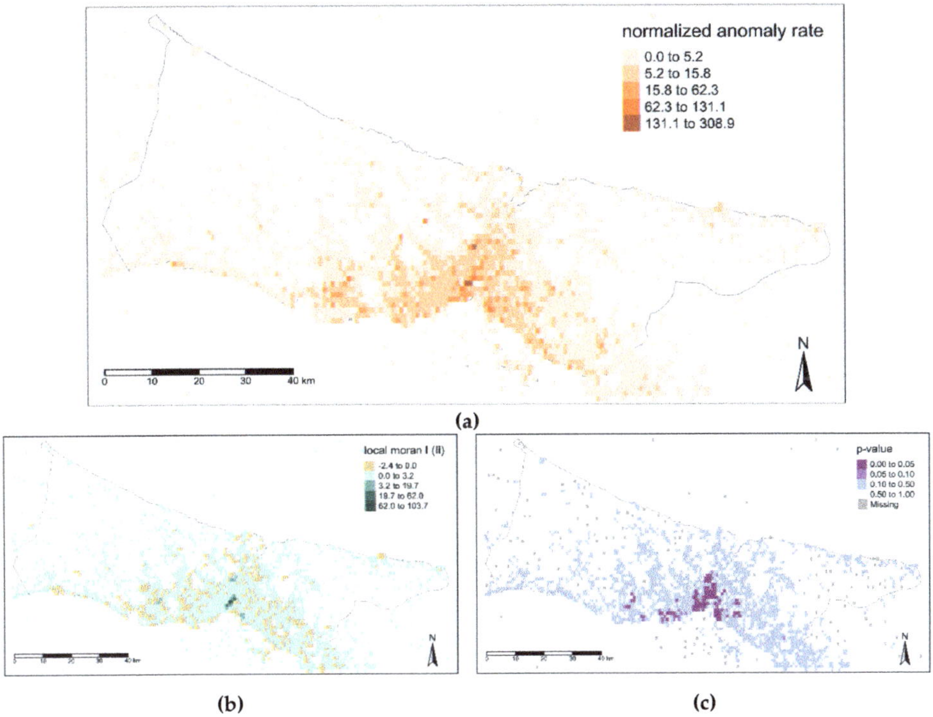

Figure 12. Anomaly tendency assessment (**a**) Overall anomaly rate; (**b**) local Moran's I; (**c**) local Moran's *p*-value.

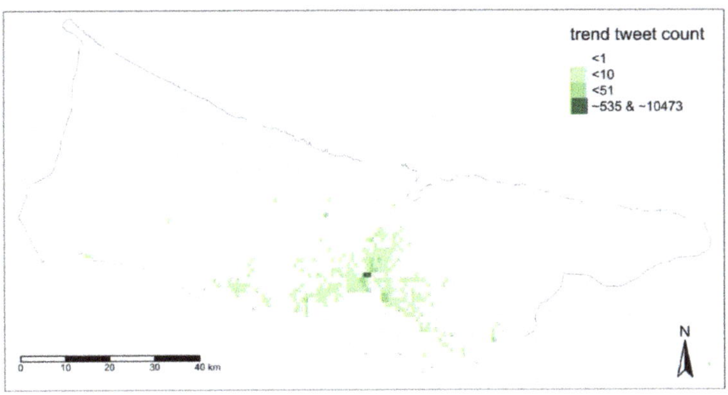

Figure 13. Average tweet count of trend values in each grid within 6 h.

Time is another aspect to detail spatial data review and citizens' footprints might vary in terms of different time levels. Therefore, the comparison between the trend map and maps belonging to different time levels was tackled at three temporal levels. The difference values defined with five classes that are below trend (least-low, low-trend), trend, above-trend (trend-high, high most). Difference values have positive and negative numbers, in addition, with the few exceptions values are not differentiated to much in these maps. With respect to this, threshold values for these classes were determined manually after exploring the data with several automatic classification techniques (such as; quantile, equal, standard deviation, kmeans, etc.). In Figure 14, the night map (a) has lower values than the trend map in the central parts while fringe parts of the urban areas have near values to trend. The difference in the second map (b) is more diverse and in some distributed areas the value is higher than the trend. In the third map for after midday time (c) and fourth map for evening time (d), the difference was reversed as the values are higher nearly in all parts of the city but dense in the central area of Istanbul (Figure 14a).

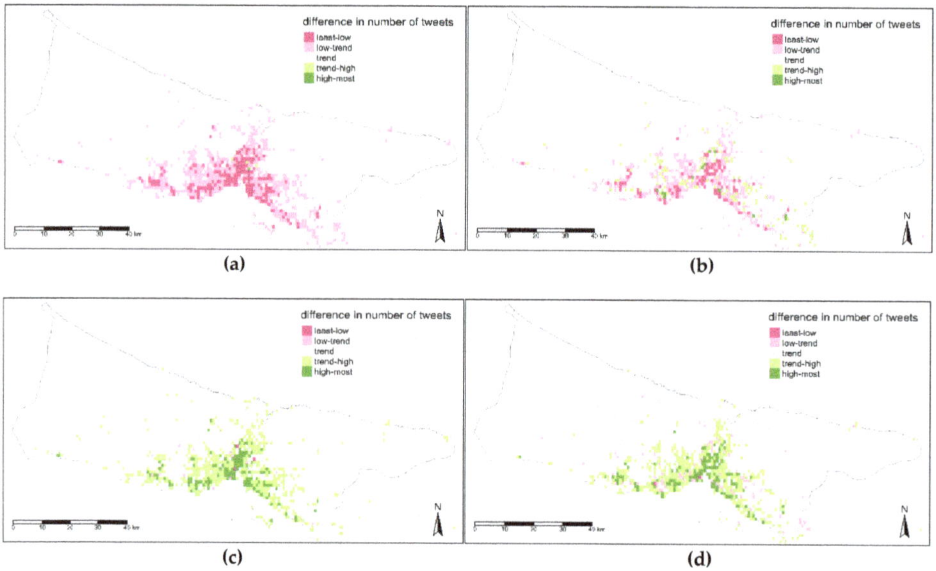

Figure 14. Difference in the number of tweets between time level 1 and trend maps (**a**) night; (**b**) before midday; (**c**) after midday; (**d**) evening.

For the second temporal level, days of the week were considered. It is explicitly seen, maps of weekdays (a, b, c, d, e) are pretty similar to each other while the weekend days (f, g) apart from them with the spots having higher values near Istanbul Strait and along seaside (Figure 15). Basing on this weekday's map, there are no direct clustered values for a region and central areas are a mix of classes above and below trend values. The most part belongs to classes that are less than the trend in weekdays, while the most and specifically the central and coastline parts have higher values than trends for weekends.

In the third time level assessment, seasons of the year were mapped (Figure 16). The central area has high value spots in winter and spring seasons while this area has less value in summer and autumn seasons (Figure 16). These seasonal plots reflect one side class for all parts of the city, that are above trend for winter and spring, below trend for summer and autumn.

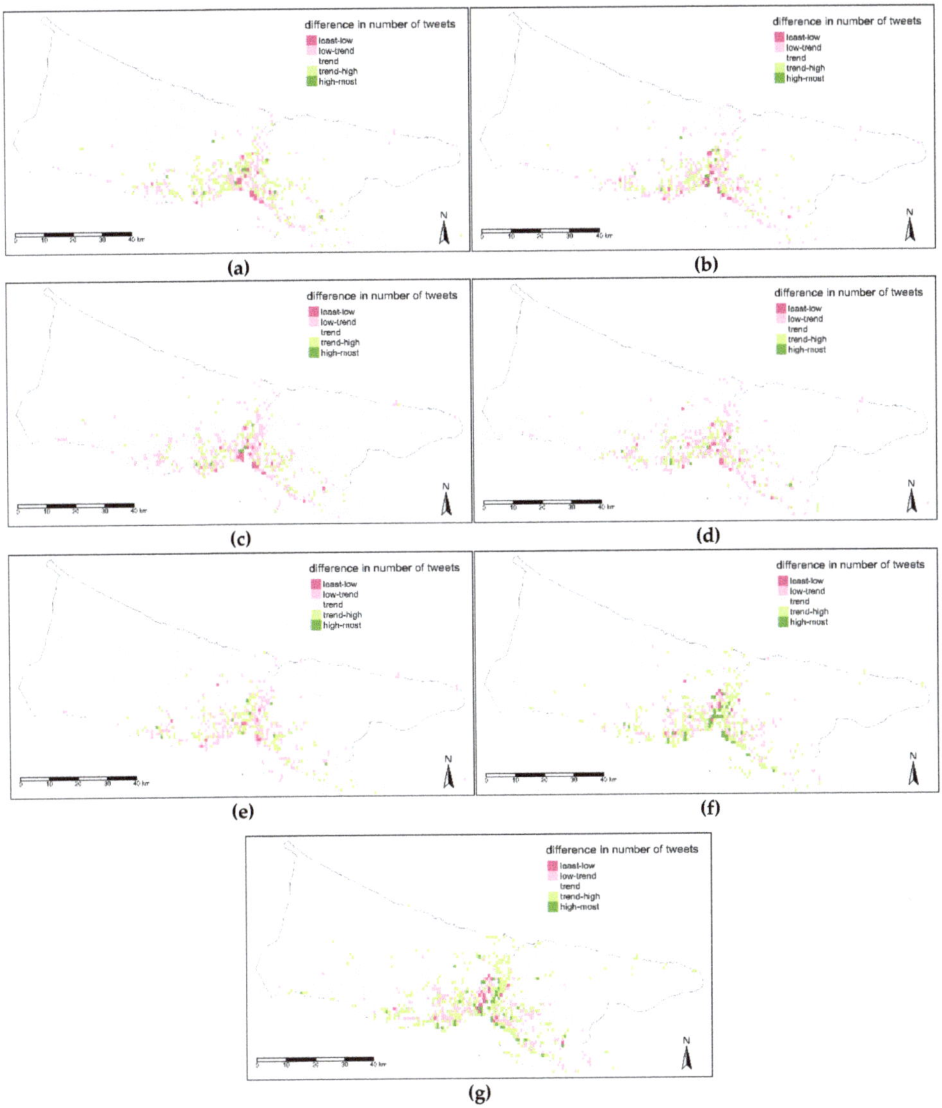

Figure 15. Spatiotemporal Bias for Temporal Level-2 (**a**) Monday; (**b**) Tuesday; (**c**) Wednesday; (**d**) Thursday; (**e**) Friday; (**f**) Saturday; (**g**) Sunday.

Global Moran's I was adopted to test the comparison maps' values spatial correlation. Though the I value is changing between −0.1 and 0.1 for each map and the p-values are above 0.1, it is not possible to say the time variance is spatially auto correlated. That means there is no significant difference between the difference values spatially although there is certain difference between the trend and time level maps as those seen in Figures 14–16.

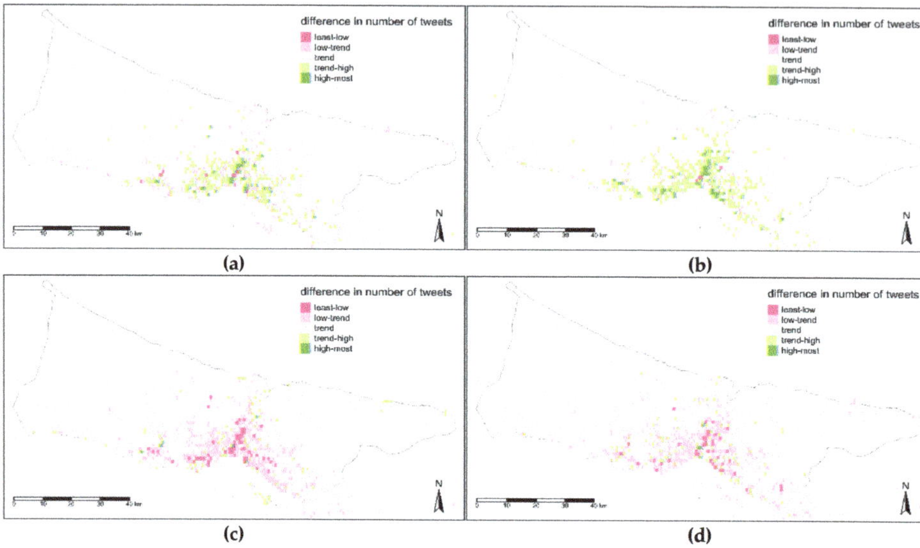

Figure 16. Spatiotemporal Bias for Temporal Level-3 (**a**) Winter; (**b**) Spring; (**c**) Summer; (**d**) Autumn.

4. Discussion

Social media is an invaluable source of data that is generated by human sensors due to its immense sensing capability and continuity [56,57]. Though it has various types of accountholders [58], the content and the spatial activity of each of them varies too [59]. There is research that provides background information to explain the activities of users on social media and categorizing them [41,60]. And some research on the credibility of users [61,62] and some others try to determine coordinated users who behave together and manipulate data content [63]. And studies claim social media is full of rumors and most part of the accountholders spreading wrong information while there are emergencies and do not correct the content even if they are informed later [64,65]. In respect to this, social media data requires to be assessed without removing any data but accepting all these deficits and considering them with the nature of itself, since it is not possible to control each user credibility in real-time without historical data or demographic information. Although data includes several issues such as credibility, rumors, representation inequalities in terms of user, it has a reference pattern in order to be assessed for monitoring systems. This study evaluated and presented general citizen footprint, most likely regional anomaly maps and spatiotemporal biases in Istanbul. These inferences as reference maps provide interpretation easiness for monitoring the city.

This study evaluates a-year SMD with the methodology given in Section 2.1. From the data revealed in this study it has been realized that to investigate the spatiotemporal change in SMD, the difference in representation levels of users should be normalized. For spatial exploration, miscellaneous representations of users are avoided with the spatio-temporal normalization technique. Anomalies that data might have due to an unusual event or coordinated users' activity, detected and replaced with the expected value. The locations, which tend to have more anomaly counts are determined as location biased spots. And the data norm displays the most represented locations by the more account holders. Besides that, the data norm is used as reference to explore spatiotemporal biases. It is obvious that data has several kinds of biases and the evaluated data can be used as a reference to discriminate any abnormality.

The results of this study can be enhanced with the finer spatial grain for lattice-based monitoring and finer time grain instead of 6 hourly. Further studies can also be developed over data content by doing several text analyses in order to find the most co-occurred word in a lattice and make a contribution to the reference map in that way.

Istanbul is the most populated city in Turkey with over 15 million citizens [66] and 3 million visitors which makes this city very important to be monitored for the sake of the living standards and responsive emergency management as well. There are several smart city projects separate conducted by local authorities; however, those projects are limited with the base map digitization or some municipality paperwork processes. Since citizens of Turkey have high potentials to generate spatial data on Twitter compared to other many countries, Twitter is eligible for citizen-based projects in Istanbul [41]. In further studies, evaluated data in this study provide the benchmarking knowledge to establish a dynamic monitoring system for Istanbul.

This study exposed four outcomes as mentioned below. The first outcome reveals that highly active users generate the majority of the data and as a general approach, removing this data within a pseudo-cleaning process conceals a large amount of data. The second one is the anomaly outcome results changes due to the diverse representation levels of the users. That is why; data normalization in terms of representation levels plays an important role in the detection of the true anomaly. The third outcome exposes that, as shown in Figure 12a, spatiotemporally normalized data represent strong spatial anomaly tendency at the urban center. The last outcome shows that the trend data is dense in the urban center and the spatiotemporal bias assessments show the data density varies in terms of the time of day, day of week and season of the year.

Twitter API is used in this study as in commonly used for other academic studies. Twitter declares that this API provides randomized 1% of public tweets in real-time [67]. There are empirical research that tests this randomization by comparing this sampled amount with the Firehose API data which provides the whole public tweets [68,69]. Studies found out there is no significant indication that the sampling of the Twitter API is biased with one exception since Twitter randomized the tweets by assigning ID for each tweet regarding the millisecond time. Because, this randomization is plausible since it exceeds a person's capability to share tweets in that quickness, unlike a bot. The used data within this study is normalized both spatially and non-spatially to avoid representation bias that can also eliminate noise due to bot accounts.

In this study, PostgreSQL with the PostGIS extension is used for data handling. PostgreSQL is an open-source relational database management system (RDBMS) that can be deployed to different environments such as a desktop, a cloud or a hybrid environment database. The storing capacity and the time cost for the processes rely on the specifications of the environment. This relational database is adequate to handle a large amount of data for basic operations (such as; insert, select and update) as in this study but could not be the best option for big data studies transactions [70]. NoSQL database like MongoDB has enhanced functionality on the big data processing performance especially the ones performed on unstructured data. SMD has unstructured content and RDBMS has issues while structuring the big amount of data. For this reason, NoSQL should be preferred while processing the big amount of unstructured text of SMD [70,71]. This work is also planned to extend with other cities' data including text mining in the context of space-time and the finer-grained temporal analyses. Therefore, in further studies a NoSQL database management system will be considered to handle such data.

There are numerous studies that conceptualize the measures of data quality in the context of VGI [72,73]. Data quality studies on VGI are mostly tackling data quality measures such as; completeness, positional accuracy and granularity on Open Street Map [11,73,74]. Generally, approaches for data quality are assessed in two class as intrinsic and extrinsic. In the intrinsic assessment, there is no use of an external reference map unlike extrinsic data quality assessment [75]. SMD was assessed in several aspects in this study in order to understand data bias, anomalies and trends. Since there is no external data use for these assessments within this study, this study provides the methodology for assessing the intrinsic data quality of the SMD.

The methodology proposed in this study can be used to extract the unbiased daily routines of the social media data of the regions for the normal days and this can be referred for the emergency or unexpected event cases to detect the change or impacts. Data assessment in this study is based on revealing the citizen footprints in SM and designed to explore anomalies, trends and bias within data.

In further studies, inferences from this study will be used to functioning a citizen-based monitoring system for Istanbul. The system design conceptually will follow the steps; tweets are collected in real-time, a number of tweets from the distinct users for each spatial grid is calculated, normalized number of tweets is assessed with the anomaly detection algorithm with regression line over trend data, detected anomalies are assessed with the most likely regional anomaly maps and the decision is made for emergency conditions. In addition, the proposed methodology is planned to work out on other big cities in order to contribute to other researchers by providing the results (reference maps) on a designed webpage for our future projects.

Author Contributions: Conceptualization, Ayse Giz Gulnerman, Himmet Karaman, and Serdar Bilgi; Data curation, Ayse Giz Gulnerman; Formal analysis, Ayse Giz Gulnerman; Funding acquisition, Himmet Karaman; Investigation, Ayse Giz Gulnerman and Direnc Pekaslan; Methodology, Ayse Giz Gulnerman, Himmet Karaman, Direnc Pekaslan and Serdar Bilgi; Project administration, Himmet Karaman; Software, Ayse Giz Gulnerman; Supervision, Himmet Karaman; Visualization, Ayse Giz Gulnerman; Writing—original draft, Ayse Giz Gulnerman and Direnc Pekaslan; Writing—review & editing, Ayse Giz Gulnerman, Himmet Karaman, Direnc Pekaslan and Serdar Bilgi All authors have read and agreed to the published version of the manuscript.

Funding: This research was funded by ISTANBUL TECHNICAL UNIVERSITY SCIENTIFIC RESEARCH PROJECTS FUNDING PROGRAM, grant number MDK-2017-40569 and SCIENTIFIC and TECHNOLOGICAL RESEARCH COUNCIL of TURKEY (TUBITAK -2214/A Grant Program), grant number 1059B141600822.

Conflicts of Interest: The authors declare no conflict of interest. The funders had no role in the design of the study; in the collection, analyses or interpretation of data; in the writing of the manuscript or in the decision to publish the results.

References

1. Ball, J. Towards a methodology for mapping 'regions for sustainability'using PPGIS. *Prog. Plan.* **2002**, *58*, 81–140. [CrossRef]
2. Hall, G.B.; Chipeniuk, R.; Feick, R.D.; Leahy, M.G.; Deparday, V. Community-based production of geographic information using open source software and Web 2.0. *Int. J. Geogr. Inf. Sci.* **2010**, *24*, 761–781. [CrossRef]
3. Sieber, R. Public participation geographic information systems: A literature review and framework. *Ann. Assoc. Am. Geogr.* **2006**, *96*, 491–507. [CrossRef]
4. Goodchild, M.F. Citizens as voluntary sensors: Spatial data infrastructure in the world of Web 2.0. *Int. J. Spat. Data Infrastruct. Res.* **2007**, *2*, 24–32.
5. Elwood, S.; Goodchild, M.F.; Sui, D.Z. Researching Volunteered Geographic Information: Spatial Data, Geographic Research and New Social Practice. *Ann. Assoc. Am. Geogr.* **2012**, *102*, 571–590. [CrossRef]
6. Hecht, B.J.; Stephens, M. A Tale of Cities: Urban Biases in Volunteered Geographic Information. *ICWSM* **2014**, *14*, 197–205.
7. Gulnerman, A.G.; Gengec, N.E.; Karaman, H. Review of Public Tweets Over Turkey Within a Pre-Determined Time. *Isprs-Int. Arch. Photogramm. Remote Sens. Spat. Inf. Sci.* **2016**, *3*, 153–159. [CrossRef]
8. Hecht, B.; Shekhar, S. *From GPS and Google Maps to Spatial Computing*, 2014 ed.; Coursera Inc.: Mountain View, CA, USA, 2014.
9. Goodchild, M. NeoGeography and the nature of geographic expertise. *J. Locat. Based Serv.* **2009**, *3*, 82–96. [CrossRef]
10. Ballatore, A.; Jokar Arsanjani, J. Placing Wikimapia: An exploratory analysis. *Int. J. Geogr. Inf. Sci.* **2018**, 1–18. [CrossRef]
11. Haklay, M. How good is volunteered geographical information? A comparative study of OpenStreetMap and Ordnance Survey datasets. *Environ. Plan. B Plan. Des.* **2010**, *37*, 682–703. [CrossRef]
12. Mooney, P.; Corcoran, P.; Winstanley, A.C. Towards quality metrics for OpenStreetMap. In Proceedings of the 18th SIGSPATIAL International Conference on Advances in Geographic Information Systems, San Jose, CA, USA, 20 November 2010; 2010; pp. 514–517.

13. Stephens, M. Gender and the GeoWeb: Divisions in the production of user-generated cartographic information. *GeoJournal* **2013**, *78*, 981–996. [CrossRef]
14. Gardner, Z.; Mooney, P. Investigating gender differences in OpenStreetMap activities in Malawi: A small case-study. In Proceedings of the AGILE Conference, Lund, Sweden, 12–15 June 2018; pp. 12–15.
15. Haklay, M.; Basiouka, S.; Antoniou, V.; Ather, A. How many volunteers does it take to map an area well? The validity of Linus' law to volunteered geographic information. *Cartogr. J.* **2010**, *47*, 315–322. [CrossRef]
16. Brown, G. A review of sampling effects and response bias in internet participatory mapping (PPGIS/PGIS/VGI). *Trans. Gis* **2017**, *21*, 39–56. [CrossRef]
17. Zooniverse. People-Powered Research. Available online: https://www.zooniverse.org/ (accessed on 18 October 2017).
18. Scistarters. Science We Can Do Together. Available online: https://scistarter (accessed on 18 October 2017).
19. Ushahidi. Read The Crowd. Available online: https://www.ushahidi.com/ (accessed on 18 October 2017).
20. Wald, D.J.; Quitoriano, V.; Worden, C.B.; Hopper, M.; Dewey, J.W. USGS "Did You Feel It? *Internet-Based Macroseismic Intensity Maps* **2012**, *54*. [CrossRef]
21. USGS. DYFI Summary Maps. Available online: https://earthquake.usgs.gov/data/dyfi/summary-maps.php (accessed on 3 September 2019).
22. Tarhan, C.; Coşkun, Z.; Zülfikar, C. Deprem Bilgi Sistemi [Earthquake Information System]. In Proceedings of the Turkey Earthquake Engineering and Seismology Conference, Hatay, Turkey, 25–27 September 2013; pp. 22–25.
23. Kocaman, S.; Anbaroglu, B.; Gokceoglu, C.; Altan, O. A review on citizen science (CitSci) applications for disaster management. *Int. Arch. Photog. Rem. Sens. Spat. Inf. Sci.* **2018**, *42*, W4. [CrossRef]
24. Statista. Most Popular Social Networks Worldwide as of October 2019, Ranked by Number of Active Users. Available online: https://www.statista.com/statistics/272014/global-social-networks-ranked-by-number-of-users/ (accessed on 30 December 2019).
25. Houston, J.B.; Hawthorne, J.; Perreault, M.F.; Park, E.H.; Goldstein Hode, M.; Halliwell, M.R.; Turner Mcgowen, S.E.; Davis, R.; Vaid, S.; McElderry, J.A.; et al. Social media and disasters: A functional framework for social media use in disaster planning, response and research. *Disasters* **2015**, *39*, 1–22. [CrossRef]
26. Sakaki, T.; Okazaki, M.; Matsuo, Y. Earthquake shakes Twitter users: Real-time event detection by social sensors. In Proceedings of the 19th International Conference on World Wide Web, Raleigh, NC, USA, 26–30 April 2010; pp. 851–860.
27. Gao, H.; Barbier, G.; Goolsby, R. Harnessing the crowdsourcing power of social media for disaster relief. *IEEE Intell. Syst.* **2011**, *26*, 10–14. [CrossRef]
28. Muralidharan, S.; Rasmussen, L.; Patterson, D.; Shin, J.-H. Hope for Haiti: An analysis of Facebook and Twitter usage during the earthquake relief efforts. *Public Relat. Rev.* **2011**, *37*, 175–177. [CrossRef]
29. Acar, A.; Muraki, Y. Twitter for crisis communication: Lessons learned from Japan's tsunami disaster. *Int. J. Web Based Communities* **2011**, *7*, 392–402. [CrossRef]
30. McClendon, S.; Robinson, A.C. Leveraging geospatially-oriented social media communications in disaster response. In Proceedings of the 9th International Conference on Information Systems for Crisis Response and Management, Vancouver, BC, Canada, 22–25 April 2012.
31. Ishino, A.; Odawara, S.; Nanba, H.; Takezawa, T. Extracting transportation information and traffic problems from tweets during a disaster. *Proc. Immm* **2012**, 91–96.
32. Iwanaga, I.S.M.; Nguyen, T.M.; Kawamura, T.; Nakagawa, H.; Tahara, Y.; Ohsuga, A. Building an earthquake evacuation ontology from twitter. In Proceedings of the Granular Computing (GrC) IEEE International Conference, Kaohsiung, Taiwan, 8–10 November 2011; pp. 306–311.
33. Bruns, A.; Liang, Y.E. Tools and methods for capturing Twitter data during natural disasters. *First Monday* **2012**, *17*, 1–8. Available online: http://eprints.qut.edu.au/49716 (accessed on 30 December 2019). [CrossRef]
34. Wang, Z.; Ye, X.; Tsou, M.H. Spatial, temporal and content analysis of Twitter for wildfire hazards. *Nat. Hazards* **2016**, *83*, 523–540. [CrossRef]
35. Mendoza, M.; Poblete, B.; Valderrama, I. Nowcasting earthquake damages with Twitter. *EPJ Data Sci.* **2019**, *8*, 3. [CrossRef]
36. Zou, L.; Lam, N.S.; Shams, S.; Cai, H.; Meyer, M.A.; Yang, S.; Lee, K.; Park, S.-J.; Reams, M.A. Social and geographical disparities in Twitter use during Hurricane Harvey. *Int. J. Digit. Earth* **2019**, *12*, 1300–1318. [CrossRef]

37. Leetaru, K.; Wang, S.; Cao, G.; Padmanabhan, A.; Shook, E. Mapping the global Twitter heartbeat: The geography of Twitter. *First Monday* **2013**, *18*. [CrossRef]
38. Middleton, S.E.; Middleton, L.; Modafferi, S. Real-Time Crisis Mapping of Natural Disasters Using Social Media. *IEEE Intell. Syst.* **2014**, *29*, 9–17. [CrossRef]
39. Karaman, H.; Şahin, M.; Elnashai, A.S.; Pineda, O. Loss assessment study for the Zeytinburnu district of Istanbul using Maeviz-Istanbul (HAZTURK). *J. Earthq. Eng.* **2008**, *12*, 187–198. [CrossRef]
40. Karaman, H.; Erden, T. Net earthquake hazard and elements at risk (NEaR) map creation for city of Istanbul via spatial multi-criteria decision analysis. *Nat. Hazards* **2014**, *73*, 685–709. [CrossRef]
41. Sloan, L.; Morgan, J. Who tweets with their location? Understanding the relationship between demographic characteristics and the use of geoservices and geotagging on Twitter. *PLoS ONE* **2015**, *10*, e0142209. [CrossRef]
42. Clement, J. Global Digital Population as of April 2019 (in Millions). Available online: https://www.statista.com/statistics/617136/digital-population-worldwide/ (accessed on 25 June 2019).
43. Basiri, A.; Haklay, M.; Foody, G.; Mooney, P. Crowdsourced geospatial data quality: Challenges and future directions. *Int. J. Geogr. Inf. Sci.* **2019**. [CrossRef]
44. Basiri, A.; Haklay, M.; Gardner, Z. The impact of biases in the crowdsourced trajectories on the output of data mining processes. In Proceedings of the AGILE Conference, Lund, Sweden, 12–15 June 2018.
45. Li, L.; Goodchild, M.F.; Xu, B. Spatial, temporal and socioeconomic patterns in the use of Twitter and Flickr. *Cartogr. Geogr. Inf. Sci.* **2013**, *40*, 61–77. [CrossRef]
46. Lansley, G.; Longley, P.A. The geography of Twitter topics in London. *Comput. Environ. Urban Syst.* **2016**, *58*, 85–96. [CrossRef]
47. Arthur, R.; Williams, H.T. The human geography of Twitter: Quantifying regional identity and inter-region communication in England and Wales. *PLoS ONE* **2019**, *14*, e0214466. [CrossRef] [PubMed]
48. Malik, M.M.; Lamba, H.; Nakos, C.; Pfeffer, J. Population bias in geotagged tweets. In Proceedings of the Ninth International AAAI Conference on Web and Social Media, Oxford, UK, 26–29 May 2015.
49. Tsou, M.-H.; Zhang, H.; Jung, C.-T. Identifying data noises, user biases and system errors in geo-tagged twitter messages (tweets). *arXiv* **2017**, arXiv:1712.02433.
50. Gengec, N.E. Geo Tweets Downloader. Available online: https://github.com/nagellette/geo-tweet-downloader (accessed on 26 August 2017).
51. Vallis, O.; Hochenbaum, J.; Kejariwal, A. AnomalyDetection: Anomaly Detection Using Seasonal Hybrid Extreme Studentized Deviate Test. R Package Version. 2014. Available online: https://rdrr.io/github/twitter/AnomalyDetection/man/AnomalyDetectionVec.html (accessed on 19 April 2020).
52. Twitter. Anomaly Detection with R. Available online: https://github.com/twitter/AnomalyDetection (accessed on 26 August 2018).
53. Hochenbaum, J.; Vallis, O.S.; Kejariwal, A. Automatic anomaly detection in the cloud via statistical learning. *arXiv* **2017**, arXiv:1704.07706.
54. Bivand, R.; Altman, M.; Anselin, L.; Assunção, R.; Berke, O.; Bernat, A.; Blanchet, G. Package 'Spdep'. 2015. Available online: https://mran.microsoft.com/snapshot/2017-08-23/web/packages/spdep/spdep.pdf (accessed on 9 December 2015).
55. Anselin, L. Local indicators of spatial association—LISA. *Geogr. Anal.* **1995**, *27*, 93–115. [CrossRef]
56. Goodchild, M.F. Citizens as sensors: The world of volunteered geography. *GeoJournal* **2007**, *69*, 211–221. [CrossRef]
57. Zhao, S.; Zhong, L.; Wickramasuriya, J.; Vasudevan, V. Human as Real-Time Sensors of Social and Physical Events: A Case Study of Twitter and Sports Games. *arXiv* **2011**, arXiv:1106.4300.
58. Zi, C.; Steven, G.; Haining, W.; Sushil, J. Who is tweeting on Twitter: Human, bot or cyborg? In *Proceedings of the 26th Annual Computer Security Applications Conference*; ACM: Austin, Texas, 2010; pp. 21–30. [CrossRef]
59. Cheng, Z.; Caverlee, J.; Lee, K. You are where you tweet: A content-based approach to geo-locating twitter users. In Proceedings of the 19th ACM International Conference on Information and Knowledge Management, Toronto, ON, Canada, 26–30 October 2010; pp. 759–768.
60. Issa, E.; Tsou, M.H.; Nara, A.; Spitzberg, B. Understanding the spatio-temporal characteristics of Twitter data with geotagged and non-geotagged content: Two case studies with the topic of flu and Ted (movie). *Ann. Gis* **2017**, *23*, 219–235. [CrossRef]

61. Gayo-Avello, D.; Metaxas, P.T.; Mustafaraj, E.; Strohmaier, M.; Schoen, H.; Gloor, P.; Castillo, C.; Mendoza, M.; Poblete, B. Predicting information credibility in time-sensitive social media. *Internet Res.* **2013**, *23*, 560–588. [CrossRef]
62. Wang, B.; Zhuang, J. Rumor response, debunking response and decision makings of misinformed Twitter users during disasters. *Nat. Hazards* **2018**, *93*, 1145–1162. [CrossRef]
63. Abbasi, M.-A.; Liu, H. Measuring user credibility in social media. In Proceedings of the International Conference on Social Computing, Behavioral-Cultural Modeling and Prediction, Washington, DC, USA, 2–5 April 2013; pp. 441–448.
64. Middleton, S.E.; Krivcovs, V. Geoparsing and Geosemantics for Social Media: Spatiotemporal Grounding of Content Propagating Rumors to Support Trust and Veracity Analysis during Breaking News. *Acm Trans. Inf. Syst.* **2016**, *34*, 1–26. [CrossRef]
65. Ma, J.; Gao, W.; Wong, K.-F. Rumor detection on twitter with tree-structured recursive neural networks. In Proceedings of the 56th Annual Meeting of the Association for Computational Linguistics, Melbourne, Australia, 20 July 2018; pp. 1980–1989.
66. Turkish Statistical Institute. Main Statistics, Population and Demography, Population Statistics, Population of Provinces by Years. Available online: http://www.turkstat.gov.tr/UstMenu.do?metod=temelist (accessed on 3 September 2019).
67. Twitter. Products for Researchers. Available online: https://developer.twitter.com/en/use-cases/academic-researchers/products-for-researchers (accessed on 29 March 2020).
68. Morstatter, F.; Pfeffer, J.; Liu, H.; Carley, K.M. Is the sample good enough? Comparing data from twitter's streaming api with twitter's firehose. In Proceedings of the Seventh International AAAI Conference on Weblogs and Social Media, Menlo Park, CA, USA, 8–11 July 2013; pp. 400–408.
69. Morstatter, F.; Pfeffer, J.; Liu, H. When is it biased? Assessing the representativeness of twitter's streaming API. In Proceedings of the 23rd International Conference on World Wide Web, Seoul, Korea, 7–11 April 2014; pp. 555–556.
70. Jung, M.; Youn, S.; Bae, J.; Choi, Y. A Study on Data Input and Output Performance Comparison of MongoDB and PostgreSQL in the Big Data Environment. In Proceedings of the 2015 8th International Conference on Database Theory and Application (DTA), Jeju Island, Korea, 25–28 November 2015; pp. 14–17.
71. Mathew, A.B.; Kumar, S.M. Analysis of data management and query handling in social networks using NoSQL databases. In Proceedings of the 2015 International Conference on Advances in Computing, Communications and Informatics (ICACCI), Kochi, India, 10–13 August 2015; pp. 800–806.
72. Senaratne, H.; Mobasheri, A.; Ali, A.L.; Capineri, C.; Haklay, M. A review of volunteered geographic information quality assessment methods. *Int. J. Geogr. Inf. Sci.* **2017**, *31*, 139–167. [CrossRef]
73. Ballatore, A.; Zipf, A. A conceptual quality framework for volunteered geographic information. In Proceedings of the International Conference on Spatial Information Theory, Colfax, NM, USA, 12–16 October 2015; pp. 89–107.
74. Mocnik, F.-B.; Mobasheri, A.; Zipf, A. Open source data mining infrastructure for exploring and analysing OpenStreetMap. *Open Geospat. DataSoftw. Stand.* **2018**, *3*, 7. [CrossRef]
75. Mocnik, F.-B.; Mobasheri, A.; Griesbaum, L.; Eckle, M.; Jacobs, C.; Klonner, C. A grounding-based ontology of data quality measures. *J. Spat. Inf. Sci.* **2018**, *2018*, 1–25. [CrossRef]

International Journal of
Geo-Information

MDPI

Article

Spatial and Temporal Patterns in Volunteer Data Contribution Activities: A Case Study of eBird

Guiming Zhang

Department of Geography and the Environment, College of Natural Sciences and Mathematics, University of Denver, CO 80208, USA; guiming.zhang@du.edu

Received: 2 September 2020; Accepted: 9 October 2020; Published: 11 October 2020

Abstract: Volunteered geographic information (VGI) has great potential to reveal spatial and temporal dynamics of geographic phenomena. However, a variety of potential biases in VGI are recognized, many of which root from volunteer data contribution activities. Examining patterns in volunteer data contribution activities helps understand the biases. Using eBird as a case study, this study investigates spatial and temporal patterns in data contribution activities of eBird contributors. eBird sampling efforts are biased in space and time. Most sampling efforts are concentrated in areas of denser populations and/or better accessibility, with the most intensively sampled areas being in proximity to big cities in developed regions of the world. Reported bird species are also spatially biased towards areas where more sampling efforts occur. Temporally, eBird sampling efforts and reported bird species are increasing over the years, with significant monthly fluctuations and notably more data reported on weekends. Such trends are driven by the expansion of eBird and characteristics of bird species and observers. The fitness of use of VGI should be assessed in the context of applications by examining spatial, temporal and other biases. Action may need to be taken to account for the biases so that robust inferences can be made from VGI observations.

Keywords: volunteered geographic information (VGI); data contribution activities; spatial and temporal patterns; biases; eBird

1. Introduction

Empowered by the ubiquitous geospatial technologies such as global navigation satellite system trackers and location-aware smart phones, many ordinary citizens are now acting as human sensors and voluntarily contributing geo-referenced ground observations regarding a broad array of natural and social phenomena. Such geospatial data contributed by citizen volunteers are collectively referred to as volunteered geographic information (VGI) [1]. The most prominent VGI initiative is OpenStreetMap (OSM) [2], a platform on which volunteers compile map data (e.g., detailed streets, roads, points of interest etc.) for much of the world. eBird, a popular citizen science project [3,4], is yet another VGI platform where birdwatchers around the world contribute and share geo-referenced birding records on a daily basis. Data from such VGI platforms has been widely used, for example, to support land management, network modeling and routing [5], and biodiversity conservation and research [3,6].

VGI has a great potential for revealing spatial and temporal dynamics of the geographic phenomena under observation [7,8]. However, VGI data quality issues have long been under scrutiny and, particularly, a variety of potential biases in VGI are recognized [3,9–15]. In order to draw robust inferences from VGI, such biases need to be understood and properly accounted for in VGI data analyses [16–18]. Some of the biases (e.g., observer and taxonomic biases) can be attributed to volunteer contributors' background (e.g., social, demographic and economic status, level of expertise); others are deeply rooted in their data contribution activities [3,19,20]. For example, individual volunteers have their own interests or motivations and often determine where and when to conduct observations at their own will, having no intent to coordinate sampling efforts with each other nor to follow any designed sampling scheme (e.g., stratified random sampling). As such, volunteer data contribution is often biased in space and time, which leads to biased spatial and temporal coverage in VGI observations. Examining spatial and temporal patterns in volunteer data contribution activities improves understanding of the spatial and temporal biases embedded in VGI. Such investigation in turn sheds light upon devising methods for bias mitigation to improve the reliability of inferences made from VGI [17,18]. It also helps identify any spatial and temporal observation gaps within VGI datasets toward which future sampling efforts can be directed.

A few studies have examined patterns in volunteer data contribution activities in various projects, and some consistent patterns exist. With respect to contributor variability, a relatively small share of volunteers often contribute most of the data whilst a large portion of contributors are ephemeral [19,21–23], which is a phenomenon of participation inequality consistently observed across online communities that is characterized by Zipf's law and the 90-10-1 rule [24–26]. Regarding temporal variability, volunteer data contributions are very uneven across time [21]. For instance, most contributions to OSM were made during the afternoon and evening hours and more contributions were made on Sundays [22]. Twitter users tweet the most around 13:00–1400 and 20:00–21:00 throughout the week while Flickr users are more active during weekends and most photos are taken during the afternoon hours [23]. In terms of spatial variability, most geographic areas have few contributors and contributions and most contributions tend to cluster in major cities with high population density [23,27]. As for identifying the pattern-shaping factors, Bittner [28] identified social biases in data contributions to OSM and Wikimapia in Jerusalem, Israel. Boakes et al. [19] revealed species abundance, ease of identification and tree height were positively related to the number of records that contributed to three biodiversity citizen science projects in the Greater London area. Based on a study using data from four citizen science projects in Denmark, Geldmann et al. [29] suggested distance to roads, human population density and land cover can be used to account for spatial bias in volunteer sampling efforts. Li et al. [23] discovered that well-educated people in the occupations of management, business, science, and arts are more likely to be involved in the generation of georeferenced tweets on Twitter and photos on Flickr. Although these studies examined patterns in volunteer data contribution through an array of lenses, few have investigated the patterns along the spatial and temporal dimensions at the same time (except [23]). More research is also in need for modeling and understanding how various factors may shape volunteer data contribution patterns.

This study aims to thoroughly examine the spatial and temporal patterns in volunteer data contribution activities among eBird contributors. eBird was launched by the Cornell Lab of Ornithology and the National Audubon Society in 2002 and has become the world's largest biodiversity-related citizen science project [3,30]. eBird data are freely accessible to anyone and have been used to support conservation decisions and help inform bird research worldwide [30]. Using the eBird mobile application or website, birders can upload information regarding when, where, and how they conduct birding and fill out a checklist of the birds seen and heard. As of 31 December 2019, over a half-million eBird contributors had collectively contributed over 50 million geo-referenced sampling events (i.e., checklists) containing more than 700 million bird observations on over 10,000 bird species across 253 countries and territories around the world [31].

The data contribution patterns in eBird have been examined through spatial or temporal profiling. For example, researchers profiled the number of submitted observations and checklists over the years 2003–2013 by month [3,6,32]. Yet, there lacks temporal profiling at finer granularity. In assessing global survey completeness of eBird data, La Sorte and Somveille [33] visualized the number of checklists, the number of species, and survey completeness (calculated by day, week, and month) based on data accumulated during 2002–2018 within equal-area hexagon cells that are 49,811 km^2 in area at the finest spatial resolution. It is a rather coarse spatial resolution to reveal spatial patterns at finer spatial scales. Many of these results (except [33]) have been outdated given the fast-growing capacity of eBird. For instance, from 1 January 2015 through 31 December 2019, over 33 million new sampling events

(~62% of the total) and 450 million new bird observations (~ 64% of the total) were submitted to eBird, and the cumulative number of contributors more than doubled [31]. The eBird website (ebird.org) provides interactive grid cell maps (~20 km resolution) showing the spatial distribution of the number of observed bird species and species relative frequency [34]. Such visualizations help understand trends in species distributions. However, they are not very useful for revealing the spatial and temporal patterns in birder's data contribution activities (i.e., sampling efforts). In summary, an up-to-date, wholistic spatial and temporal profiling of the eBird data at finer spatial and temporal resolutions, and modeling effects of factors in shaping sampling efforts are much needed to better understand the status quo of data contribution patterns in VGI projects such as eBird and beyond.

This study reports a comprehensive profiling of eBird data to discover the spatial and temporal patterns in volunteer data contribution activities. Discovering the patterns helps understand spatial and temporal biases and thus informs better data use. It can also reveal spatial and temporal gaps in existing sampling efforts; birders may direct future birding efforts to under-observed regions and/or time periods to improve eBird data coverage. Besides, the effects of environmental and cultural factors in shaping the spatial pattern of volunteer sampling efforts is explored in this study through spatially explicit modeling. The modeling provisions quantitative information on spatial variation of sampling efforts, which may be incorporated into other modeling process (e.g., species distribution modeling) for explicitly accounting for spatial bias to improve modeling performance [29].

2. Materials and Methods

2.1. eBird Data

eBird data (version: July 2020) were requested and downloaded from the eBird website (ebird.org), and records with an observation date before or on 31 December 2019 were used for analysis in this study. The dataset contains sampling event data and bird observation data [31]. Essentially, sampling event data have records regarding where (latitude, longitude), when (observation date) and by whom (identified by observer id) each birding session (identified by sampling event identifier) was conducted. These records reflect birder's sampling efforts. Bird observation data contain information on the observed bird species (identified by species scientific name) and their count estimates, among others, during each birding session. A bird observation can be related to a sampling event base on a common sampling event identifier present in both records.

eBird data were pre-processed, parsed and loaded into PostgreSQL/PostGIS, a free and open-source object-relational database with geospatial capabilities (postgis.net). As of 31 December 2019, a cumulative total of 548,365 eBird contributors (observers) had contributed 53,837,394 sampling events containing 716,876,356 bird observations on 10,379 bird species in 253 countries and territories around the world. The distribution of eBird data across the countries and territories is highly skewed (Table 1). The countries and territories with the 10 largest numbers of sampling events account for 89.9% of sampling events and 84.6% of sampling locations world-wide.

Table 1. Top 10 countries and territories most intensively sampled by eBird contributors.

Country/Territory	Sampling Events		Sampling Locations		Observers	Species
	n	%	n	%		
United States	36,540,720	67.9%	5,166,510	61.6%	417,288	1444
Canada	6,304,830	11.7%	915,075	10.9%	59,733	761
Australia	1,392,746	2.6%	218,756	2.6%	11,176	876
India	1,005,541	1.9%	219,430	2.6%	18,824	1536
Spain	757,638	1.4%	148,401	1.8%	9484	687
United Kingdom	714,712	1.3%	123,621	1.5%	11,898	756
Mexico	488,476	0.9%	108,411	1.3%	15,813	1140
Taiwan	453,852	0.8%	75,321	0.9%	3421	742
Costa Rica	383,191	0.7%	68,171	0.8%	12,855	930
Portugal	369,303	0.7%	59,118	0.7%	4170	620

2.2. Visualizing Spatial and Temporal Patterns

The spatial and temporal patterns in data contribution activities of eBird contributors were examined by visualizing results of spatial and temporal queries and analyses on the eBird data. SQL (standard query language) queries were used to obtain summary statistics regarding sampling events, observers, and reported bird species. The first two statistics reflect sampling efforts of eBird contributors, whilst the third indicates observed diversity of bird species.

The above summary statistics were aggregated and mapped on a grid of 0.25° latitude × 0.25° longitude cells (~28 km × 28 km at the equator) across the globe to visualize spatial patterns. Temporal patterns were visualized by aggregating and plotting the summary statistics over time periods of various granularities (year, month, day of week). These visualizations together provide a wholistic view of the spatial and temporal patterns in the volunteer data contribution activities of eBird contributors.

2.3. Modeling Sampling Efforts

The above spatial and temporal visualizations, although useful to uncover spatial and temporal patterns in volunteer data contribution activities, provide little insights on the underlying drivers shaping the patterns. To this end, spatial modeling was conducted to identify and quantify the effects of environmental and cultural factors in shaping the current spatial patterns in sampling efforts of eBird contributors. The Maxent approach (Section 2.3.3) was adopted to model the spatial pattern in eBird sampling efforts based on sampling locations in the most recent full year of 2019 (Section 2.3.1) and covariate data characterizing environmental and cultural factors (Section 2.3.2).

2.3.1. Sampling Efforts

There were 2,131,692 geographically unique eBird sampling locations world-wide in 2019 (Figure 1). These sampling locations represent volunteer sampling efforts in a most recent full-year cycle. Note that locations on the sea represent observations made from ships.

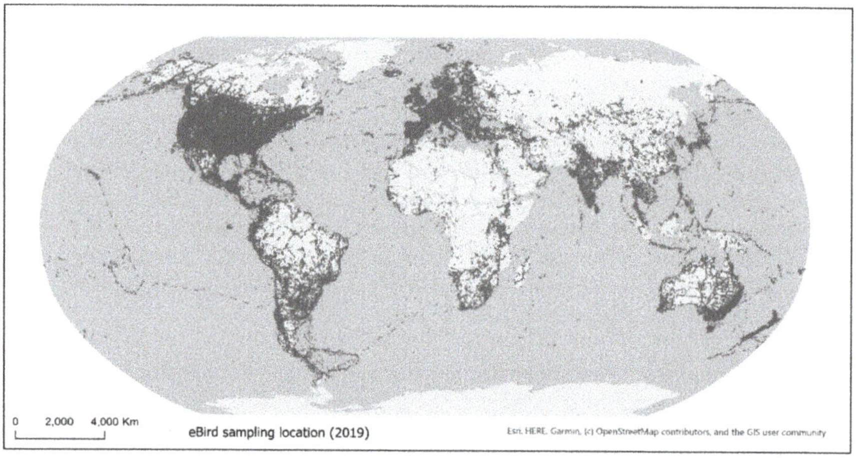

Figure 1. Spatial distribution of eBird sampling locations in 2019.

2.3.2. Covariates

A set of five covariates representing environmental and cultural factors was used for modeling the spatial pattern in eBird sampling efforts. According to analyses of data from four citizen science projects in Denmark [29], land cover, population density and road density often are the major variables that determine spatial bias in citizen science. These covariates are indicators of vegetation/land use condition, human activity intensity and infrastructure. In addition, the country-level United

Nations Human Development Index (HDI), a summary measure of average achievement in key dimensions of human development including life expectancy, years of education and gross national income per capita [35], was used in modeling because birding as a recreational activity is more often conducted by highly educated citizens with higher annual income [36]. Although the covariates are often correlated, each of them does contain a certain amount of independent information. Moreover, the Maxent modeling method used in this study (Section 2.3.3) does not require uncorrelated variables to achieve good model performance.

Besides, given that eBird is a global project with contributors from all over the world submitting data through either the eBird mobile app or website, contributor's knowledge of the written language in which the app or website is available could also play a role in determining the large-scale spatial bias in eBird data. Specifically, as of August 2015, the eBird mobile app is only available in five languages including Spanish, French, Chinese (Traditional), German, and English [37], although more languages have been added since then. The language in which the app and website is available may impact where and who would use them for contributing data to eBird. Therefore, a country-level official language map was used as an additional cultural variable in modeling the spatial pattern in eBird sampling efforts. Note that although spoken languages often have ambiguous geographic boundaries, official languages have much more clearly delineated boundaries (e.g., political boundaries). Moreover, official languages, often including both the spoken and written components, are widely taught in a country's school system.

A consensus world land cover dataset compiled by the EarthEnv project [38] was downloaded from here [39]. Population density map projected for 2020 produced by NASA's Socioeconomic Data and Applications Center [40] was downloaded from here [41]. Road density data compiled by the Global Roads Inventory Project [42] were downloaded from here [43]. The road density was for all roads (highways, primary roads, secondary roads, tertiary roads, and local roads). The most recent release of 2018 HDI data [35] with HDI values for all United Nations member countries [35] were downloaded from here [44]. The country-level official language map compiled by CIA World Factbook, University of Groningen was downloaded through a web feature service here [45]. It contains the first, second and third (if any) official languages of each country. In this study, countries were categorized based on the ordered list of official languages. Land cover, population density, and road density data are in raster format at a spatial resolution of 30 arc seconds (about 1 km at the equator). The vector-format HDI map and language map were rasterized to the same spatial resolution as the other covariates (Figure 2).

Frequency distributions of the sampling locations on the covariates were plotted against their respective background frequency distributions (distributions of all covariate values across the world) (Figure 3). The background distribution of the official language covariate was computed as per-language percentages of the 2019 world population at the country level (population data were obtained here [46]). The background distribution of HDI was computed as frequency distribution of the country-level HDI values weighted by the population of each country in 2019. Relative frequencies of land cover type, population density and road density were computed as area percentage. The frequency distributions show that the sampling intensity of eBird contributors is higher in areas of cultivated and managed vegetation and in urban/built-up areas. Although only about 40% of the sampling locations are in areas with population density above 100 persons/km^2, this is a high percentage considering the background distribution. About 75% of the sampling locations are in areas with road density greater than 250 m/km. Approximately 65% of the sampling locations are in countries where English is the official language (e.g., U.S. and U.K.) and another 10% in countries where Spanish is the official language. Finally, 80% of the sampling locations are in well-developed countries with HDI greater than 0.9. All indicate biases in sampling efforts along dimensions of the environmental and cultural factors.

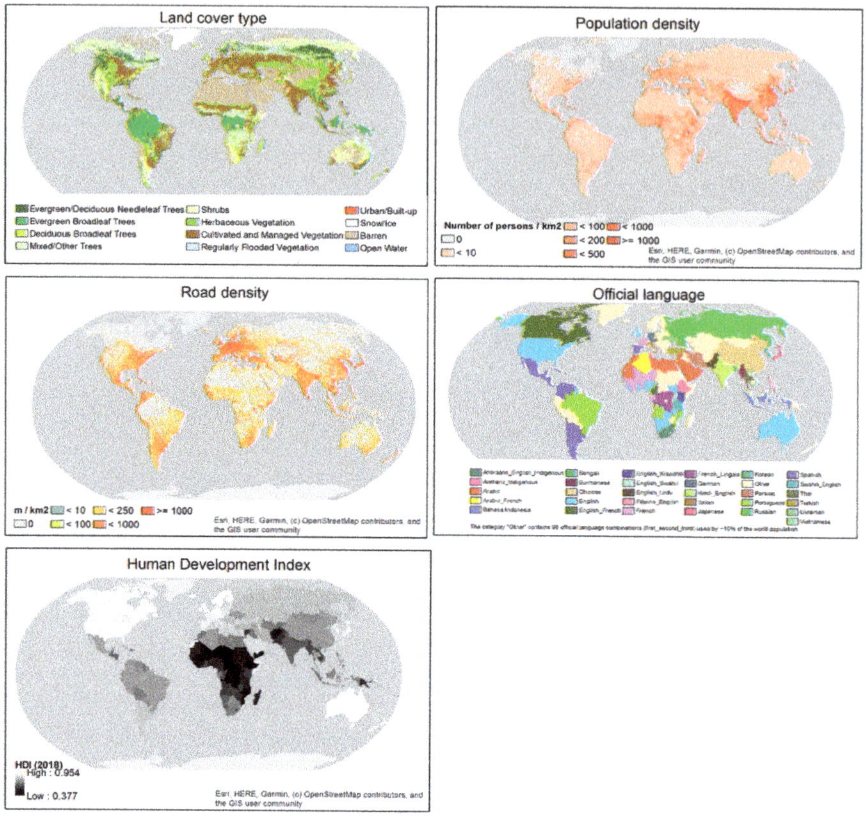

Figure 2. Covariates used for modeling the spatial pattern in sampling efforts of eBird contributors.

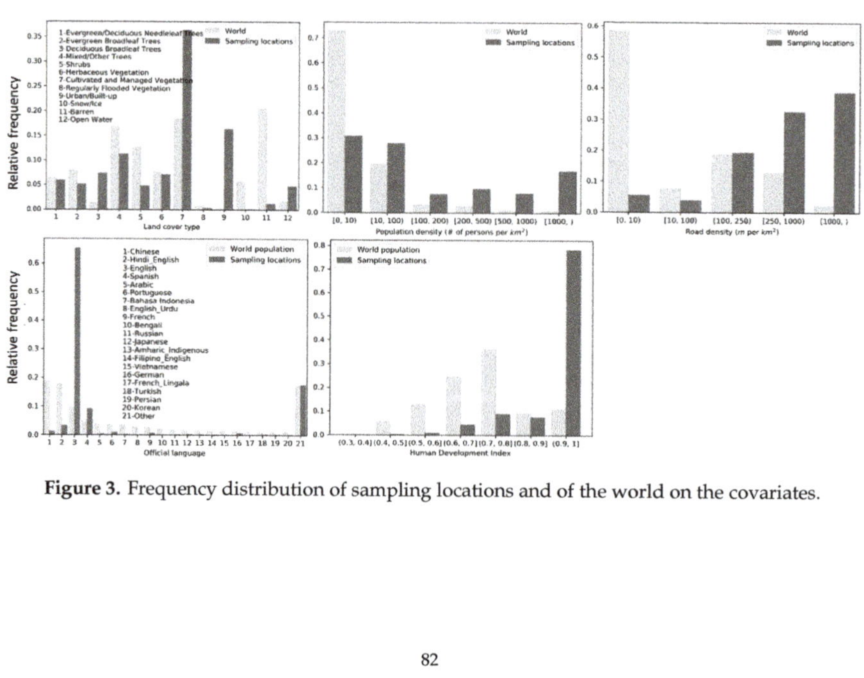

Figure 3. Frequency distribution of sampling locations and of the world on the covariates.

2.3.3. Modeling Method

The Maxent (maximum entropy) approach [47] was adopted to model the spatial pattern in eBird sampling efforts. Maxent is a general-purpose machine-learning method for making predictions or inferences from incomplete information and it has been widely used in various application domains, for example, modeling species distribution based on species "presence-only" data (e.g., occurrence locations) [47] and modeling geographic distribution of tourists from the locations tourists visited [48]. Maxent is well suited for modeling eBird sampling efforts as birding locations are also "presence-only" data. A conceptual overview of the Maxent method is provided below (readers interested in the mathematical details are referred to [47]).

Maxent estimates a probability distribution over a geographic area consisting of discrete raster cells (probability surface) based on two inputs: localities indicating the occurrences of a target event and covariate data layers characterizing the environmental factors that affect the event's occurrences. The probability of the event occurring at a cell is a function of the in-situ environmental conditions. The probability distribution is determined following the maximum entropy principle. That is, the distribution should be as close to a uniform distribution as possible while conforming to constraints embedded in the event occurrence localities. For example, expectation of the distribution on environmental variables should be close to the empirical averages observed at the occurrence localities. Maxent has been widely applied in various domains such as for modeling species geographic distribution [47] and for predicting geographic distribution of tourists [48].

In this study, eBird sampling locations (i.e., occurrence localities) and raster data layer of the four environmental and cultural covariates were input to Maxent to estimate the probability of a location being sampled by birders (sampling probability). The Maxent software version 3.4.0 [49] was used in this study. Most parameter settings of Maxent were kept to the defaults (e.g., auto feature, cloglog output format) as they have been fine-tuned on a large dataset and are supposed to work well in general [50]. Changes were made on four parameters. First, samples (eBird sampling locations) were not added to background as the author observed adding samples to background would greatly degrade model performance because the large number of sampling locations when added to background would severely bias background. Second, the number of background points was set to 500,000 (the default is 10,000) given the large study area (i.e., world continents and islands) from which random background points are selected. Third, the maximum number of iterations was changed from the default 500 iterations to 2,000 iterations to ensure the optimization procedure converges. Fourth, Maxent by default removes sample locations that are within the same raster cell of the covariates. This default setting was changed such that only duplicate locations with identical geographic coordinates were removed. Since the duplicates are removed by the model, information regarding repeated visits of the same location was not considered in the modeling process. The model thus effectively models the probability of the occurrence of at least one sampling event at a location, which is different to modeling the total spatial bias in eBird sampling efforts, as some sites/locations have many sampling events. After removing out-of-extent locations and duplicate locations, the number of eBird sampling locations was reduced to $n = 1{,}920{,}182$. Half of the locations ($n = 960{,}091$) were used for model training and the other half were used as test data for evaluating model performance.

Maxent model performance was evaluated by computing AUC (the area under the curve) of the predicted sampling probability map based on sampling locations in the test data and randomly selected background locations [50]. The AUC, a threshold-independent model performance measure, is the probability that the predicted sampling probability at a randomly chosen location will be higher than that at a randomly chosen background location [47]. The AUC ranges from 0 to 1, with AUC = 1 indicating perfect model performance, AUC = 0.5 indicating performance comparable to a random model, and AUC < 0.5 indicating worse-than-random model performance.

3. Results

3.1. Spatial and Temporal Patterns

3.1.1. Sampling Events

Sampling events conducted by eBird contributors were highly biased over the geographic space (Figure 4). Much of the world has not yet been sampled by eBirders. Among the grid cells that have been sampled, half have fewer than nine sampling events within a cell. Sampled areas are mostly developed regions of the world with better accessibility (e.g., higher road density). The most intensively sampled areas are in proximity to big cities of the world.

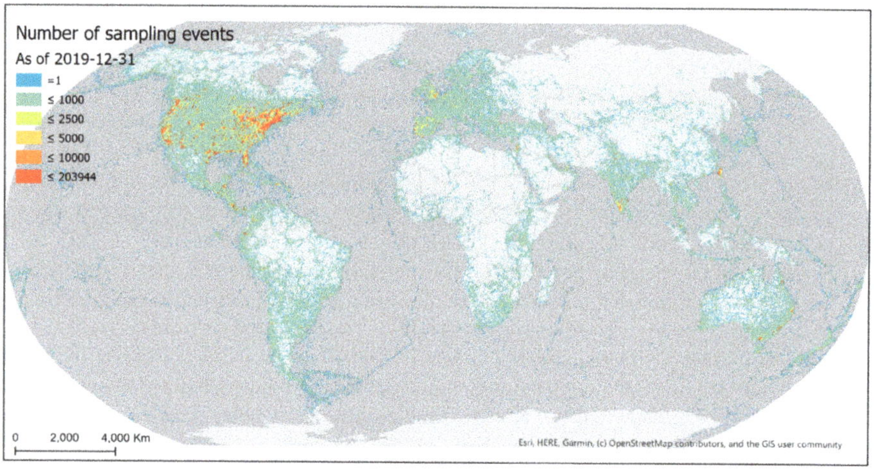

Figure 4. The number of cumulative eBird sampling events as of 31 December 2019 mapped over 0.25° latitude × 0.25° longitude grid cells. Intervals were determined loosely following quartile classification method.

On a yearly basis, the total number of sampling events has been increasing exponentially (i.e., at a faster pace) since 2002 (Figure 5). Over the months of the years 2002–2019, the number of sampling events in the northern hemisphere increased starting from March or April and peaked in May (Figure 6). Sampling efforts then significantly decreased and reached the lowest in July or August. In the southern hemisphere, sampling events were the fewest in May or June and the most in October or November. The overall number of sampling events across the world followed a monthly trend similar to that in the northern hemisphere. Over the days of the week through the years (Figure 7), there were more sampling events on weekends than on weekdays.

The number of species reported across sampling events is skewed (Figure 8). On average, 13 species were reported in each sampling event. Yet, no more than nine species were reported in half of the sampling events. About 15.5% of the sampling events reported only a single species, 36.1% reported 2–10 species, 37.5% reported 10–30 species, and 10.6% reported 30 species or more.

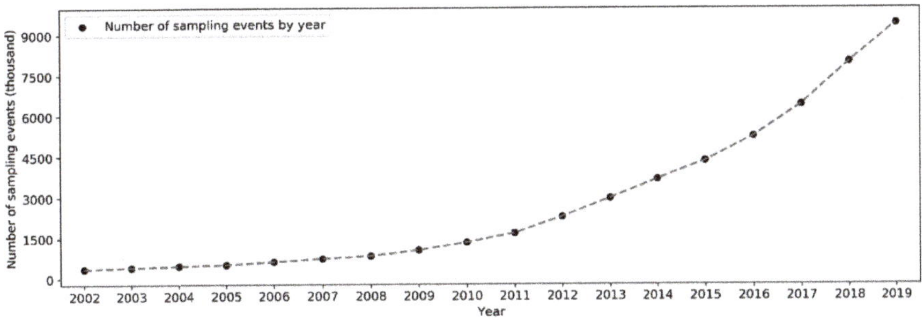

Figure 5. Number of sampling events in each year (2002–2019).

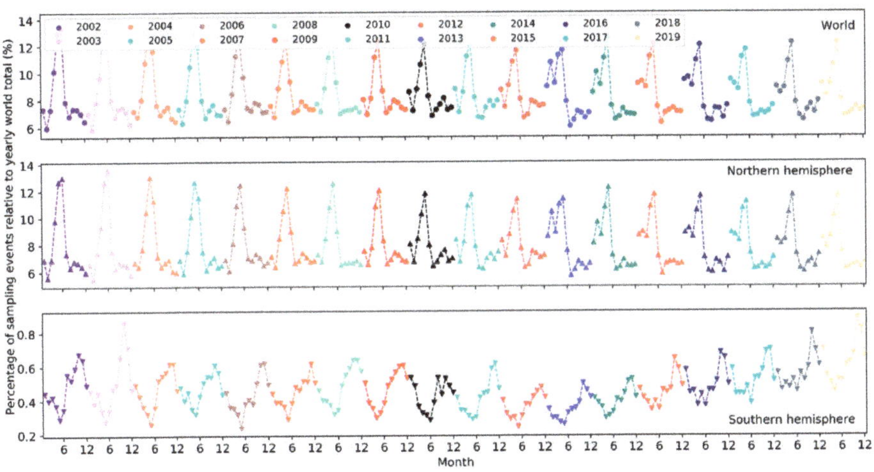

Figure 6. Percentage of sampling events in each month relative to the yearly total number of events.

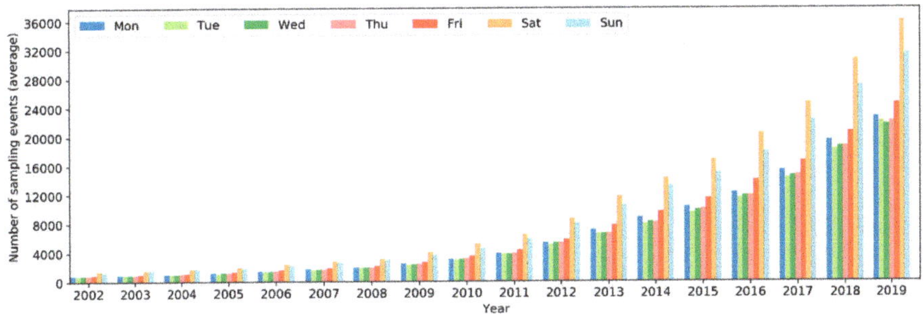

Figure 7. Average number of sampling events on each day of the week over the years.

There are sampling events associated with very large number of species. After checking records in the database, it was found there are $n = 29$ events with 300 or more observed species, and $n = 480$ events with 200 or more species (statistics obtained based on bird observations reviewed and approved by eBird). Most ($n = 349$) of the events have associated trip comments providing contextual information, although some are not in English. Overall, many of these events with large number of species are not 'regular' birding events. For example, some events are (1) compilations of birding records over prolonged birding periods (e.g., days or weeks), (2) records imported from publications, (3) birding events involving groups of birders who submitted records in a single record, (4) special birding events such as field guide, and (5) Big Day birding events where birders aimed to find as many birds as possible in a single day. Nonetheless, such birding events are not expected to have a significant impact on the reported statistics in this article (e.g., average number of species per event), given the very large sample size (n = 53,837,394 events in total).

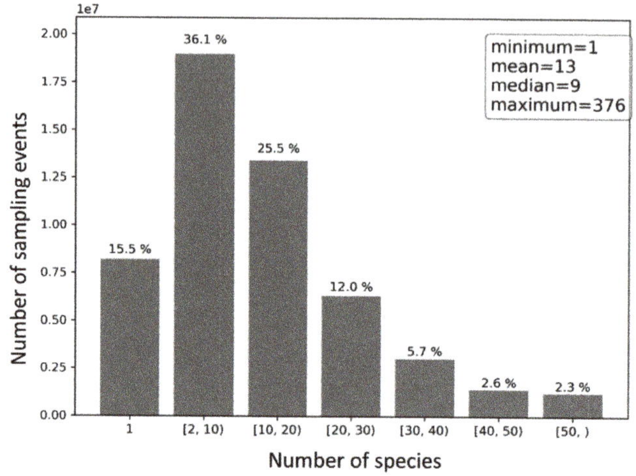

Figure 8. Frequency distribution of sampling events by the number of reported species.

3.1.2. Observers

The spatial distribution pattern of eBird contributors (observers) over the 0.25° latitude × 0.25° longitude grid cells (Figure 9) was similar to that of the sampling events. Among the grid cells that have been covered by observers, half of the cells have fewer than four contributing observers.

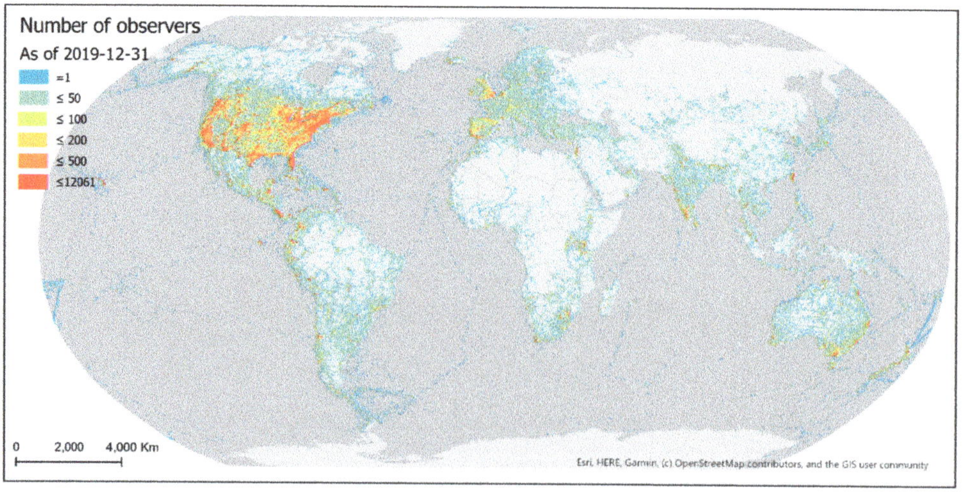

Figure 9. The cumulative number of observers as of 31 December 2019 mapped over 0.25° latitude × 0.25° longitude grid cells. Intervals were determined loosely following quartile classification method.

The number of active observers (observers who contributed at least one sampling event) has been increasing exponentially since 2002 (Figure 10). Notably, there was a significant boost in the number of observers in 2013. Over the months of the years (Figure 11), the highest peak in the number of active observers in the northern hemisphere occurred often in May, but since 2013 there was another peak in February. In the southern hemisphere, birders were most active in October or November, but since 2015 there was another peak in May (see Section 4.2.2 for possible explanations). The overall number of active observers across the world followed a monthly trend similar to that in the northern hemisphere. Over the days of the week (Figure 12), there were more active observers on weekends than on weekdays.

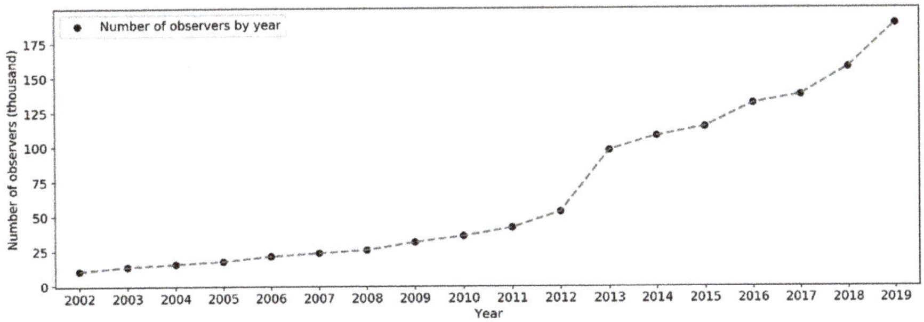

Figure 10. Number of active observers in each year (2002–2019).

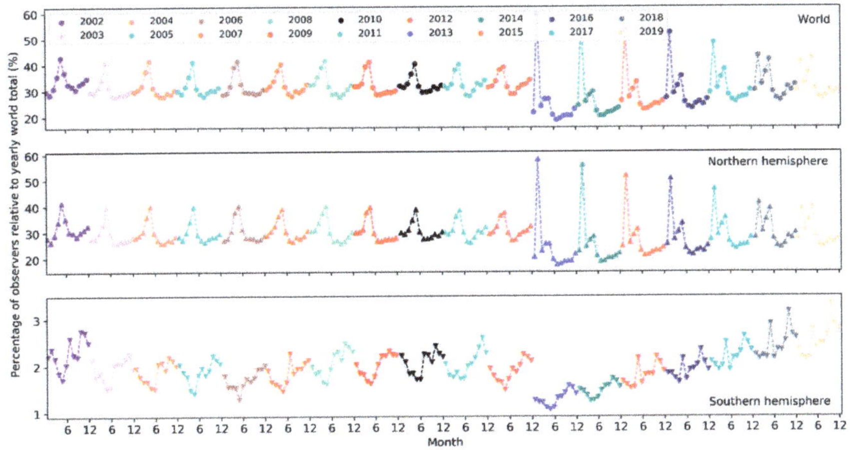

Figure 11. Percentage of observers in each month relative to the yearly total number of observers.

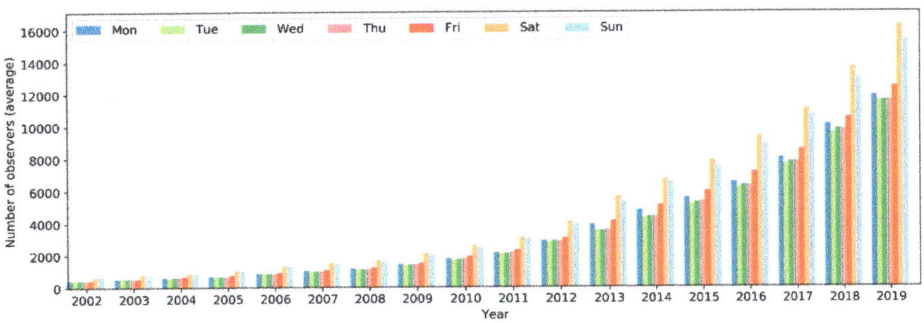

Figure 12. Average number of active observers on each day of the week over the years.

Observers were grouped by (1) the year in which they submitted the first observation to eBird, (2) the year in which they submitted the last observation, (3) the number of years between the first and last observations, and (4) the number of days with observations (Figure 13). Grouping results of (1) and (2) reflect the number of observers entering and exiting eBird each year (except for 2019), respectively. Over the years, the number of entering and exiting observers both increased approximately exponentially. There was a sharp rise in the number of observers entering eBird in 2013, followed by a slight decrease and plateau during 2014–2017, and a modest increase in 2018 and another significant leap in 2019. Grouping results of (3) and (4) reveal the temporal span of birding activities of the observers in units of year and day, respectively. About 67.9% of the observers contributed data only in a single year, 9.7% contributed in two consecutive years, 5.2% contributed in three consecutive years, and 17.2% contributed in four or more consecutive years. Speaking of the number of days with observation, 40.6% of the observers contributed observations on a single day, 36.3% contributed on 2–10 days, and 23.1% contributed on 10 or more days. Such patterns in contributors span over time are consistent with the life cycle of contributors in collaborative online communities [51].

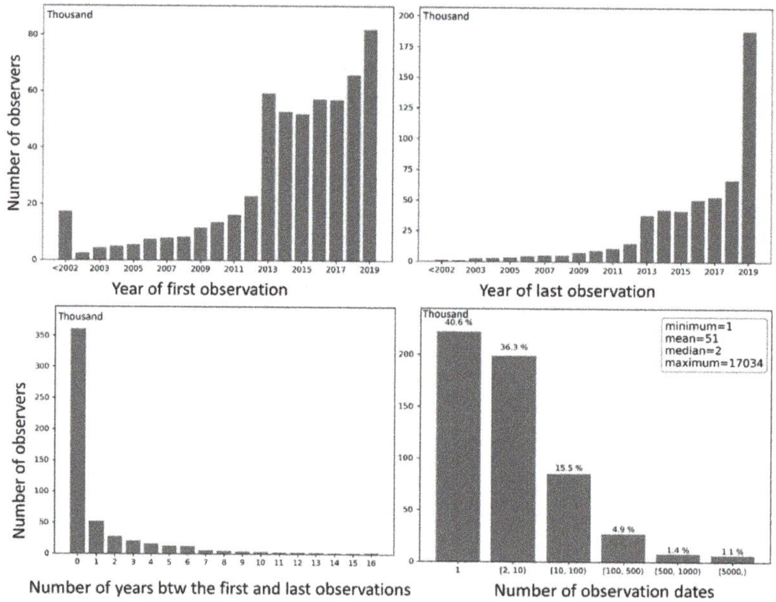

Figure 13. Frequency distribution of observers by year of first observation (**upper left**), year of last observation (**upper right**), number of years between the first and last observations (**lower left**), and number of active dates (**lower right**).

On average, each observer contributed 98 sampling events, sampled 32 locations, and reported 84 species (Figure 14). Nonetheless, half of the observers contributed no more than three sampling events, sampled only one location, and reported no more than 17 species. About 37% of the observers contributed just one sampling events, 53.8% contributed 2–100 sampling events, and 9.2% contributed 100 or more sampling events. Approximately 52.9% of the observers sampled only one location, 41.4% sampled 2–100 locations, and 5.7% sampled 100 or more locations. Roughly 13.5% of the observers reported only one species, 67.2% reported 2–100 species, and 19.3% reported 100 or more species.

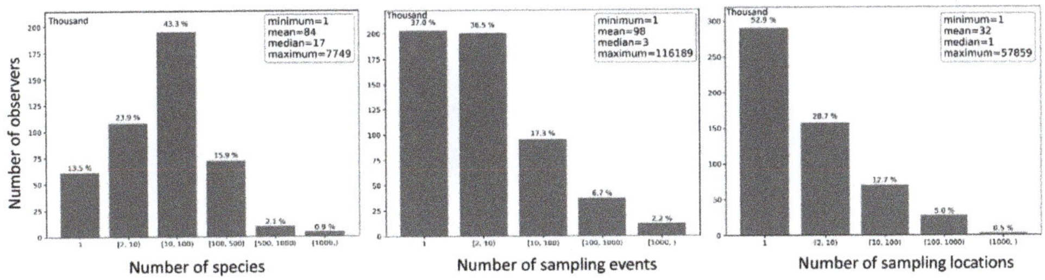

Figure 14. Frequency distribution of observers by number of reported species (**left**), number of sampling events (**center**) and number of sampling locations (**right**).

3.1.3. Bird Species

The distribution of bird species reported by eBird contributors is also highly skewed and spatially biased (Figure 15). Many parts of the world still do not have any bird species reported because of the lack of sampling efforts in those areas (Section 3.1.1). Among the grid cells with bird observations, half of them have a number of reported bird species below 42.

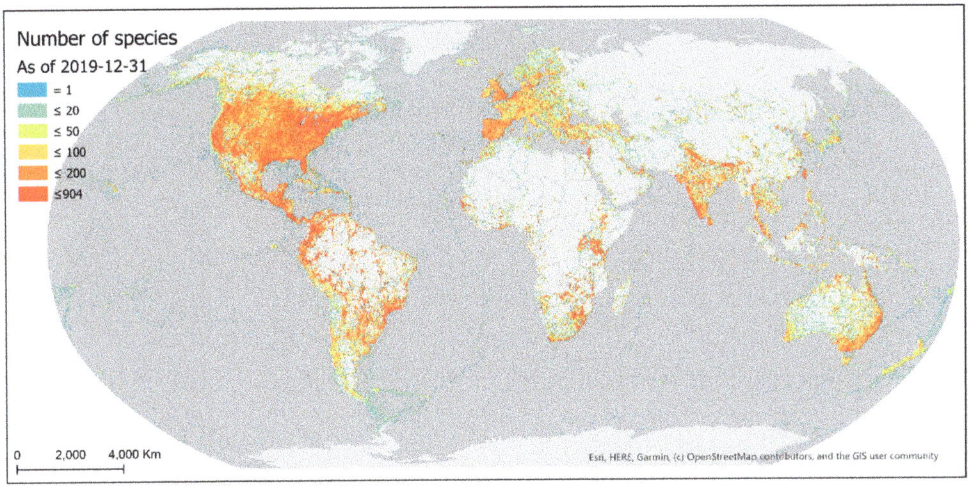

Figure 15. The cumulative number of species reported to eBird as of 31 December 2019 mapped over 0.25° latitude × 0.25° longitude grid cells. Intervals were determined loosely following quartile classification method.

The total number of species reported to eBird in a single year has been increasing in a linear fashion since 2002 but plateaued starting 2017 (Figure 16). From 2002 to 2019, the number of reported bird species increased from 8,740 to 10,053 (~15% increase). There was a sharp rise in 2005 and yet another jump around 2010. A larger number of species were observed in November–March in the northern hemisphere whilst in July–November in the southern hemisphere (Figure 17). The overall number of reported species often peaked in October or November. Over the days of the week (Figure 18), a larger number of species were reported on weekends than on weekdays.

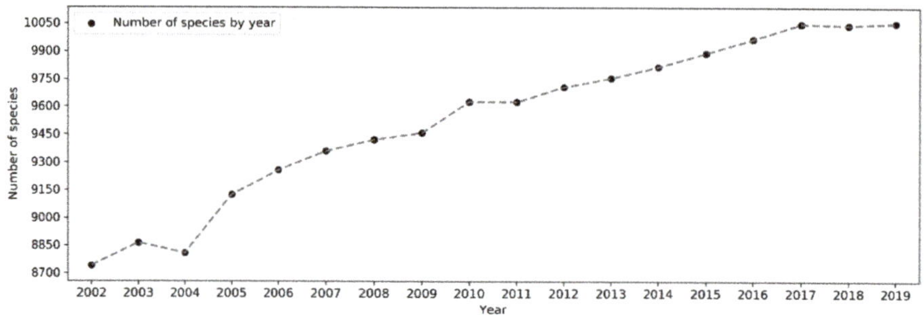

Figure 16. Number of species reported in each year (2002–2019).

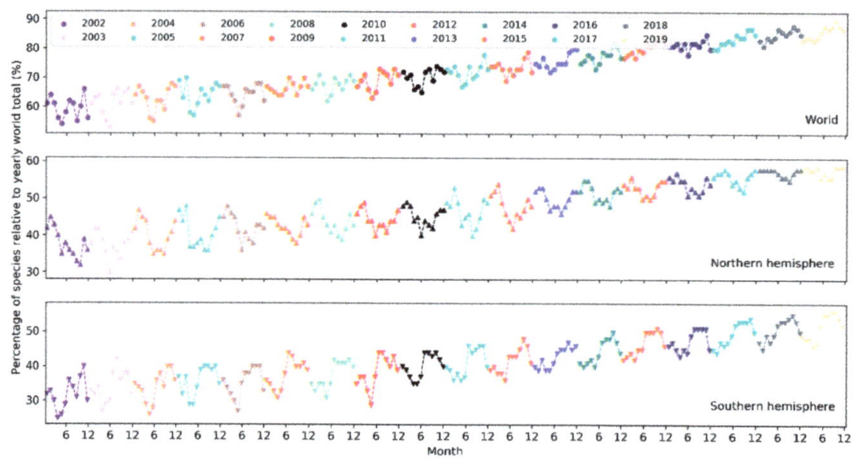

Figure 17. Percentage of species reported in each month relative to the yearly total number of species.

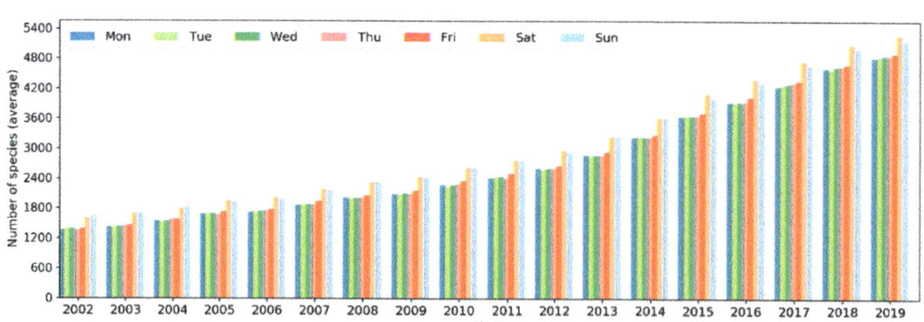

Figure 18. Average number of species reported on each day of the week over the years.

For each bird species present in the eBird database, the number of observers who reported the species and the number of sampling events in which the species was reported were counted (Figure 19). Half of the species were reported by no more than 671 observers and in no more than 1942 sampling events; on average, each bird species was reported by 3617 observers and in 66,037 sampling events, suggesting overall highly repetitive species observations. Only 0.4% of the species were reported by a single observer and 0.3% in a single sampling event, whilst about 83.7% of the species were reported by over 100 observers and 89.2% in over 100 sampling events.

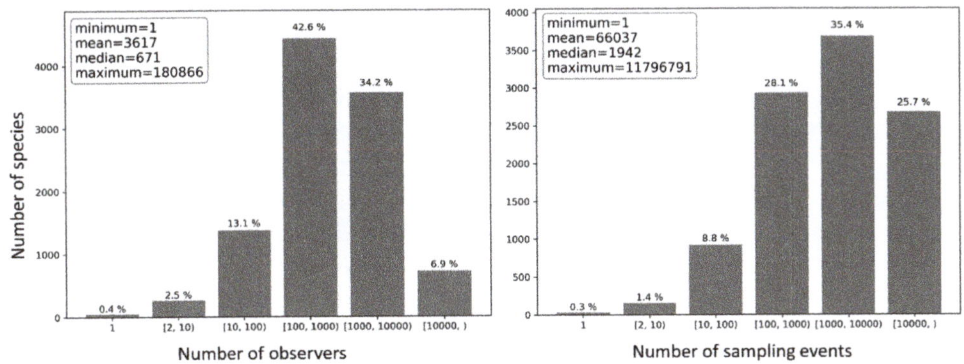

Figure 19. Distribution of the number of species by the number of observers who reported the same bird species (**left**) and by the number of sampling events in which the same species was reported (**right**).

3.2. Modeling Sampling Efforts

3.2.1. Analysis of Variable Importance

According to estimates of percent contributions of the environmental and cultural variables to the model provided in Maxent output (Table 2), road density and official language seem to be the two most important factors that contributed to the Maxent model. Land cover and HDI have much less contribution, and population density has very little contribution to the model. Permutation importance for each variable is the resulting drop in training AUC (normalized to percentages) when the values of that variable on training sampling and background locations are randomly permuted and the model is trained on the permuted data. Training AUC would drop the most when HDI values are permuted, followed by official language. Permutation on the other three variables would result in little drop in training AUC. It suggests that HDI and official language are important controlling factors determining birders' sampling locations at large spatial scales (e.g., country-level).

Table 2. Relative contributions of the environmental and cultural variables to the Maxent model.

Variable	Percent Contribution (%)	Permutation Importance (%)
Road density	45.8	0
Official language	31.5	31
Land cover	9.6	0
HDI	9.4	69
Population density	3.8	0

Moreover, based on the results of the jackknife test of variable importance (Figure 20), the variable with highest test AUC when used in isolation is road density, which therefore appears to have the most useful information by itself. By this standard, HDI and official language have slightly less useful information, and land cover and population have the least useful information. The variable that decreases test AUC the most when it is omitted is official language, which therefore appears to have the most information that is not present in the other variables. Nevertheless, the decreases in test AUC when omitting each variable is rather slight, meaning that using any four variables would result in model performance (as measured by the test AUC) very close to that of all five variables. Yet, variable contribution and variable importance should be interpreted with caution when the predictor variables are correlated. In this case, for example, there exist positive correlations between population density and road density at the raster cell level (*Pearson*'s $r = 0.48, p < 0.001$) and between HDI and mean road density at the country level (*Pearson*'s $r = 0.27, p < 0.001$).

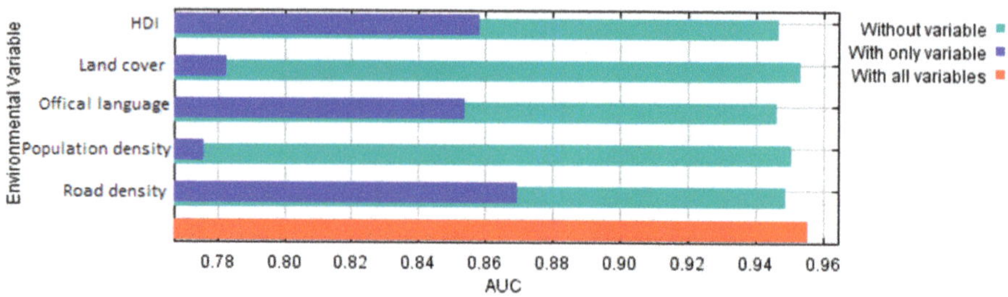

Figure 20. Jackknife test of variable importance to the Maxent model based on test AUC.

3.2.2. Modeled Sampling Probability

The map of sampling probability of eBird contributors (i.e., probability of the occurrence of at least one sampling event) modeled and predicted using the Maxent method (with all five covariates) is shown in Figure 21. The AUC computed for the probability map based on the held-out test data is 0.955, indicating an excellent model performance. That is, the model generally predicts higher sampling probability values at sampling locations in the test data than at randomly selected background locations. Across the globe, English and Spanish-speaking countries have the highest modeled sampling probability. Other European countries and some countries in Asia also have higher modeled sampling probability. Interestingly, most countries with high modeled sampling probability are highly developed countries in the world. There exists much spatial variation in the modeled sampling probability within each country. For example, higher sampling probability was modeled in southern parts of Canada and the east half and west coast of the United States, particularly in the vicinity of big cities. In Australia, the highest sampling probability was found on the east and southeast coasts, especially in the vicinity of big cities. These areas and/or cities are where most population reside and most infrastructure (e.g., roads) have been built.

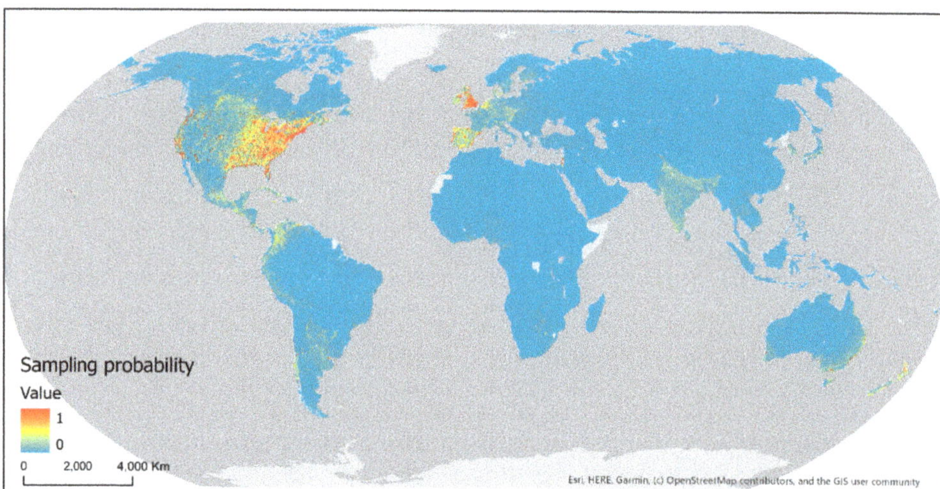

Figure 21. Map of sampling probability of eBird contributors modeled and predicted using Maxent.

4. Discussion

4.1. Spatial Patterns in eBird Data

Existing eBird sampling efforts (sampling events and observers) were mostly concentrated in areas of denser populations and/or better accessibility (e.g., higher road density), with the most intensively sampled areas being in proximity to big cities in developed regions of the world (Figures 4 and 9). Such spatial bias highlights significant disparities in the birding activities of eBird contributors across developed and under-developed regions. Due to the spatially biased sampling efforts, the number of reported bird species was also spatially biased towards areas where more sampling efforts occur (Figure 15). Despite the extensive geographic coverage of eBird data, there are gaps in the sampling efforts of eBird contributors. Many parts of the world still have not been sampled, such as central Africa, the Amazon, and Siberia.

Some reasons underlying the spatial patterns in sampling efforts of eBird contributors are related to characteristics of the social and physical environment. Maxent modeling of sampling efforts with the selected covariates was intended to unveil some of such effects. As revealed by the results of variable importance analysis (Section 3.2.1), among the factors considered, official language, road density and HDI could help explain much of the spatial variability in sampling efforts of eBird contributors across the world. Official language and HDI seem to determine spatial bias in sampling efforts at large spatial scales (e.g., country-level). English and Spanish-speaking countries have been sampled more intensively by eBird contributors (Figure 3). The eBird website and mobile app are available in only a few languages. Language advantage may have encouraged participation of certain birder populations, but it is also a barrier that may have prevented other birders from contributing to eBird. A large proportion (~ 80%) of the sampling locations are found in well-developed countries with high HDI values (above 0.9) (Figure 3), indicating that birdwatching as a recreational activity is more often conducted by citizens in well-developed countries. Road density may determine spatial bias in sampling efforts at smaller spatial scales (e.g., within countries). A large proportion of sampling locations are in areas of high road density. In contrast, areas with little or no sampling efforts (e.g., central Africa, the Amazons, and Siberia) are geographic areas with very limited accessibility.

However, one should not expect the covariates considered in the modeling to explain all spatial variabilities in eBird sampling efforts, as in fact many more factors may play a role (and thus modeling using other covariates may lead to more meaningful results). For example, the limited number of eBird observations in some countries can be explained also by the use of other local platforms for birdwatching, such as the official birding exchange platform in Switzerland [52], which is available also in Italian, French and German languages or the more general iNaturalist platform [53]. In developed European countries, the lack of observations to eBird is due to the use of such alternative platforms, instead of a simple language barrier. Whereas in African countries there are different reasons other than language and infrastructure for low reporting on eBird, e.g., civil and military conflicts [54].

Other reasons underlying local spatial patterns in the observations may be related to the characteristics of the observation target, i.e., bird species rarity and richness, etc. (see [55] and references therein). Birders go birdwatching with the hope to see birds. Therefore, where birding activities are conducted depends on where birds occur. If birders have prior knowledge of where (geographic area) birds prefer, they will be inclined to look for birds in such areas. The eBird mobile app and website, based on existing observations in the database, produce a "hotspots" map showing areas of rich bird species and species maps showing relative frequency of individual species. Many eBird contributors would probably use such maps as guides when deciding where to watch birds, which may help improve birding efficiency (e.g., see more birds in a birding session). However, this may also reinforce existing spatial bias in sampling efforts because the "hotspots" map is based on data resulted from spatially-biased sampling efforts.

4.2. Temporal Patterns in eBird Data

Possible reasons underlying the temporal patterns in eBird data across the years, months, and days of the week are discussed in the following three sub-sections.

4.2.1. Patterns across the Years

Over the years, sampling events and observers have been increasing exponentially since 2002 (Figures 5 and 10). This is a trend consistent across many large-scale online communities, and such rises are generally related to information and communication technology advancements, infrastructures development, etc. [22,56]. Specifically to eBird, several developments in the history of the eBird project may have contributed to the accelerated rise of volunteer data contribution activities, for example, the release of birder-engaging tools in mid-2005 [32]; expansion of eBird to include New Zealand in 2008 and later to cover the whole world in 2010 [57]; the released mobile app called "BirdLog", which was the first and only app that made it possible for birders to record and submit information to eBird in the field [58]; the release of the eBird mobile app [58] and the Merlin mobile app for bird identification in 2014 [59]; and the release of the eBird mobile app supporting multiple languages in 2015 [37]. The significant boost in the number of observers from 2012–2013 (Figures 10 and 13) may be attributed to the release of the "BirdLog" mobile app in 2012.

Despite the exponential increase in sampling efforts, the number of species reported to eBird increased only in a linear fashion over the years (Figure 16), suggesting that the marginal increase in the number of new species reported to eBird is disproportionately less significant compared to the huge amount of additional sampling efforts brought by new observers. The significant increases in the number of reported species in 2005 and in 2010 may be due to the release of birder-engaging tools in 2005 and the expansion of eBird to a global coverage in 2010, respectively.

4.2.2. Patterns across the Months

Over the months, temporal trends in sampling events, observers and reported bird species differ across the northern and southern hemispheres (Figures 6, 11 and 17). In the northern hemisphere, the number of sampling events, observers and reported bird species all peaked in May. This may be because in this hemisphere the breeding season of many birds begin in May and many birders go for birdwatching more intensively during that season (e.g., the North American Breeding Bird Survey). There was another higher peak in February among observers since 2013. It may be explained by (1) many birders across the world participate in the Great (Global) Backyard Bird Count (GBBC) every February [60]; (2) the "BirdLog" mobile birding app released in 2012 made recording and submitting data directly in the field possible [58], which may have helped engage more birders in GBBC. More species were reported in and around winter months during November–March. Given that sampling efforts in terms of the number of active birders and the number of sampling events were not as intensive in winter months, the larger number of reported bird species may be an indicator of highly "efficient" winter birding activities of a smaller group of skilled birders who could identify many bird species. For example, Christmas Bird Count occurs December 14 to January 5 every year, mostly in U.S. and Canada. There were no December or January spikes in the number of active birders nor in the number of sampling events (Figures 6 and 11). It may be that the birders who participated in Christmas Bird Count may already have been active in other periods. Only 0.2–0.5% of the sampling events in each year during 2002–2019 mentioned "Christmas Bird Count" (or its variants) in the trip comments. However, CBC participants may have contributed to the large number of species reported in December and January (Figure 17).

Peaks in the southern hemisphere fell into different months than in the north but generally followed similar seasonal treads. For example, sampling efforts (sampling events, observers) increased starting from May or June until October or November (Figures 6 and 11), a period encompassing the breeding season of birds in the southern hemisphere. More species were also reported over this period

(Figure 17). The number of active birders often peak in October or November, but since 2015 there was another peak in May and the October peak became much higher than the numbers in November and December, which may be due to the yearly Global Big Day event in May [61] and the October Big Day event [62] organized by eBird to engage more birders in birding. Yet, since most of the contributors are ephemeral one-time contributors submitting only a single record, such events (e.g., Big Day, Great Backyard Bird) may only boost participation over the limited timeframes of the events. Increases in the overall participation are more related to other developments (e.g., the release of "BirdLog" in 2012).

The timing of bird migration may also help explain the monthly fluctuations. Many birders often wait for migration seasons to go out and look for migrant birds. In the northern hemisphere, for example, the increases in the number of sampling events, active birders and species in March–April and in September–October correspond to the typical Spring migration arrival and Fall departure timeline, respectively.

Globally, temporal patterns in sampling efforts over the months are similar to those in the northern hemisphere, which had an order of magnitude more sampling events and observers compared to the southern hemisphere, and thus, was driving the patterns in the overall sampling efforts across the globe. The number of reported species has a comparable magnitude across the two hemispheres but followed "opposite" trends over the months due to the opposite seasonality in the two hemispheres.

4.2.3. Patterns across the Days of the Week

Over the days of a week, more observers were active on weekends than on weekdays, reporting a larger number of sampling events and more species. This pattern is consistent across the years (Figures 7, 12 and 18). It could be attributed to the fact that birdwatching is just a hobby to many people, some of whom may take a break from routine life and work responsibilities on weekends and thus can spend more time on birding.

However, the level of expertise of the birders who were active on weekends and the quality of their data submissions was notably different from those active on weekdays (Figure 22). The average number of species per observer, an indicator of the level of birding expertise, was lower among birders active on weekends. Moreover, the number of incomplete checklists (i.e., not all species were identified and reported) and unapproved observations per observer, two indicators of data quality, were both higher among birders active on weekends. The evidence suggests birders who were active on weekdays were of an overall higher level of expertise and data contributed by them were of higher quality. This might be due to the fac that a larger number of novice birders tend to go birding only on weekends while expert birder may conduct birding on weekdays besides weekends.

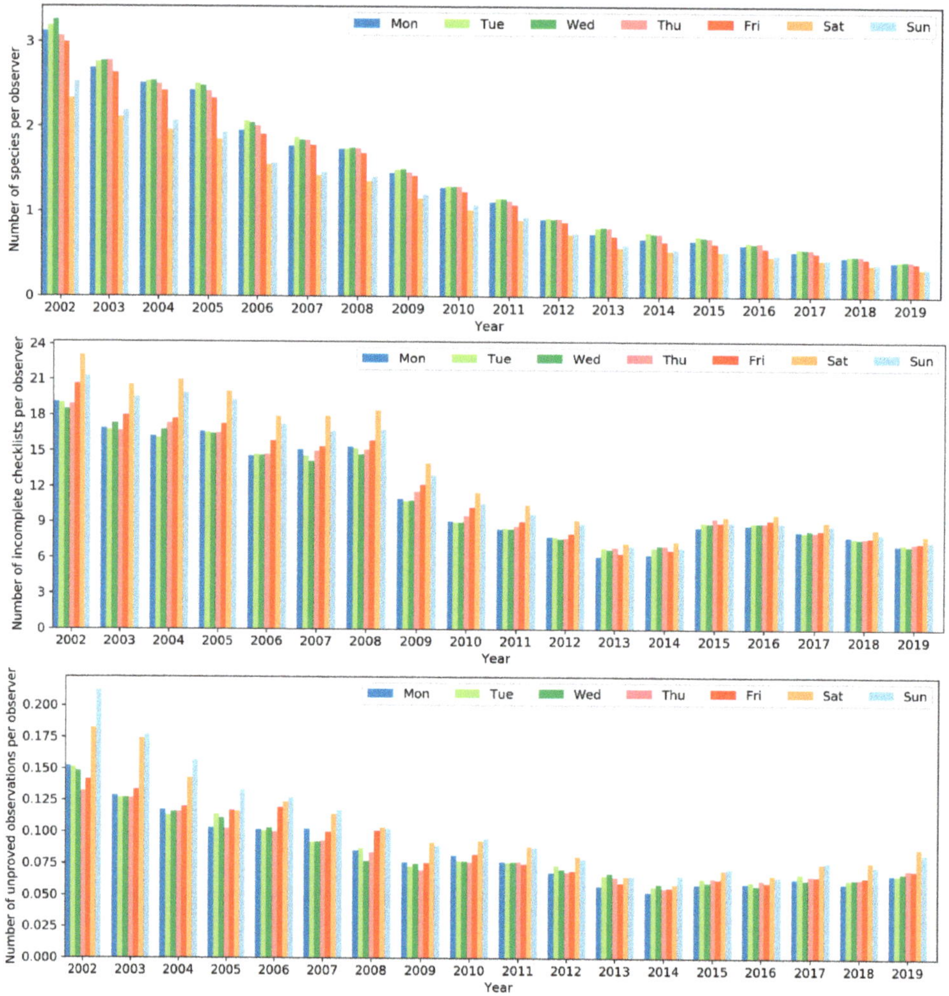

Figure 22. Number of species (**top**), incomplete checklists (**center**), and unapproved observations (**bottom**) per observer across the days of the week.

4.3. Biases in VGI and Their Implications

Several forms of biases exist in VGI due to the characteristics of volunteer data contribution activities. When utilizing VGI, one should assess the fitness of use of VGI in the context of particular applications by examining the extent of the biases, their potential impacts, and possible methods to account for the biases.

4.3.1. Spatial Bias

Spatial bias is often intrinsic to the sampling efforts of volunteer data contributors. Unlike traditional geographic sampling where sampling locations are designed following a certain spatial sampling scheme (e.g., stratified random sampling), individual volunteers choose sampling locations (i.e., where to conduct observations) largely at their own will in ad hoc or opportunistic manners, without coordinating sampling efforts with other volunteers [18]. As a result, volunteer sampling efforts often concentrate in more accessible areas (Figure 4) and such sampling efforts are subject to spatial bias. In some cases, volunteers do make their decisions regarding where to sample based on where others have sampled. For example, eBird contributors may consult the "hotspots" map and species distribution maps provided by eBird [34] and accordingly select "hotspots" (areas with larger number of reported bird species) for birding. This, however, may reinforce existing spatial bias in sampling efforts, as areas with higher sampling intensity get sampled repeatedly (over-sampled) while areas with lower sampling intensity remain under-sampled.

Geographic data contributed by volunteers, when using geographic samples for analysis and modeling, need to be representative so properties of the underlying population can be inferred from the sample with satisfactory accuracy. The representativeness of geographic samples collected through traditional spatial sampling protocols is often ensured by following a rigorous sampling scheme such that sampling locations properly cover the environmental gradients in the geographic area of interest [63]. However, due to spatial bias in volunteer sampling efforts, the sampling locations may not have a good coverage over the environmental gradients, which impedes the representativeness of volunteer-contributed geographic samples [64]. For instance, more occurrence locations of a bird species in urban areas as reflected in the eBird database do not necessarily mean the species actually prefer urban habitats; it may be that there were simply not sufficient sampling efforts in non-urban habitats to discover the species. When the occurrence locations are used to predict species distribution, e.g., through species distribution modeling, such spatial bias needs to be accounted for in order to improve modeling and prediction accuracy [8,16,17,55,65–67].

4.3.2. Temporal Bias

Temporal bias also exists in volunteer sampling efforts, potentially at multiple temporal granularities. As profiling of the eBird data shows, sampling efforts and reported bird species are increasing over the years, with significant monthly fluctuations and notably more contributions on weekends. When observations from VGI are used to analyze temporal changes of geographic phenomena, such temporal bias should be accounted for [15]. For instance, the increasing number of bird species reported to eBird (Figure 16) does not mean increasing bird diversity on Earth. The increase is basically attributed to the increase in sampling efforts. As another example, although in the northern hemisphere there are a larger number of bird species reported to eBird in May than in September, one cannot definitively conclude that more bird species are present in May; there are simply not as many sampling efforts in September, and thus, the comparison would not be meaningful without disentangling the effects of the uneven sampling efforts. In fact, when eBird observations are used for modeling and predicting species geographic distribution, temporal variations in sampling efforts are often controlled for by selecting only observations within certain periods of roughly uniform sampling efforts [8,17].

4.3.3. Contributor Bias

Volunteer contributors are of various levels of skill in contributing data, and skill level of the same contributor may change over time. For example, the varied levels of expertise among eBird contributors may well be reflected in the number of active days they report observations (Figure 13) and in the number of reported sampling events, sampling locations and bird species (Figure 14). Many observers are ephemeral one-time contributors submitting only a single record. Only a small portion of the contributor are experts who tend to actively contribute large quantities of data over the long term [51]. In fact, when counting birds, there exist both between-observer differences [68] and within-observer differences (i.e., a change in ability to count birds of a given species after an observer's first year experience) [69].

Such observer bias may need to be accounted for when using VGI data in analyses. For instance, [70] reveal that for low-density populations, using data contributed by novice and experienced observers together may lead to erroneous site occupancy models. Other studies have found that observation skills of volunteer contributors can be estimated using species accumulation curves [71] and incorporating observer quality as a covariable to account for observer differences [68]; removing an observer's first year of observation [69] improves population trends estimation, and incorporating estimates of observer expertise in occupancy model improves species distributions from citizen science data [72].

4.3.4. Observation Bias

Bias also exists regarding the observation target. While some targets are easy to observe and identify, others may be more challenging. As a result, volunteer observations may be biased towards the easy targets in data volume. Common species may be reported repeatedly in many sampling events and by many observers, while only few records of rare and elusive species may be reported. Another complication is that observers may have their respective preferences on what to observe and report. Expert birders may be only interested in reporting rare species while ignoring common species. Such observation bias in VGI may deserve treatment in certain VGI applications. In fact, eBird let users report whether a checklist (i.e., sampling event) includes all species they could detect and identify ("complete" checklist). This makes it possible to filter the checklists to use only complete ones in analysis and modeling and thus enables analysts to move away from the reporting preference issue mentioned above [73]. eBird encourages observers to submit complete checklists, and a high proportion (~75%) of the submitted checklists in eBird database are complete.

5. Conclusions

Using eBird as an example, this study explores spatial and temporal patterns in volunteer data contribution activities. The sampling efforts of eBird contributors are biased in space and time. Most sampling efforts are concentrated in areas of denser populations and/or better accessibility, with the most intensively sampled areas being in proximity to big cities in developed regions of the world. Due to the spatially biased sampling efforts, reported bird species are also spatially biased towards areas where more sampling efforts occur. Temporally, eBird sampling efforts and reported bird species are increasing over the years, with significant monthly fluctuations and notably more data reported on weekends. Such trends are driven by continued development of the eBird project and by characteristics of both the bird species (e.g., breeding season) and the observers (e.g., more birding on weekends). Other forms of biases also exist in volunteer data contribution activities (e.g., contributor bias, observation bias). The fitness of use of VGI in the context of particular applications should be evaluated by examining the extent of the spatial, temporal and other biases and their potential impacts. In many cases, the biases need to be accounted for such that reliable inferences can be made from VGI observations.

Funding: This research was funded by Microsoft AI for Earth—Azure Compute Credit Grants.

Acknowledgments: This study was supported by the Faculty Start-up Funds and the Faculty Research Fund at the University of Denver. The author thanks the many eBird contributors for their efforts of contributing birding records to eBird, the Cornell Lab of Ornithology for making eBird data open and freely available, and Alison Johnston (Research Associate at eBird) and the anonymous reviewers for their constructive comments that helped improve this manuscript.

Conflicts of Interest: The authors declare no conflict of interest.

References

1. Goodchild, M.F. Citizens as sensors: The world of volunteered geography. *GeoJournal* **2007**, *69*, 211–221. [CrossRef]
2. Haklay, M.; Weber, P. OpenStreetMap: User-generated street maps. *IEEE Pervasive Comput.* **2008**, *7*, 12–18. [CrossRef]
3. Sullivan, B.L.; Wood, C.L.; Iliff, M.J.; Bonney, R.E.; Fink, D.; Kelling, S. eBird: A citizen-based bird observation network in the biological sciences. *Biol. Conserv.* **2009**, *142*, 2282–2292. [CrossRef]
4. Wood, C.; Sullivan, B.; Iliff, M.; Fink, D.; Kelling, S. eBird: Engaging birders in science and conservation. *PLoS Biol.* **2011**, *9*, e1001220. [CrossRef] [PubMed]

5. Arsanjani, J.J.; Zipf, A.; Mooney, P.; Helbich, M. *OpenStreetMap in GIScience: Experiences, Research, and Applications*; Springer: Berlin/Heidelberg, Germany, 2015; ISBN 3319142801.
6. Sullivan, B.L.; Aycrigg, J.L.; Barry, J.H.; Bonney, R.E.; Bruns, N.; Cooper, C.B.; Damoulas, T.; Dhondt, A.A.; Dietterich, T.; Farnsworth, A.; et al. The eBird enterprise: An integrated approach to development and application of citizen science. *Biol. Conserv.* **2014**, *169*, 31–40. [CrossRef]
7. Sachdeva, S.; McCaffrey, S.; Locke, D. Social media approaches to modeling wildfire smoke dispersion: Spatiotemporal and social scientific investigations. *Inf. Commun. Soc.* **2017**, *20*, 1146–1161. [CrossRef]
8. Fink, D.; Hochachka, W.M.; Zuckerberg, B.; Winkler, D.W.; Shaby, B.; Munson, M.A.; Hooker, G.; Riedewald, M.; Sheldon, D.; Kelling, S. Spatiotemporal exploratory models for broad-scale survey data. *Ecol. Appl.* **2010**, *20*, 2131–2147. [CrossRef] [PubMed]
9. Malik, M.M.; Lamba, H.; Nakos, C.; Pfeffer, J. Population Bias in Geotagged Tweets. In Proceedings of the Nineth International AAAI Conference on Web and Social Media, Oxford, UK, 26–29 May 2015; pp. 18–27.
10. Brown, G. A review of sampling effects and response bias in Internet participatory mapping (PPGIS/PGIS/VGI). *Trans. GIS* **2017**, *21*, 39–56. [CrossRef]
11. Hecht, B.; Stephens, M. A tale of cities: Urban biases in volunteered geographic information. In Proceedings of the Eighth International Conference on Web and Social Media (ICWSM), Ann Arbor, MI, USA, 1–4 June 2014; pp. 197–205.
12. Zhang, Y.; Li, X.; Wang, A.; Bao, T.; Tian, S. Density and diversity of OpenStreetMap road networks in China. *J. Urban Manag.* **2015**, *4*, 135–146. [CrossRef]
13. Yang, A.; Fan, H.; Jing, N.; Sun, Y.; Zipf, A. Temporal Analysis on Contribution Inequality in OpenStreetMap: A Comparative Study for Four Countries. *ISPRS Int. J. Geo-Inf.* **2016**, *5*, 5. [CrossRef]
14. Basiri, A.; Haklay, M.; Foody, G.; Mooney, P. Crowdsourced geospatial data quality: Challenges and future directions. *Int. J. Geogr. Inf. Sci.* **2019**, *33*, 1588–1593. [CrossRef]
15. Boakes, E.H.; McGowan, P.J.K.; Fuller, R.A.; Ding, C.; Clark, N.E.; O'Connor, K.; Mace, G.M. Distorted views of biodiversity: Spatial and temporal bias in species occurrence data. *PLoS Biol.* **2010**, *8*, e1000385. [CrossRef] [PubMed]
16. Zhang, G. Enhancing VGI application semantics by accounting for spatial bias. *Big Earth Data* **2019**, *3*, 255–268. [CrossRef]
17. Zhang, G.; Zhu, A.-X. A representativeness directed approach to spatial bias mitigation in VGI for predictive mapping. *Int. J. Geogr. Inf. Sci.* **2019**, *33*, 1873–1893. [CrossRef]
18. Zhu, A.-X.; Zhang, G.; Wang, W.; Xiao, W.; Huang, Z.-P.; Dunzhu, G.-S.; Ren, G.; Qin, C.-Z.; Yang, L.; Pei, T.; et al. A citizen data-based approach to predictive mapping of spatial variation of natural phenomena. *Int. J. Geogr. Inf. Sci.* **2015**, *29*, 1864–1886. [CrossRef]
19. Boakes, E.H.; Gliozzo, G.; Seymour, V.; Harvey, M.; Smith, C.; Roy, D.B.; Haklay, M. Patterns of contribution to citizen science biodiversity projects increase understanding of volunteers' recording behaviour. *Sci. Rep.* **2016**, *6*, 1–11. [CrossRef] [PubMed]
20. Antoniou, V.; Skopeliti, A. Measures and indicators of VGI quality: An overview. *ISPRS Ann. Photogramm. Remote Sens. Spat. Inf. Sci.* **2015**, *II-3/W5*, 345–351. [CrossRef]
21. Sauermanna, H.; Franzonib, C. Crowd science user contribution patterns and their implications. *Proc. Natl. Acad. Sci. USA* **2015**, *112*, 679–684. [CrossRef]
22. Neis, P.; Zipf, A. Analyzing the contributor activity of a volunteered geographic information project-The case of OpenStreetMap. *ISPRS Int. J. Geo-Inf.* **2012**, *1*, 146–165. [CrossRef]
23. Li, L.; Goodchild, M.F.; Xu, B. Spatial, temporal, and socioeconomic patterns in the use of Twitter and Flickr. *Cartogr. Geogr. Inf. Sci.* **2013**, *40*, 61–77. [CrossRef]
24. Nielsen, J. The 90-9-1 Rule for Participation Inequality in Social Media and Online Communities. 2006. Available online: https://www.nngroup.com/articles/participation-inequality (accessed on 1 October 2020).
25. Haklay, M.E. *Why is Participation Inequality Important?* Ubiquity Press: London, UK, 2016.
26. Carron-Arthur, B.; Cunningham, J.A.; Griffiths, K.M. Describing the distribution of engagement in an Internet support group by post frequency: A comparison of the 90-9-1 Principle and Zipf's Law. *Internet Interv.* **2014**, *1*, 165–168. [CrossRef]
27. Girres, J.-F.; Touya, G. Quality assessment of the French OpenStreetMap dataset. *Trans. GIS* **2010**, *14*, 435–459. [CrossRef]

28. Bittner, C. Diversity in volunteered geographic information: Comparing OpenStreetMap and Wikimapia in Jerusalem. *GeoJournal* **2017**, *82*, 887–906. [CrossRef]
29. Geldmann, J.; Heilmann-Clausen, J.; Holm, T.E.; Levinsky, I.; Markussen, B.; Olsen, K.; Rahbek, C.; Tøttrup, A.P. What determines spatial bias in citizen science? Exploring four recording schemes with different proficiency requirements. *Divers. Distrib.* **2016**, *22*, 1139–1149. [CrossRef]
30. Audubon. Cornell Lab of Orithnology about eBird. Available online: https://ebird.org/about (accessed on 17 September 2019).
31. eBird. eBird Basic Dataset Metadata (v1.12). 2019. Available online: https://ebird.org/science/download-ebird-data-products (accessed on 17 September 2019).
32. Kelling, S.; Lagoze, C.; Wong, W.-K.; Yu, J.; Damoulas, T.; Gerbracht, J.; Fink, D.; Gomes, C. eBird: A Human/Computer Learning Network to Improve Biodiversity Conservation and Research. *AI Mag.* **2013**, *34*, 10–20. [CrossRef]
33. La Sorte, F.A.; Somveille, M. Survey completeness of a global citizen-science database of bird occurrence. *Ecography* **2020**, *43*, 34–43. [CrossRef]
34. eBird. Explore Hotspots-eBird. Available online: https://ebird.org/hotspots (accessed on 17 September 2019).
35. UNDP. *Human Development Indices and Indicators: 2018 Statistical Update*; UNDP: New York, NY, USA, 2018.
36. USFWS. *Birding in the United States: A Demographic and Economic Analysis Addendum to the 2011 National Survey of Fishing, Hunting, and Wildlife-Associated Recreation*; U.S. Fish and Wildlife Service: Washington, DC, USA, 2013. Available online: https://digitalmedia.fws.gov/digital/collection/document/id/1874/ (accessed on 1 October 2020).
37. eBird. Mobile Now Available in 5 Languages. Available online: https://ebird.org/news/mobiletranslation/ (accessed on 20 March 2020).
38. Tuanmu, M.N.; Jetz, W. A global 1-km consensus land-cover product for biodiversity and ecosystem modelling. *Glob. Ecol. Biogeogr.* **2014**, *23*, 1031–1045. [CrossRef]
39. Global 1-km Consensus Land Cover. Available online: http://www.earthenv.org/landcover (accessed on 15 June 2020).
40. Center for International Earth Science Information Network—CIESIN—Columbia University. Gridded Population of the World, Version 4 (GPWv4): Population Density, Revision 11. 2018. Available online: https://data.nasa.gov/dataset/Gridded-Population-of-the-World-Version-4-GPWv4-Po/w4yu-b8bh (accessed on 15 June 2020).
41. Population Density, v4.11 (2000, 2005, 2010, 2015, 2020). Available online: https://sedac.ciesin.columbia.edu/data/set/gpw-v4-population-density-rev11 (accessed on 15 June 2020).
42. Meijer, J.R.; Huijbregts, M.A.J.; Schotten, K.C.G.J.; Schipper, A.M. Global patterns of current and future road infrastructure. *Environ. Res. Lett.* **2018**, *13*, 064006. [CrossRef]
43. GRIP Global Roads Database. Available online: https://www.globio.info/download-grip-dataset (accessed on 15 June 2020).
44. Human Development Data (1990–2018). Available online: http://hdr.undp.org/en/data (accessed on 1 October 2020).
45. An Overview of All the Official Languages Spoken per Country. Available online: http://www.arcgis.com/home/item.html?id=5c6ec52c374249a781aede5802994c95 (accessed on 15 June 2020).
46. 2020 World Population by Country. Available online: https://worldpopulationreview.com/ (accessed on 15 June 2020).
47. Phillips, S.J.; Anderson, R.P.; Schapire, R.E. Maximum entropy modeling of species geographic distributions. *Ecol. Model.* **2006**, *190*, 231–259. [CrossRef]
48. Yan, Y.; Kuo, C.; Feng, C.; Huang, W.; Fan, H. Coupling maximum entropy modeling with geotagged social media data to determine the geographic distribution of tourists. *Int. J. Geogr. Inf. Sci.* **2018**, *32*, 1699–1736. [CrossRef]
49. Phillips, S.J.; Dudík, M.; Schapire, R.E. Maxent Software for Modeling Species Niches and Distributions, (Version 3.4.1). 2019. Available online: https://biodiversityinformatics.amnh.org/open_source/maxent (accessed on 1 March 2019).
50. Phillips, S.J.; Dudík, M. Modeling of species distributions with Maxent: New extensions and a comprehensive evaluation. *Ecography* **2008**, *31*, 161–175. [CrossRef]

51. Bégin, D.; Devillers, R.; Roche, S. The life cycle of contributors in collaborative online communities-the case of OpenStreetMap. *Int. J. Geogr. Inf. Sci.* **2018**, *32*, 1611–1630. [CrossRef]
52. Welcome to ornitho.ch. Available online: https://www.ornitho.ch/ (accessed on 1 October 2020).
53. iNaturalist. Available online: https://www.inaturalist.org/ (accessed on 1 October 2020).
54. Conflict Is Still Africa's Biggest Challenge in 2020. Available online: https://reliefweb.int/report/world/conflict-still-africa-s-biggest-challenge-2020 (accessed on 1 October 2020).
55. Johnston, A.; Moran, N.; Musgrove, A.; Fink, D.; Baillie, S.R. Estimating species distributions from spatially biased citizen science data. *Ecol. Model.* **2020**, *422*, 108927. [CrossRef]
56. Newman, G.; Graham, J.; Crall, A.; Laituri, M. The art and science of multi-scale citizen science support. *Ecol. Inform.* **2011**, *6*, 217–227. [CrossRef]
57. Wikipedia eBird. Available online: https://en.wikipedia.org/wiki/EBird (accessed on 17 July 2020).
58. eBird Mobile App for iOS Now Available! Available online: https://ebird.org/news/ebird_mobile_ios1 (accessed on 1 October 2020).
59. Cornell Lab of Orinithology Merlin. Available online: https://merlin.allaboutbirds.org/the-story/ (accessed on 21 July 2020).
60. Great (Global) Backyard Bird Count This Weekend! Available online: https://ebird.org/news/great-global-backyard-bird-count-this-weekend/ (accessed on 3 October 2020).
61. Global Big Day—9 May 2020. Available online: https://ebird.org/news/global-big-day-9-may-2020 (accessed on 21 July 2020).
62. October Big Day—19 October 2019. Available online: https://ebird.org/news/october-big-day-19-october-2019 (accessed on 21 July 2020).
63. Jensen, R.R.; Shumway, J.M. Sampling our world. In *Research Methods in Geography: A Critical Introduction*; Gomez, B., Jones, J.P., III, Eds.; John Wiley & Sons: Hoboken, NJ, USA, 2010; pp. 77–90.
64. Zhang, G.; Zhu, A.-X. The representativeness and spatial bias of volunteered geographic information: A review. *Ann. GIS* **2018**, *24*, 151–162. [CrossRef]
65. Pardo, I.; Pata, M.P.; Gómez, D.; García, M.B. A novel method to handle the effect of uneven sampling effort in biodiversity databases. *PLoS ONE* **2013**, *8*, e52786. [CrossRef]
66. Stolar, J.; Nielsen, S.E. Accounting for spatially biased sampling effort in presence-only species distribution modelling. *Divers. Distrib.* **2015**, *21*, 595–608. [CrossRef]
67. Robinson, O.J.; Ruiz-Gutierrez, V.; Fink, D. Correcting for bias in distribution modelling for rare species using citizen science data. *Divers. Distrib.* **2018**, *24*, 460–472. [CrossRef]
68. Sauer, J.R.; Peterjohn, B.G.; Link, W.A. Observer differences in the North American Breeding Bird Survey. *Auk* **1994**, *111*, 50–62. [CrossRef]
69. Kendall, W.L.; Peterjohn, B.G.; Sauer, J.R.; Url, S. First-time observer effects in the North American Breeding Bird Survey. *Auk* **1996**, *113*, 823–829. [CrossRef]
70. Fitzpatrick, M.C.; Preisser, E.L.; Ellison, A.M.; Elkinton, J.S. Observer bias and the detection of low-density populations. *Ecol. Appl.* **2009**, *19*, 1673–1679. [CrossRef] [PubMed]
71. Kelling, S.; Johnston, A.; Hochachka, W.M.; Iliff, M.; Fink, D.; Gerbracht, J.; Lagoze, C.; La Sorte, F.A.; Moore, T.; Wiggins, A.; et al. Can Observation Skills of Citizen Scientists Be Estimated Using Species Accumulation Curves? *PLoS ONE* **2015**, *10*, e0139600. [CrossRef] [PubMed]
72. Johnston, A.; Fink, D.; Hochachka, W.M.; Kelling, S. Estimates of observer expertise improve species distributions from citizen science data. *Methods Ecol. Evol.* **2018**, *9*, 88–97. [CrossRef]
73. Johnston, A.; Hochachka, W.; Strimas-Mackey, M.; Ruiz Gutierrez, V.; Robinson, O.; Auer, T.; Kelling, S.; Fink, D. Analytical guidelines to increase the value of citizen science data: Using eBird data to estimate species occurrence. *bioRxiv* **2020**. Available online: https://www.biorxiv.org/content/10.1101/574392v3.full.pdf (accessed on 15 June 2020).

International Journal of *Geo-Information*

MDPI

Article

Role of Maximum Entropy and Citizen Science to Study Habitat Suitability of Jacobin Cuckoo in Different Climate Change Scenarios

Priyanka Singh [1,*], Sameer Saran [1] and Sultan Kocaman [2]

1 Geoinformatics Department, Indian Institute of Remote Sensing (ISRO), 4-Kalidas Road, Dehradun 248001, India; sameer@iirs.gov.in

2 Department of Geomatics Engineering, Hacettepe University, Ankara 06800, Turkey; sultankocaman@hacettepe.edu.tr

* Correspondence: priyanka.iirs@gmail.com; Tel.: +91-7011828901

Abstract: Recent advancements in spatial modelling and mapping methods have opened up new horizons for monitoring the migration of bird species, which have been altered due to the climate change. The rise of citizen science has also aided the spatiotemporal data collection with associated attributes. The biodiversity data from citizen observatories can be employed in machine learning algorithms for predicting suitable environmental conditions for species' survival and their future migration behaviours. In this study, different environmental variables effective in birds' migrations were analysed, and their habitat suitability was assessed for future understanding of their responses in different climate change scenarios. The Jacobin cuckoo (*Clamator jacobinus*) was selected as the subject species, since their arrival to India has been traditionally considered as a sign for the start of the Indian monsoon season. For suitability predictions in current and future scenarios, maximum entropy (Maxent) modelling was carried out with environmental variables and species occurrences observed in India and Africa. For modelling, the correlation test was performed on the environmental variables (bioclimatic, precipitation, minimum temperature, maximum temperature, precipitation, wind and elevation). The results showed that precipitation-related variables played a significant role in suitability, and through reclassified habitat suitability maps, it was observed that the suitable areas of India and Africa might decrease in future climatic scenarios (SSPs 2.6, 4.5, 7.0 and 8.5) of 2030 and 2050. In addition, the suitability and unsuitability areas were calculated (in km^2) to observe the subtle changes in the ecosystem. Such climate change studies can support biodiversity research and improve the agricultural economy.

Keywords: citizen science; machine learning; Indian monsoon; Jacobin cuckoo; Maxent; species distribution model; habitat suitability; range expansion; WorldClim; CMIP

Citation: Singh, P.; Saran, S.; Kocaman, S. Role of Maximum Entropy and Citizen Science to Study Habitat Suitability of Jacobin Cuckoo in Different Climate Change Scenarios. *ISPRS Int. J. Geo-Inf.* **2021**, *10*, 463. https://doi.org/10.3390/ijgi10070463

Academic Editors: Maria Antonia Brovelli and Wolfgang Kainz

Received: 31 May 2021
Accepted: 2 July 2021
Published: 6 July 2021

1. Introduction

The exponential change in the climate has directly affected the spatial distribution of species and communities in ecosystems, which is an essential requirement to understand the functions and processes of the ecosystem. As such, species movement in response to climatic shifts could be projected from species distribution models (SDMs), which provide an empirical way to assess the climatic impacts for the changes of species habitats (for example, reference [1]). A habitat is defined as a particular location where species live and reproduce with certain characteristics, behaviour, interactions and population patterns [2]. A favourable habitat that is significant for the survival of a species is called habitat suitability [3] and has importance in ecological research through habitat suitability modelling, which can help in conservation and protection plans. Several studies have investigated the suitability of species' habitats using maximum entropy (Maxent) [4] for evaluating the species range using geolocation data; for example, references [5–12]. The

Maxent model has gained popularity in the literature for the modelling of species' spatial distributions, and related studies have received over 5000 citations in the Web of Science Core Collection, mostly used by distribution modelers (ca. >60%) [13]. In recent decades, countless studies have been carried out on the suitability of species' habitats using the maximum entropy method for evaluating species ranges using presence-only data [14–19]. Such studies help to derive useful guidelines on the parameterisations of the model, such as the minimum sample size and data sample requirement, selection of random samples from voluminous datasets and the determination of subsampling process for range predictions per species and per sample size [14–16,18,19].

Although such data-intensive modelling approaches can help in identifying the major factors behind species range expansion [20], the occurrences of species, which poses a significant contribution to the model, should preferably be recorded across spatial (species range) and temporal (time of observation) contexts. However, it is intrinsically a problematic and costly task to record geographically varied near-real-time observations, because such activities need the continuous monitoring of species movements, so the species are tagged with tracking devices without causing any harm to them. Therefore, these real-time or near-real-time observations through a volunteering approach for data collection could help in quantifying the species fitness at a large spatial scale and informing about the changes in climatic patterns. The species properties can also be obtained from floras, the literature, herbaria and museums as theoretical data to model the habitat suitability of a species [21]. However, the main challenge remains the spatial uncertainty, which may be sourced from incorrect geotagging or wrong datum information [21,22]. Until recently, the species data were collected and recorded as a textual description in the forms of names and places [22], and digitising the textual information also causes substantial errors and brings spatial uncertainty in the order of several kilometres [23]. Various techniques have been developed to estimate and to document the location uncertainty among species' occurrence records in order to eliminate high errors prior to suitability modelling [22,24].

Trained and untrained volunteers have helped in the data collection processes as a citizen science approach, which may provide robust and rigorous data with qualitative and quantitative attributes. The data collected using citizen science approaches have been applied to ecological niche models in recent years to mitigate the gaps in the quantity and quality of data, which also improved the approximation of the metric of interest [25]. Species distribution modelling for particular species requires a sufficient number of occurrences distributed across its extant [26–28]. Citizen science is a broad concept that can be understood in different forms [29], from highly systematic protocols to opportunistic surveys with no sampling designs [29,30].

In this paper, a ML-based maximum entropy (Maxent) algorithm was applied to Jacobin cuckoo (pied cuckoo or pied-crested cuckoo) occurrences with environmental variables, such as to evaluate the potential habitat suitability in Africa and in India. For the modelling procedure, the birds' occurrences were first divided into three time periods—June–September, which refers to India's southwest monsoon, October–December, which refers to India's northwest monsoon and Africa's wet season, and January–May—to predict the suitability site of the Jacobin cuckoo, as most parts of India have winter and summer seasons. This approach was also used to predict the change in monsoon patterns by modelling how this bird's favourable habitats will shift under different climate change scenarios. For this future prediction modelling, the obtained modelling results of current suitability with the existing environmental variables and species occurrences was projected with future models to observe the probable climatic changes in 2030 (an average of 2021–2040) and 2050 (an average of 2041–2060). In addition, the areas of suitability and unsuitability sites were also calculated to analyse the increase or decrease in the ecological system in response to changes in the monsoon patterns. The first ever Indian monsoon climate change study in terms of Jacobin cuckoos' migration was performed by Singh and Saran [31], in which the geographic occurrences of the Jacobin cuckoo with 19 current bioclimatic variables were modelled using the ML=based Maxent model in R software. This trained model was then projected with future bioclimatic variables under the RCP8.5 scenario of the Coupled Model Intercomparison Project (CMIP5) to assess the predicted changes in the suitability of Pied cuckoos' habitats by 2050. Specifically, the current and future bioclimatic variables were used at a resolution of 2.5 arc seconds (~4.5 km at the equator) of latitude and longitude. The above-mentioned study signified that the major environmental variables that affect the suitability of the Jacobin cuckoo were isothermality (16.8%), the mean temperature of the warmest quarter (15.7%), annual precipitation, precipitation of the warmest quarter (13.6%) and precipitation of the wettest month (11.3%) during the Indian summer monsoon season, i.e., June–October. As per the current suitability predictions, the states of Southern India—Andhra Pradesh, Goa, Karnataka, Kerala, Maharashtra and Tamil Nadu—and Northern India—Uttarakhand and Himachal Pradesh—showed high, as well as medium, habitat suitability, and the western states (e.g., Gujarat) displayed medium suitability; southern Africa was found unsuitable for this bird, because in the June–October months, a dry and hot climate is experienced there, which is not a favourable habitat. However, according to the future suitability prediction of this bird, the Jacobin cuckoo range contraction could happen in all parts of India except the southern parts of Tamil Nadu due to increased greenhouse gas emissions and a decrease in precipitation of the warmest quarter. In addition, the quantiles (5% and 95%) of the relevant environmental variables were calculated to observe the changes in climates of now and 2050 with respect to Indian monsoon seasons.

2. Citizen Science as a Biodiversity Research Method

The act of engaging volunteers in scientific tasks has proliferated in the past few decades with offered, more pressing opportunities for participants to deliver advanced approaches and make meaningful insights into their collected data. The activity of effectively utilising crowdsourcing, along with the Internet and mobile applications, over large geographic regions is known as citizen science. Citizen science "is a process where concerned citizens, government agencies, industry, academia, community groups, and local institutions collaborate to monitor, track and respond to issues of common community concern" [32] and "where citizens and stakeholders are included in the management of resources" [33,34]. Citizen science involves both professionals and non-professionals participating in both scientific thinking and data collection [33,35] with the support of technological advancements, such as smart phones with location services, camera, accelerometer, etc. [36]. However, based on its nature of engagement and utility in diverse domains, citizen science may have different conceptual definitions and meanings.

According to the nature of engagement, the galaxy of citizen science is categorised into four levels—crowdsourcing, distributed intelligence, participatory science and extreme [37]. Crowdsourcing is the most basic level, where the general public can contribute to science by processing and analysing collected data. The next level is distributed intelligence, in which citizens learn new skills before becoming involved in data collection and interpretation activities. The third level is participatory science, where citizens are involved with research groups for defining problems and data collection. The last level is extreme, where citizens are equipped with full control to define problems, collect data and performing analyses on it.

As per its utility in various projects with different aims, Wiggins and Crownston [38] classified citizen science projects into five mutually exclusive and exhaustive types-action, education, conservation, investigation and virtual projects. The various action projects address local issues with the joint collaboration of citizens and scientists/researchers—for example, references [39–49]-and education projects help in improving the knowledge of citizens as part of the curriculum [50–55]. The conservation projects focus on the management of natural resources—for example, reference [19]—investigation projects emphasise the study of citizen's observations combined with different parameters to answer scientific questions [56,57] and virtual projects involve remote citizen science activities [58–61].

The above classification schemes can be demonstrated with the example of Project PigeonWatch, which is one of the citizen science projects at the Cornell Lab of Ornithology (CLO) and the National Audubon Society that engage many volunteers of all ages and professions throughout the world to collect hands-on data to study and analyse pigeon colour variations. On the basis of the above checklist, this project can be characterised as an "investigation" project, and the utilised approach is "crowdsourcing".

Amidst various citizen science projects, 72% of the projects relate to the discipline of biology [62], and due to such dominancy in this area, the study and research on the diversity and distribution of species [63] advance the rapid need for biodiversity monitoring, conservation planning and ecological research. Many citizen science programs have been realised over the span of years or even decades and are still being carried out to study the patterns of nature on a large spatial scale by collecting data on different locations and habitats of species. The way of collecting such information on the species' locations, habitats and other related information [63–65] by enlisting the public in scientific activities is now considered the best practice. It is not necessarily true that the scientific output is always benefitted by robust strategies and inferences from highly recognised and peer-reviewed scientific publications; rather, gathering information through public participation could be a better source of scientific information to answer specific questions [66,67]. Higher credit may be given to Cornell University's Lab of Ornithology, which laid the foundation for volunteer participation in biodiversity observations, monitoring and research [52]. However, there are many other organisations and research groups that have designed various citizen science programs to collect geographically well-distributed and dense data with rigorous spatial sampling, such as Species Mapping through the Indian Bioresource Information Network (IBIN) portal [68], bioblitz [69], the shell polymorphism survey [70], the water quality survey [71] and breeding bird surveys [72]. Such diverse datasets compel the aggregation of observation data from different sources for conventional research, but the major concerns even after data aggregation are data quality [73] and techniques for combining diverse datasets into different schemas [74]. Therefore, apposite planning is required for managing the voluminous dataset integration into a uniform schema with data quality check infrastructure for handling observational biases, "false absences" that yield to inadequate sightings [75] and uneven data distributions [76]. These challenges were addressed by a global concerted effort [77] that began in 2004 and has now resulted into the largest single gateway to observation-based datasets, known as the Global Biodiversity Information Facility (GBIF).

The GBIF is an intergovernmental organisation that provides "an Internet accessible, interoperable network of biodiversity databases and information technology tools" [78] as a "cornerstone resource" [79], with a "mission to make the world's biodiversity data freely and universally available via the Internet" [79]. Currently, the GBIF portal provides open access to more than 160 million biodiversity occurrences and taxa data from 1641 institutions and volunteered surveying data around the globe. Therefore, the GBIF has become an authentic repository where various organisations/institutes share their data with quality and in large quantity, which are essential for modelling and decision-making purposes. Edwards et al. [77] performed a spatial validation of the third-largest flowering plant family, the Leguminosa, using its taxa and distribution data from the GBIF portal to evaluate the quality and coverage of its geographic occurrences. Similar reviews could be seen by Graham et al. [21] and Suarez and Tsutsui [80] for additional uses of museum specimen data, which facilitated biodiversity policy and decision-making process [80]. Amongst the various other advantages, GBIF data can be used for biodiversity assessments [81], taxonomic revisions [82], compiling red lists of threatened species [83] and habitat suitability modelling [31,84–87]. The latter is one of the prominent examples of climate change studies in which citizen science-based observations from the GBIF are being increasingly used [88–95]. In this paper, different climate change scenarios combined with the GBIF's observed occurrences of the monsoon favourable bird, *Clamator jacobinus*, are modelled using the Maxent approach to study the contemporary and future habitat suitability of this bird so that the variations in the Indian monsoon season can be examined.

3. Materials and Methods

3.1. The Jacobin (Pied) Cuckoo Species

As per Indian belief, the arrival of this partially migrated bird, the Jacobin cuckoo (*Clamator jacobinus*) (Figure 1), also known as "Chatak" in India, heralds the onset of the Indian monsoon [96]. During the summer, the bird flies from Africa to India for breeding, crossing the Arabian Sea and the Indian Ocean, as shown in Figure 2. The Jacobin Cuckoo belongs to the cuckoo order of small terrestrial birds with black and white soft plumage and long-wings with a spiffy crest on the head that quenches its thirst with raindrops. The species is also known as a brood parasite, i.e., instead of making its own nest, it lays its eggs in the nest of other birds, particularly Jungle babbler (*Turdoides striata*). This bird of an arboreal nature mostly sits on tall trees but often forages for food in low bushes and, occasionally, on the ground. It prefers well-wooded areas, forests and bushes in semi-arid regions. As widely known, the Jacobin cuckoo maintains their suitability in India in two ways. There is a population that is sighted as residents year-round in the southern part of the country. The Jacobin cuckoo is also sighted in the central and northern parts of India along with summer monsoon winds from just before the monsoon to early winter, i.e., May–August. The reason behind choosing the Jacobin cuckoo was that the arrival time of this bird is directly linked with the monsoon because it only drinks rainwater drops as it pours down and does not utilise any other water sources, such as collected rain waters, rivers, etc., to quench its thirst.

Figure 1. Image of *Clamator jacobinus* (Jacobin cuckoo or Pied cuckoo) (Source: Pied Cuckoo Macaulay Library ML32455551).

3.2. The Species Distribution Data and the Preprocessing

The distribution data was obtained from the GBIF repository that collated geographic records of this bird from surveys, museums, human observations and other data sources. Then, those occurrences recorded through "human observation" were selected for this study, because this research was focused on demonstrating the use of citizen science data for habitat suitability modelling [97]. The institutes/organisations contributed this bird's data in the "human observation" category through various citizen science programs. These institutes/organisations are the Cornell Lab of Ornithology, FitzPatrick Institute of African Ornithology, South African National Biodiversity Institute, iNaturalist.org, Observation.org, Xeno-canto Foundation for Nature Sounds, naturgucker.de, Kenya Wildlife Service, India Biodiversity Portal and A Rocha Kenya. However, the dataset contained repeated latitude and longitude values, as well as null values (NA: not available). By using a data cleaning algorithm in R, records with NA values and duplicated locations were removed. Since the

Jacobin cuckoos are known for their close association with the onset of monsoon season in India, the compiled GPS records of human observations from 1991 to 2020 were divided into the following monthly sets:

i. Based on the period of the southwest monsoon season that typically lasts from June to September, the geographic occurrences of the Jacobin cuckoo were filtered for these months starting from the year 1991 to 2020. In this period, the whole country receives more than 75% of its rainfall [98].
ii. The second input set was filtered using the months of northeast monsoon season, i.e., October–December. This monsoon season is also known as post-monsoon or winter monsoon season, in which the country receives about 60% of its annual rainfall in the coastal areas and about 40% in the interior areas [99]. Additionally, the rainy season starts from October and lasts until April–June in Africa, where the conditions are mostly suitable for its residency.
iii. The third and final set contains the data of months January–May that denote the mid-rainy period and end of the rainy season in Africa and India, respectively.

Hence, the abovementioned sets of geographic occurrences were combined with environmental datasets to understand their potential suitability ranges, environmental parameters and altered climatic variations in different climate change scenarios.

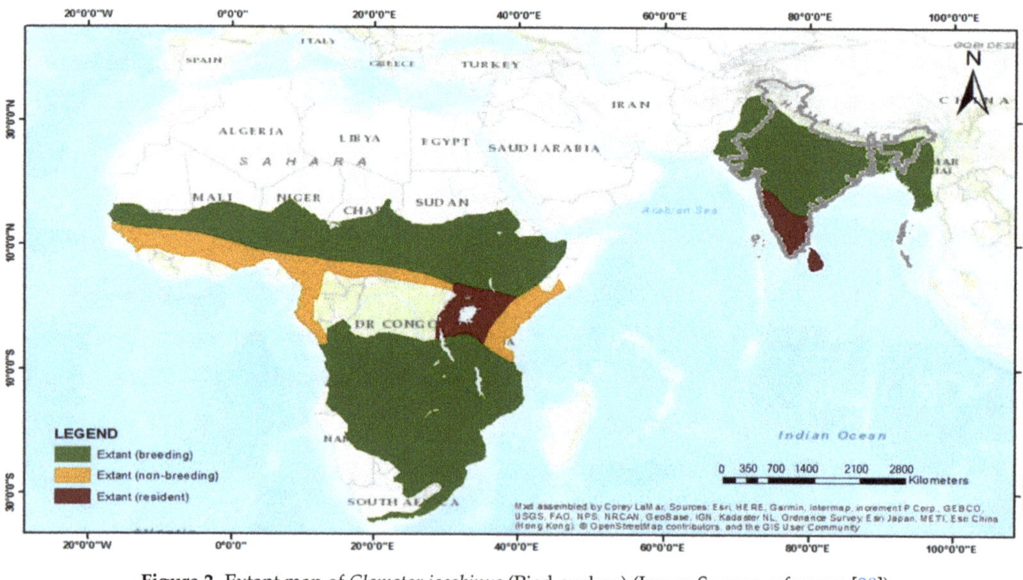

Figure 2. Extant map of *Clamator jacobinus* (Pied cuckoo) (Image Source: reference [30]).

3.3. Environmental Data

3.3.1. Selection of Environmental Variables

This section discusses the selection of environmental data that are assumed to ecologically influence mobile species like birds, particularly the Jacobin cuckoo distribution. These include bioclimatic variables, minimum temperature, maximum temperature, precipitation, elevation and wind at the spatial resolution of 2.5 arc-min from WorldClim. For the present climatic conditions, the bioclimatic variables, which were averaged for the years 1970–2000, were obtained from WorldClim version 2.1, the latest version of climate data launched in January 2020 [100]. As per this study, modelling was carried out for three different time periods; therefore, climatic variables such as precipitation, wind, minimum temperature and maximum temperature were taken and screened for the given three sets.

For future sets, 19 bioclimatic variables (Table 1) for the near-future (2021–2040) and remote-future (2041–2060) projections of the species distribution maps at 2.5 arc-min were obtained from WorldClim [101]. For future climate scenarios, the CIMP's climate data of the CNRM-ESM2-1 [102] global climate model (GCM) for four Shared Socioeconomic Pathways (SSPs): 126, 245, 370 and 585 were obtained from WorldClim's database, which was spatially downscaled and calibrated to reduce the bias.

Table 1. Environmental variables used in the habitat suitability modelling process.

Bioclimatic Variable Code	Bioclimatic Variable Name	Unit
BIO1	Annual Mean Temperature	°C
BIO2	Mean Diurnal Range (Mean of monthly (max temp–min temp))	°C
BIO3	Isothermality (BIO2/BIO7)	dimensionless
BIO4	Temperature Seasonality (Standard Deviation)	°C
BIO5	Max Temperature of Warmest Month	°C
BIO6	Min Temperature of Coldest Month	°C
BIO7	Temperature Annual Range (BIO5-BIO6)	°C
BIO8	Mean Temperature of Wettest Quarter	°C
BIO9	Mean Temperature of Driest Quarter	°C
BIO10	Mean Temperature of Warmest Quarter	°C
BIO11	Mean Temperature of Coldest Quarter	°C
BIO12	Annual Precipitation	mm
BIO13	Precipitation of Wettest Month	mm
BIO14	Precipitation of Driest Month	mm
BIO15	Precipitation Seasonality (Coefficient of Variation)	fraction
BIO16	Precipitation of Wettest Quarter	mm
BIO17	Precipitation of Driest Quarter	mm
BIO18	Precipitation of Warmest Quarter	mm
BIO19	Precipitation of Coldest Quarter	mm

The 2013 IPCC (Intergovernmental Panel on Climate Change) fifth assessment report (AR5) generated climate models from CMIP5, and the 2021 IPCC sixth assessment report (AR6) presented CMIP6 with the 10 Earth system models (ESMs) [103]. CNRM-ESM2-1 is one of the ESMs that contains interactive earth system components such as aerosols, atmospheric chemistry and the land and ocean carbon cycles. In CMIP6, sufficient amounts of data on Carbon Brief were included to analyse the future emission scenarios, such as past and future warming and climate sensitivity since CMIP5. The IPCC AR5 introduced four Representative Concentration Pathways (RCPs) that examined future greenhouse gas emissions in different climate change scenarios: RCP2.6, RCP4.5, RCP6.0 and RCP8.5. These scenarios were updated with the Shared Socioeconomic Pathways (SSPs) scenarios in CMIP6–SSP1-2.6, SSP2-4.5, SSP3-7.0, SSP4-6.0 and SSP5-8.5. SSP1 is a world of sustainability-focused growth and equality. SSP2 is known as the "middle of the road", where the historical patterns are followed, SSP3 lies right in the middle of the range of the baseline outcomes produced by ESMs, SSP4 is a more optimistic world that fails to ordain any climate policies and SSP5 depicts the worst-case scenario. These SSPs could examine the demographic and economic factors, as well as how societal picks will have an impact on greenhouse gas emissions. While in RCPs, the socioeconomic factors are not included, but only the pathways are set to examine the greenhouse gas concentrations and the amount of warming that could occur by the end of the century. In this paper, the SSPs 1–2.6, 2–4.5, 3–7.0 and 5–8.5 were used.

3.3.2. Screening of Environmental Variables

The correlation test between the environment variables for each of the three seasonal sets was carried out to retain the ecologically relevant variables in the species' suitability. Spearman's correlation coefficients [104] were applied on the variables set, where, if the variables had a Spearman correlation < 0.7, those variables were not highly correlated. Then, the Variance Inflation Factor (VIF) was calculated in R software for each remaining variable using the vifcor function of R package usdm [105] and eliminated the environmental variables whose VIF values > 3, because the smaller VIF values hold low correlations. The resulting VIF values were <3 [106], and therefore, no further variables were eliminated. This correlation test among the environmental variables was performed for India and Africa separately, and these screened environmental variables were then used in the Maxent model to predict the habitat suitability in the abovementioned study areas.

3.4. The Spatial Distribution Model with Maxent

ML-based Maxent modelling [107] is the most popular and a well-established habitat suitability modelling approach [108–116] that predicts probable distributions based on species' occurrences and environmental variables. The advantage of using Maxent is that it uses presence-only data and provides a predictive map within the study area. This works on the principle of maximum entropy that estimates the probability distribution of species' habitats with no constraints and assumes that each feature has the same mean value in the approximated distribution as the species occurrences.

In this study, the maximum entropy algorithm with bird's occurrences and screened predictor variables was modelled to predict the potential suitable habitats and analyse the relative importance of different bioclimatic factors of each point of occurrence for the Jacobin cuckoo. This method was applied on all the three sets of time periods, so that the habitat suitability analysis could be performed to validate the belief that the Jacobin cuckoo are the harbinger of the Indian monsoon and analyse the suitable climates and range of this bird in India, as well as in Africa, during the selected periods. The jackknife test was applied to recognise the importance of the environmental variables. The species occurrences were split as training (75% of the total occurrences) and test (25% of the total occurrences) data for the models' calibration and assessment, respectively. The response curves; jackknife and other features such as linear, quadratic, product, threshold and hinge were set as true parameters in the habitat suitability model. The other model parameters were used as follows:

i. "replicates = 10" tells the model about the number of replicates that the model executes for cross-validating, bootstrapping or doing sampling with replacement runs;
ii. "lq2lqptthreshold = 80" is the number of samples at which the product and threshold features start being used;
iii. "l2lqthreshold = 10" is the number of samples at which the quadratic features start being used and
iv. "hingethreshold = 15" is the number of samples at which the hinge features start being used.

The predictive performance of the generated model was then assessed by calculating the Area Under the Receiver Operator Curve (AUC) of the receiver operating characteristic (ROC) plot, which ranges between 0 (no discrimination) to 1 (perfect discrimination) [116]. The process of evaluating the model's predictive performance using AUC involves the process of setting thresholds on the model's prediction by generating various levels of false positive rates and then assessing the true positive rate as a false positive rate function. Here, the false positive rate referred to the prediction of a presence for those places where the species is absent, and the true positive rate is the successful prediction of a presence. The AUC range from 0.7 to 0.8 is acceptable, 0.8 to 0.9 is excellent and above 0.9 is an outstanding performance [117]. The dominant environmental variables in determining the species' probable distribution were assessed through the jackknife test (also called "leave-one-out") that gives the permutation importance against the environmental variables [110]. The species response curves were generated by the model to examine how the likelihood of species' occurrences responds to the variations in the changing environmental conditions.

Then, the future climatic variations (2021–2040 and 2041–2060) were also modelled to estimate how the species will respond to changes in ecological systems, as their favourable habitats will shift under different climate change scenarios (i.e., SSP1-2.6, SSP2-4.5, SSP3-7.0 and SSP5-8.5).

The predicted habitat suitability maps were then reclassified into convenient classes that represented the threshold limits that differentiated the unsuitable and suitable habitats. The reclassified classes of habitat suitability were: the unsuitable conditions with a lower threshold and the suitable conditions that were further categorised into three classes: low, medium and highly suitable. This threshold helped in interpreting the ecological significance by identifying areas that were at least suitable as similar to those areas where the species has been recorded.

4. Results

4.1. Selection of Environmental Variables

To detect the correlations among environmental variables, the Spearman's correlation coefficient threshold was set to 0.7, and then, the vifcor function was performed. From the visual assessment, some variables showed an intercorrelation that was then eliminated if their vif values was assigned as less than 3. This was the case for the minimum temperature, maximum temperature, precipitation, elevation, wind and bioclimatic variables. The final selected set of variables were then used to predict the suitability of the Jacobin cuckoo, given in Table 2.

Table 2. Environmental variables selected after the correlation test.

Time Periods	India	Africa
June–September	bio14, bio15, bio18, bio19, bio2, bio3, bio8, wind7	bio13, bio14, bio19, bio2, bio3, elevation, wind6, tmax9
October–December	bio14, bio15, bio18, bio19, bio2, bio3, wind12, prec12	bio14, bio15, bio19, bio8, bio9, wind 12
January–May	bio14, bio15, bio18, bio19, bio2, bio3, wind2, prec1	bio14, bio15, bio19, bio2, elev

4.2. Performance Evaluation Results of the Maxent Model

After executing the Maxent model on the species' occurrences and environmental variables, its predictive accuracy was evaluated by using AUC plots. As shown in Table 2, environmental variables were selected for India and Africa separately in three different time periods; therefore, the Maxent model was executed by separating the species occurrences of India and Africa into three different time periods. Additionally, the minimum temperature, maximum temperature and precipitation used in the Maxent model were taken according to these three time periods.

The AUC values for the Pied cuckoos' suitability prediction model given in Table 3 depicted that the model's prediction was very good, so that it could effectively predict the species distribution under the current and future climate scenarios.

Table 3. AUC values of the Maxent model's performance.

Time Periods	India	Africa
June–September	0.885	0.908
October–December	0.91	0.947
January–May	0.958	0.908

4.3. Variable Importance and Contribution

Tables 4 and 5 depict the heuristic estimate of the percentage contribution and the permutation importance of the environmental variables used in the Maxent model for three different time periods with species occurrence data from India and Africa. These two tables helped us to interpret the most influential environmental variables that played a significant role in the Jacobin cuckoo's habitat suitability in India and Africa. It is observed in Table 4 that bio2, bio3, bio14, bio15, bio18, bio19 and wind are common for all the three time periods in India, whereas, in Africa (Table 5), bio 14 and bio 19 are common in all three time periods. In India, wind and precipitation play a minor role, whereas, in Africa, wind and elevation hold major contributions in suitability modelling. Therefore, this section concluded that the environmental variables related to precipitation play a significant role in the distribution of the Jacobin cuckoo and are essentially required in its potential suitable habitats.

Table 4. Percent contribution and permutation importance of the environmental variables in terms of India's suitability.

Time Periods	Selected Environmental Variables	Percent Contribution	Permutation Importance
June–September	bio8 (Mean Temperature of Wettest Quarter)	52.4	40.9
	bio3 (Isothermality)	13.9	11
	bio2 (Mean Diurnal Range)	9.2	14.4
	bio18 (Precipitation of Warmest Quarter)	7.8	5.3
	wind7 (wind of July month)	6.8	9.6
	bio19 (Precipitation of Coldest Quarter)	5.2	10
	bio15 (Precipitation Seasonality)	4.5	7
	bio14 (Precipitation of Driest Month)	0.3	1.7
October–December	bio3 (Isothermality)	69.7	54.7
	bio18 (Precipitation of Warmest Quarter)	9.8	12.4
	bio2 (Mean Diurnal Range)	9.4	17.9
	wind12 (wind of December month)	3.8	5.7
	prec12 (precipitation of December month)	2.6	2.8
	bio15 (Precipitation Seasonality)	2.3	4.5
	bio19 (Precipitation of Coldest Quarter)	2	1.6
	bio14 (Precipitation of Driest Month)	0.5	0.5
January–May	bio3 (Isothermality)	56.2	49.7
	bio19 (Precipitation of Coldest Quarter)	21.7	5.1
	bio2 (Mean Diurnal Range)	13.2	22.8
	prec1 (precipitation of January month)	3.5	4.7
	bio18 (Precipitation of Warmest Quarter)	3.4	9.6
	wind2 (wind of February month)	1.3	6.3
	bio15 (Precipitation Seasonality)	0.7	1.5
	bio14 (Precipitation of Driest Month)	0.1	0.4

Table 5. Percent contribution and permutation importance of the environmental variables in terms of Africa's suitability.

Time Periods	Selected Environmental Variables	Percent Contribution	Permutation Importance
June–September	bio14 (Precipitation of Driest Month)	31.2	26.9
	bio13 (Precipitation of Wettest Month)	23.5	24.3
	wind6 (wind of June month)	15.4	19.2
	bio19 (Precipitation of Coldest Quarter)	11.1	7.2
	tmax9 (Maximum Temperature of September)	8.9	12.6
	bio2 (Mean Diurnal Range)	5.7	3.4
	elev (Elevation)	2.1	4.8
	bio3 (Isothermality)	2	1.6
October–December	bio9 (Mean Temperature of Driest Quarter)	51	30.1
	bio14 (Precipitation of Driest Month)	24.8	20.9
	wind12 (wind of December month)	8.9	18
	bio8 (Mean Temperature of Wettest Quarter)	8.1	20.9
	bio15 (Precipitation Seasonality)	5.5	8.1
	bio19 (Precipitation of Coldest Quarter)	1.7	2
January–May	bio19 (Precipitation of Coldest Quarter)	33.9	38.6
	bio15 (Precipitation Seasonality)	24.8	16.1
	bio14 (Precipitation of Driest Month)	24.6	37.9
	bio2 (Mean Diurnal Range)	15.4	6.3
	elev (Elevation)	1.4	1

4.4. Predicted Habitat Suitability Map of the Jacobin Cuckoo

Using the influential bioclimatic factors in the species distribution, the habitat suitability prediction was performed under the current and future climatic scenarios to estimate the changes in the ecological systems and how the species will respond to the changes in different climatic variations. This section discusses the spatial characteristics of the utilised distribution data, i.e., India's southwest and northeast monsoon seasons and Africa's rainy season, especially in Southern Africa.

4.4.1. Current Habitat Suitability

June–September

The months of June–September are known as India's southwest monsoon period, in which all of India receives more than 75% of its rainfall [118], whereas these months bring the dry season in Central, South and East Africa. Therefore, the recorded occurrences of June-September (separately for India and Africa) with screened environmental variables were supplied to the Maxent model, which resulted in the dominant bioclimatic variables, as well as the prediction of the current habitat suitability of this bird. After evaluating the model's performance, the output was then used to project the future habitat suitability of the Jacobin cuckoo under different climate change scenarios.

The species habitat suitability map shown in Figure 3 depicts that the areas covered with grey colour represent no suitability for this bird, whereas the yellow and brown colours represent good and low habitat suitability of the species, respectively. The species occurrences are plotted with red points on the map, and the habitat suitability ranges can be seen from the probability scale, which depicts the bird's residency. Accordingly, Africa has very low suitability during the June-September period because of its dry season, which might not be a favourable climate for the suitability of Pied cuckoos. However, the model predicted the good and high suitability of these birds in all of India as compared to Africa, because India receives 75% of its rainfall then, and thus, this could be one of the main reasons of their partial migration to India, so that they can get their suitable climatic factors. The results shown in Figure 3 were computed on occurrences from 1991 to 2020 and the climatic variables (tmin, tmax, prec and wind) of the June–September months.

In India, the major suitability predictions under the current scenario can be seen in Northern, Western and Southern India, but Eastern, Central and North-eastern India have shown very low or no suitability predictions. The model predicted the average suitability range of the Jacobin cuckoo in the following Indian states—Uttarakhand and Uttar Pradesh and in Madhya Pradesh of Central India. The reason behind its migration to these parts of India could be because of the wettest features due to the highest rainfall and the maximum number of rivers. A good suitability range was predicted in Western India, such as in Gujarat (sites near to the Gulf of Kachcha and the Gulf of Khambat) and in Maharashtra (sites surrounded by the Arabian Sea) and, also, in Southern India, such as in Western Ghats. The highest suitability of this bird was predicted in a few areas of South India, such as Southern Tamil Nadu, which is Rameswaram, Dhanushkodi and Thoothukodi. Therefore, there is a probability that this bird likes to stay at the wettest sites, which receive the highest amount of rainfall. In Africa, no habitat suitability was predicted for Jacobin cuckoo, because this time period is the dry season, which might force the Jacobin cuckoo's migration to India.

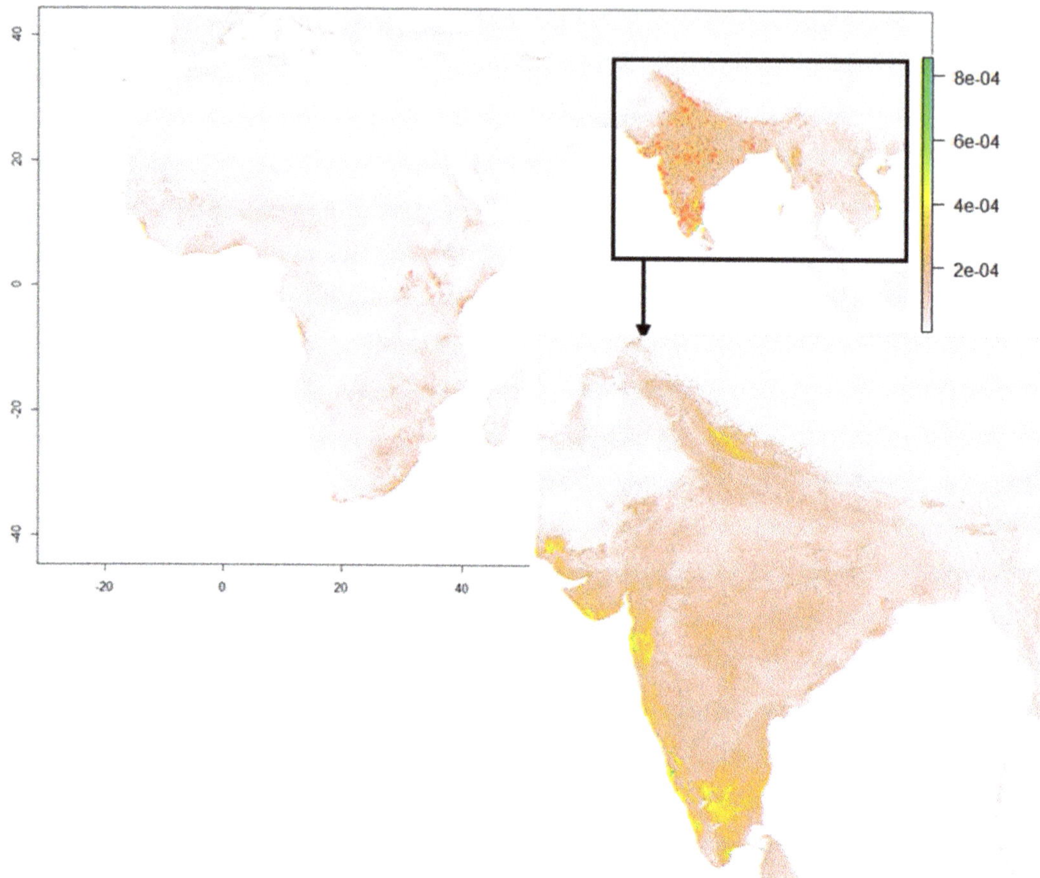

Figure 3. Habitat suitability prediction of June-September for the Pied cuckoo range.

The change in climate, particularly the Indian southwest monsoon patterns, were analysed with respect to the Jacobin cuckoo in the past 30 years by separating the datasets into two subsets—1991–2005 and 2006–2020. For this, the climatic variables were used for these yearly subsets, and their results are illustrated in Figures 4 and 5, respectively. In Figure 4, there is no suitability predicted in Africa in 1991–2006 during June–September because of its dry weather, which is unsuitable for the bird, but, in India, an adequate habitat suitability can be seen in the eastern parts and northern parts but not the southern and western parts, due to the bird's migration to India in search of a wet climate. When the suitability results of Figure 4 are compared with Figure 5, a decline is observed in the Indian monsoon rainfall in the north and east for the past 15 years in June–September. Additionally, this study on climate change was completely dependent on sightings of the Jacobin cuckoo, so one of the reasons of less suitability in Southern and Western India would be less sightings of the bird due to the unawareness of crowdsourcing. As per Google Trends search records from 2004 to the present (Figure 6), the public interest in the term "crowdsourcing" in India started in 2007.

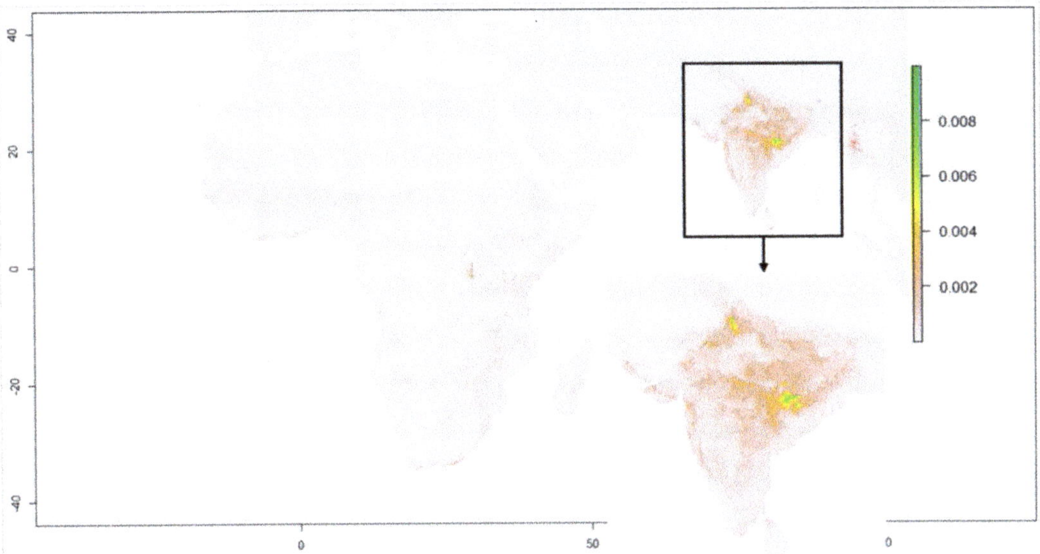

Figure 4. Habitat suitability modelling in 1991–2005 for the June–September period.

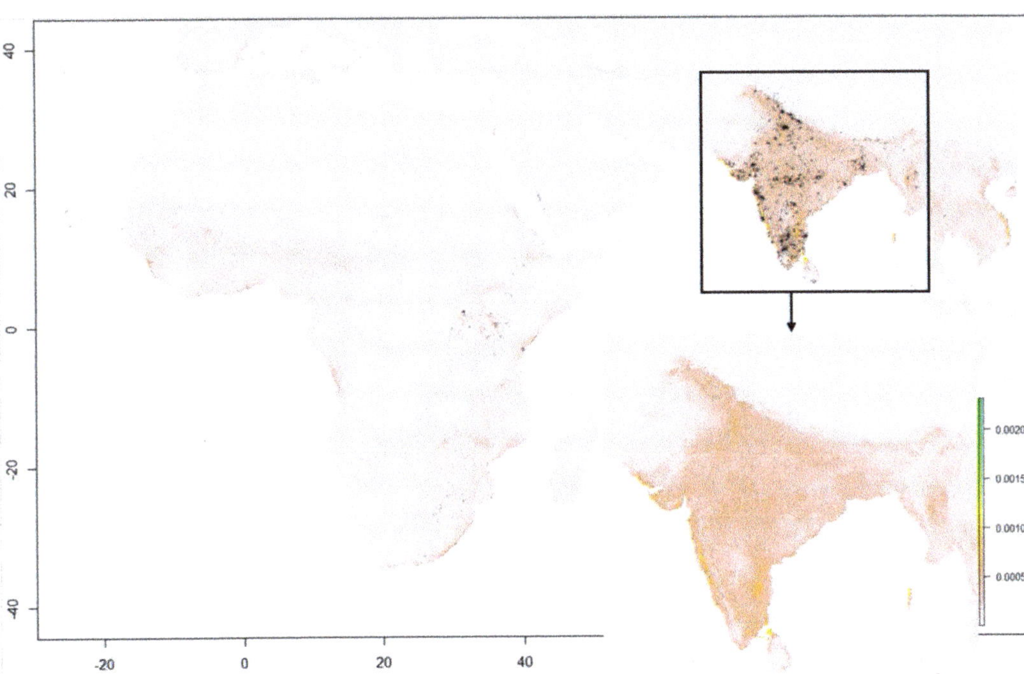

Figure 5. Habitat suitability modelling in 2006–2020 for the June-September period.

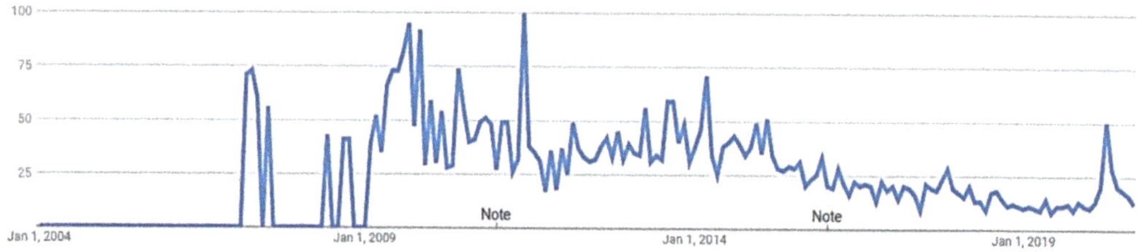

Figure 6. Google trends of "crowdsourcing" based on searches in India.

October–December and January–May

The months of October, November and December are known for the northeast monsoon or winter monsoon in India, as its direction is set from the northeast to southwest of India, and the beginning of rainy and wet season in Africa, which typically lasts until April. As such, this set of months includes monsoons of two different countries. Therefore, the results discussed in this section are divided into two sub-sections: one for October–December and another for January–May.

The northeast monsoon season from October to December in India brings rain to Andhra Pradesh mainly in the coastal regions of Kerala, Puducherry, Rayalaseema, South Karnataka and Tamil Nadu. As compared to the southwest monsoon, this monsoon period gives only 11% of the annual rainfall in India, but in Tamil Nadu, this season gives almost half of its annual rainfall. The habitat suitability map given in Figure 7 depicts that, in India, the highest habitat suitability of the Jacobin cuckoo is predicted in Tamil Nadu, good in Andhra Pradesh and Karnataka and low in Kerala. Considering the model's predictions on the bird's habitat suitability, which is correlated with the patterns of the northeast monsoon (i.e., in October, November and December), it can be linked to the belief or fact that the bird's movements are completely linked with the monsoon rains of the northeast monsoon period.

After verifying and exploring the links of the Indian monsoon arrival with the bird's sightings, subsequently, the same study was carried out to investigate the major climatic factors that could be the cause behind the bird's return journey to Africa. Figures 7 and 8 depict that, when the rainy and the wet season starts in October, the Pied cuckoos, residents of Africa, might return to their native lands and reside there until April. Therefore, the sightings of the Jacobin cuckoo were observed in the provinces of Africa, except the Western Cape. However, the highest suitability (green colour) is predicted in the coastal areas of the Eastern Cape and Kwazulu-Natal Provinces. Therefore, the research on the habitat suitability of the Jacobin cuckoo proved the correlation between sighting of the Jacobin cuckoo and the arrival of the monsoon season in India and also gives rise to ancestral tales or traditional beliefs that these Jacobin cuckoos have the magical ability of summoning rain wherever they go, such as all over the Indian subcontinent, as well as in Africa.

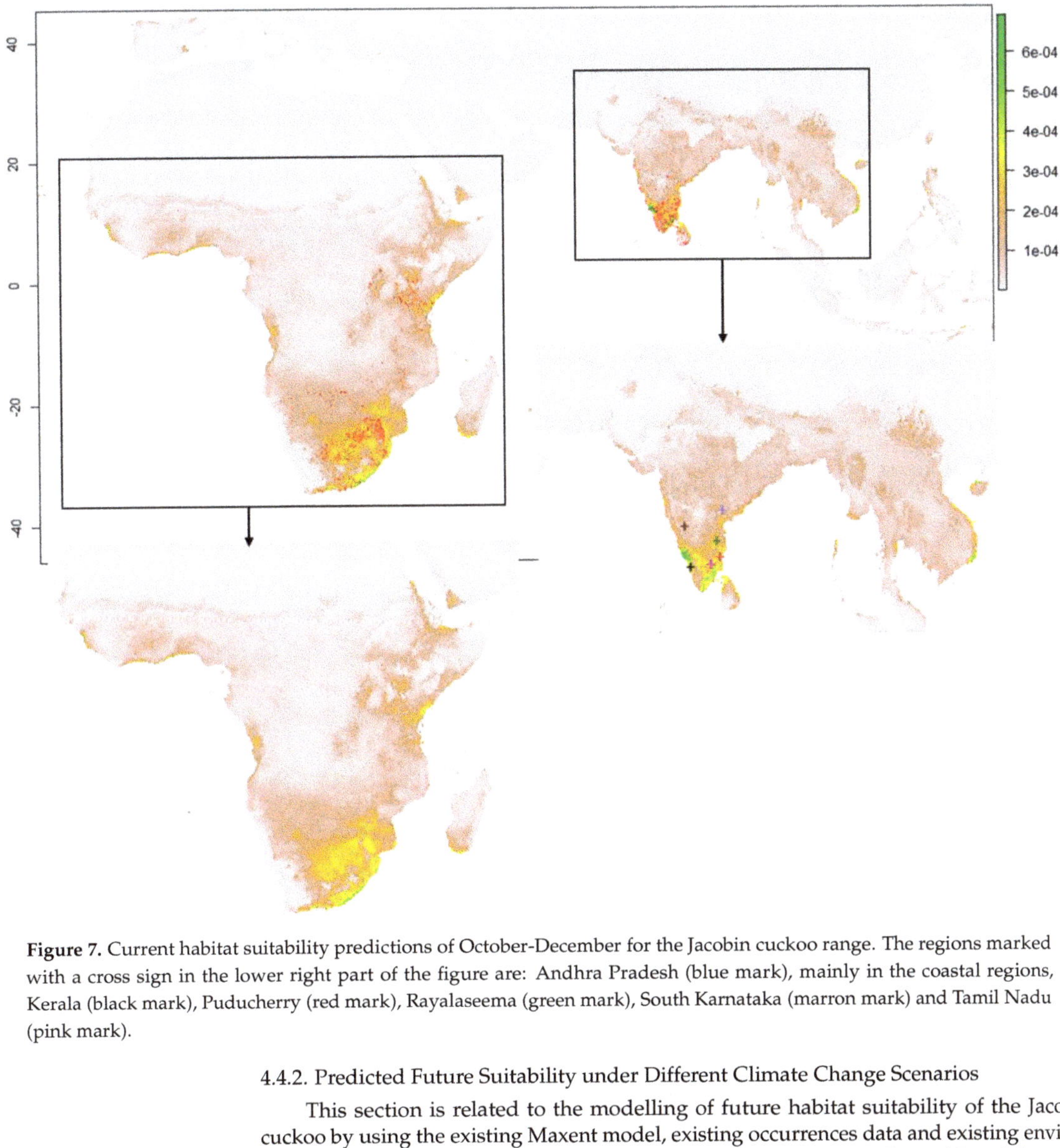

Figure 7. Current habitat suitability predictions of October-December for the Jacobin cuckoo range. The regions marked with a cross sign in the lower right part of the figure are: Andhra Pradesh (blue mark), mainly in the coastal regions, Kerala (black mark), Puducherry (red mark), Rayalaseema (green mark), South Karnataka (marron mark) and Tamil Nadu (pink mark).

4.4.2. Predicted Future Suitability under Different Climate Change Scenarios

This section is related to the modelling of future habitat suitability of the Jacobin cuckoo by using the existing Maxent model, existing occurrences data and existing environmental layers and then projecting them with future environmental variables of the years 2030 and 2050. In addition, the probable increase or decrease in suitable and unsuitable habitats in the current and future years were also estimated for the sites occupied by the species, which can be used in various climate change studies later.

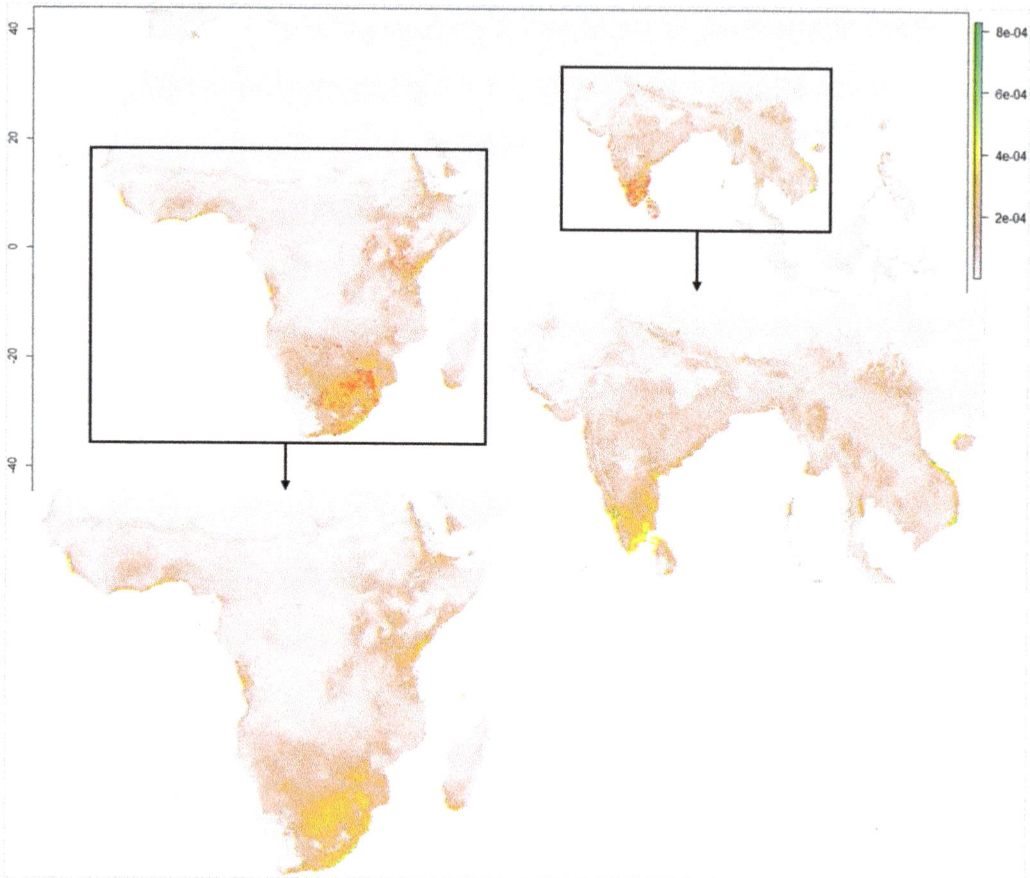

Figure 8. Current habitat suitability prediction of January-May for the Pied cuckoo range.

June–September

The resulting suitability maps was then generated using the selected environmental variables that are given in Figure 9. The figure depicts that the probable habitat suitability conditions of the Jacobin cuckoo have are relatively medium in SSPs 2.6, 4.5 and 7.0 of 2030. As compared to the current predictions in Figure 5, declines are observed in different climatic scenarios of the future years, particularly in SSP 8.5, in which the pixels of good suitability disappeared. This might be due to the estimation of higher CO^2 emissions and the increase in global warming. Table 6 depicts that there is a decline in suitable areas and an increase in unsuitable areas as compared to the current ones.

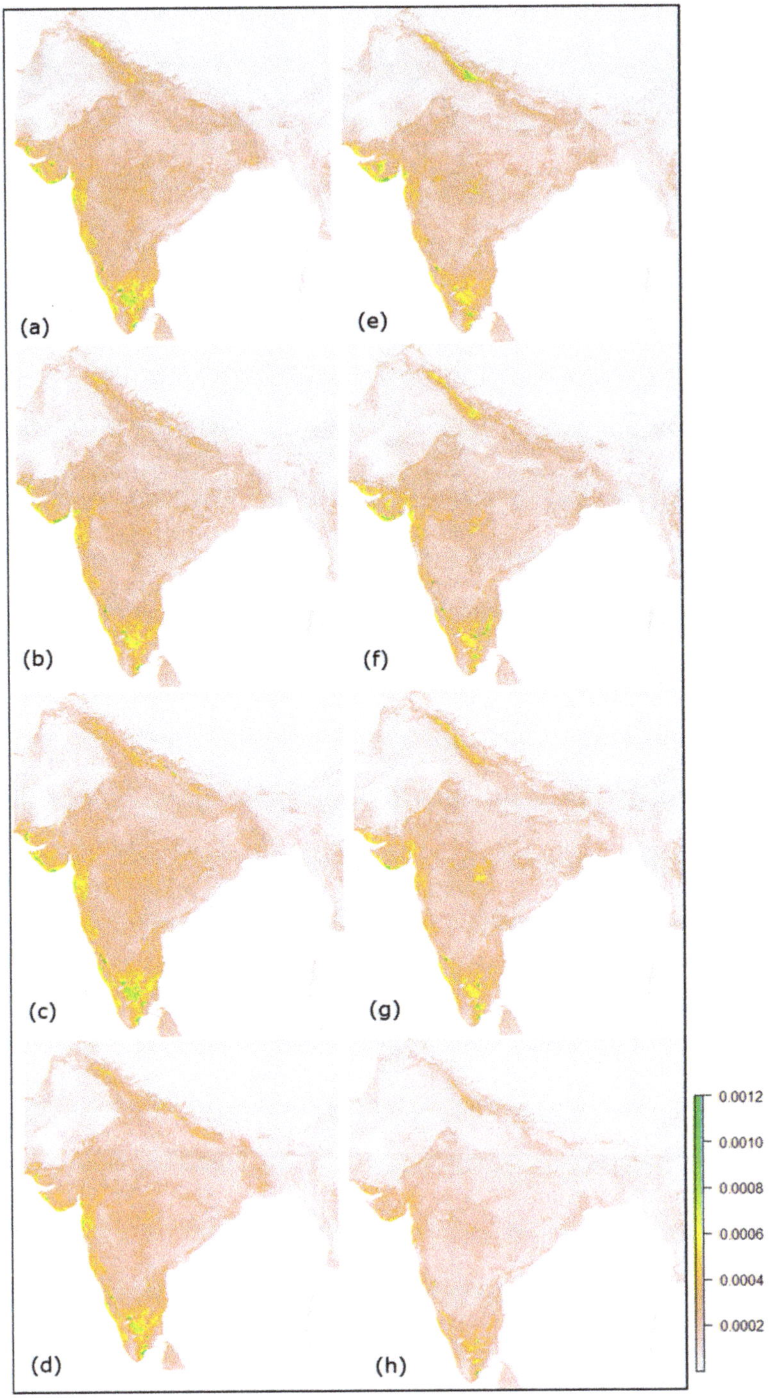

Figure 9. Predicted future habitat suitability maps of the Jacobin cuckoo during June–September of India for the 2030 SSPs 2.6, 4.5, 7.0 and 8.5 (**a**–**d**) and 2050 SSPs 2.6, 4.5, 7.0 and 8.5 (**e**–**h**).

Table 6. Predicted suitable areas under the current and future climatic conditions during June–September (India).

	Climatic Scenarios	Unsuitability (km^2)	Suitability (km^2)
Current	–	3,903,906.3	1,554,570.5
2030	SSP 2.6	3,917,983.9	1,533,619.6
	SSP 4.5	3,956,603.1	1,492,705.1
	SSP 7.0	3,925,725.7	1,530,059
	SSP 8.5	4,083,028.8	1,371,543.6
2050	SSP 2.6	3,935,078.1	1,514,768.8
	SSP 4.5	4,012,498.1	1,436,972
	SSP 7.0	4,170,800.4	1,281,060.6
	SSP 8.5	4,249,557.4	1,205,204.4

October–December and January–May

This section discusses on how the distribution of potential habitats will shift under different climate change scenarios. According to the model's future predictions for the months of October-December in India (Figure 10) and Africa (Figure 11), the habitat suitability of the Jacobin cuckoo is highly related to the current scenario (Figure 7) under SSPs 2.6 and 4.5 of 2030. However, the suitability conditions under different scenarios such as 7.0 and 8.5 of 2030 and, in 2050, under all climatic scenarios, the probability of occurrence of the Jacobin cuckoo is predicted as quite low when compared with the current one. This can be observed in Tables 7 and 8, which represent the increase in unsuitable and decrease in suitable areas of the Jacobin cuckoo in India and Africa, respectively.

Table 7. Predicted suitable areas under the current and future climatic conditions during October–December (India).

	Climatic Scenarios	Unsuitability (km^2)	Suitability (km^2)
Current	–	4,662,162.9	269,311.2
2030	SSP 2.6	4,772,917.3	198,217
	SSP 4.5	4,773,967.4	197,233.2
	SSP 7.0	4,758,425.3	200,135.6
	SSP 8.5	4,806,362.7	185,111.4
2050	SSP 2.6	4,772,249.1	1,990,134.3
	SSP 4.5	4,804,739.2	183,147.1
	SSP 7.0	4,842,675.4	165,650.5
	SSP 8.5	4,836,618.7	163,666.7

Table 8. Predicted suitable areas under the current and future climatic conditions during October–December (Africa).

	Climatic Scenarios	Unsuitability (km^2)	Suitability (km^2)
Current	–	25,328,121	1,416,705.3
2030	SSP 2.6	25,421,822.1	1,320,241.8
	SSP 4.5	25,402,520.3	1,337,374.5
	SSP 7.0	25,396,134.9	1,344,209.5
	SSP 8.5	25,417,724.7	1,321,503.3
2050	SSP 2.6	25,549,750.9	1,193,954.8
	SSP 4.5	25,598,589.5	1,144,656.8
	SSP 7.0	25,611,813.3	1,132,636.1
	SSP 8.5	25,682,361.4	1,065,214.8

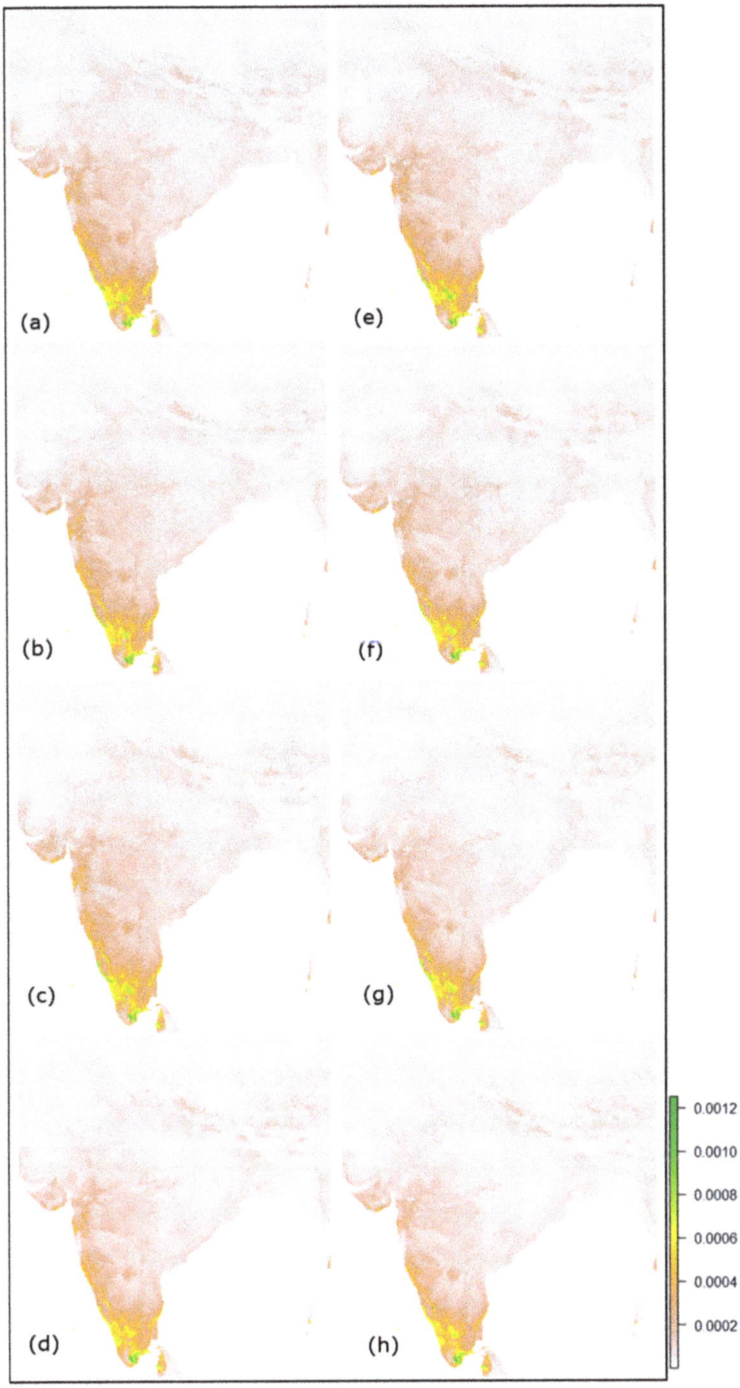

Figure 10. Predicted future suitability maps for Jacobin cuckoos from October to December: (**a**–**d**) and (**e**–**h**) represent the SSPs 2.6, 4.5, 7.0 and 8.5 for 2030 and 2050, respectively.

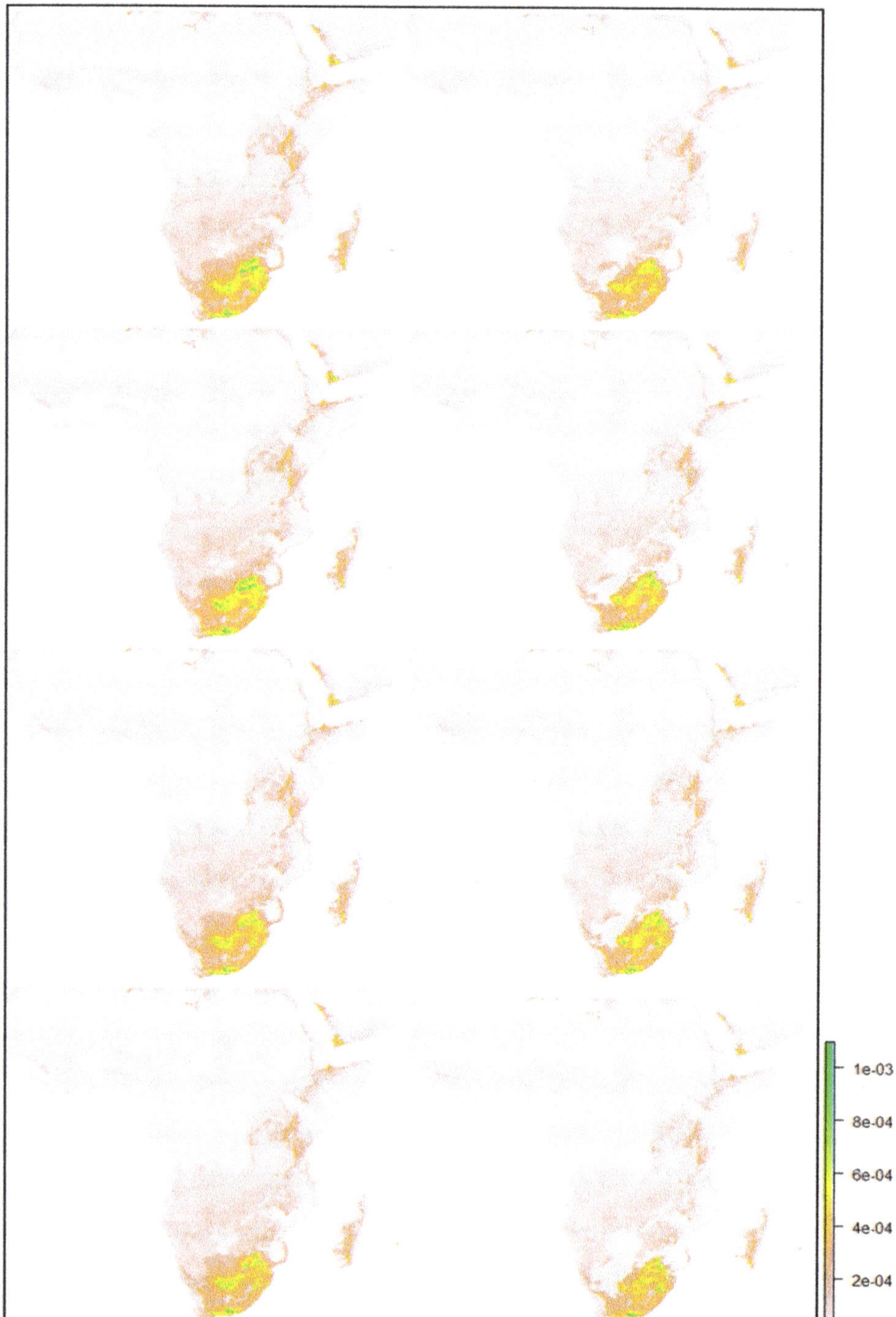

Figure 11. Predicted future suitability maps for Jacobin cuckoos from October to December in Africa: (**a–d**) and (**e–h**) represent the SSPs 2.6, 4.5, 7.0 and 8.5 for 2030 and 2050, respectively.

Other future suitability predictions for the January–May months of 2030 and 2050 through the Maxent model are shown in Figures 12 and 13 for India and Africa, respectively, which provides the probability that the bird's suitability might become low in 2050 under all climate scenarios. Such a decline in habitat suitability of the bird during these months indicates that, in the future, India, as well as Southern Africa, might receive less rain and more dryness, which will result in a decline of the Jacobin cuckoo's suitability in India (Table 9) and Africa (Table 10). Although incongruities may exist between various climate modelling approaches [119], the strategy of assessing the current suitability and predicting the future changes in the distributions of diverse species, which are influenced by different climatic patterns, is still recognised as an important research area.

Table 9. Predicted suitable areas under the current and future climatic conditions during January–May (India).

	Climatic Scenarios	Unsuitability (km^2)	Suitability (km^2)
Current	–	4,568,739.6	128,897
2030	SSP 2.6	4,585,901.9	114,579.5
	SSP 4.5	4,583,793.4	115,614
	SSP 7.0	4,584,234.3	115,404
	SSP 8.5	4,594,489.4	100,791.9
2050	SSP 2.6	4,573,664.1	121,099.9
	SSP 4.5	4,581,033.1	117,698.1
	SSP 7.0	4,593,339.3	103,510.2
	SSP 8.5	4,600,834.1	94,180.3

Table 10. Predicted suitable areas under the current and future climatic conditions during January–May (Africa).

	Climatic Scenarios	Unsuitability (km^2)	Suitability (km^2)
Current	–	15,137,454	1,205,065.2
2030	SSP 2.6	15,271,158.6	1,150,279.7
	SSP 4.5	15,242,352.7	1,143,199.7
	SSP 7.0	15,275,949	1,170,872.3
	SSP 8.5	15,385,782.4	1,140,142.1
2050	SSP 2.6	15,271,495.7	1,206,615
	SSP 4.5	15,279,882.5	1,115,013
	SSP 7.0	15,200,979.4	1,199,596.5
	SSP 8.5	15,389,258.9	1,125,855.3

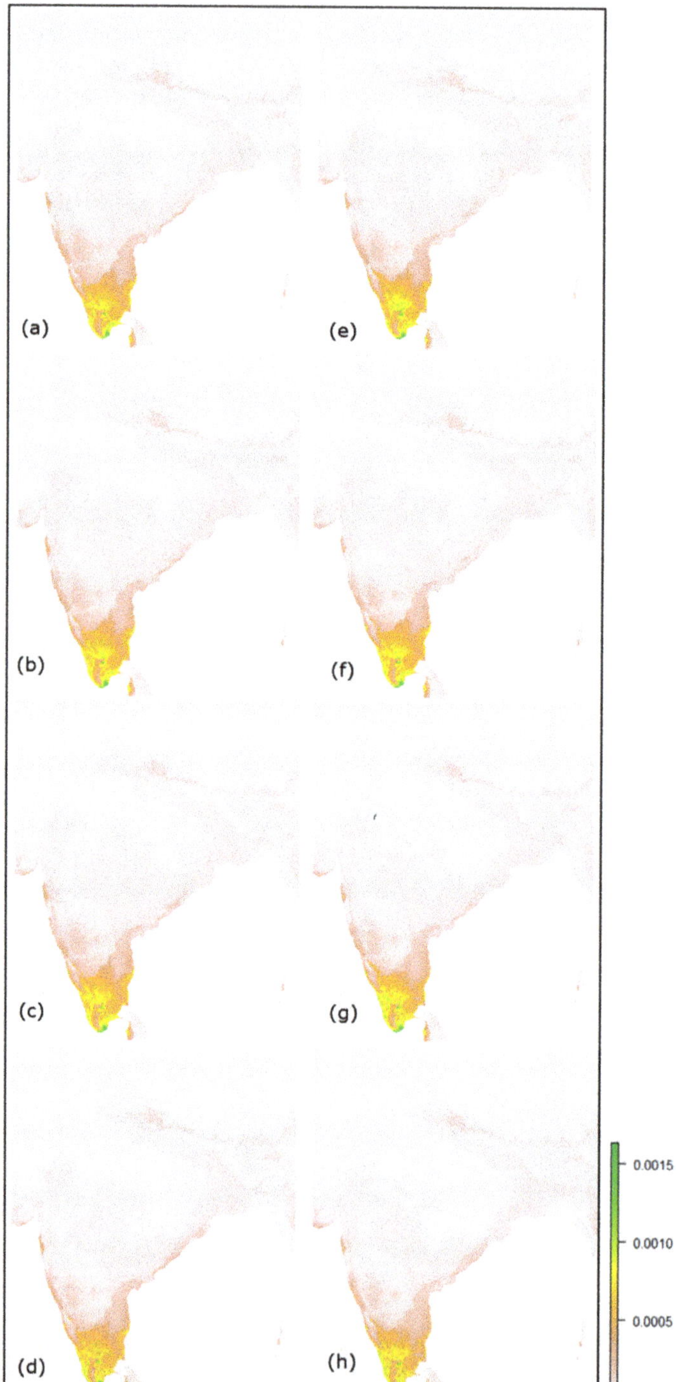

Figure 12. Predicted future suitability maps for January-May in India: (**a–d**) and (**e–h**) represent the SSPs 2.6, 4.5, 7.0 and 8.5 for 2030 and 2050, respectively.

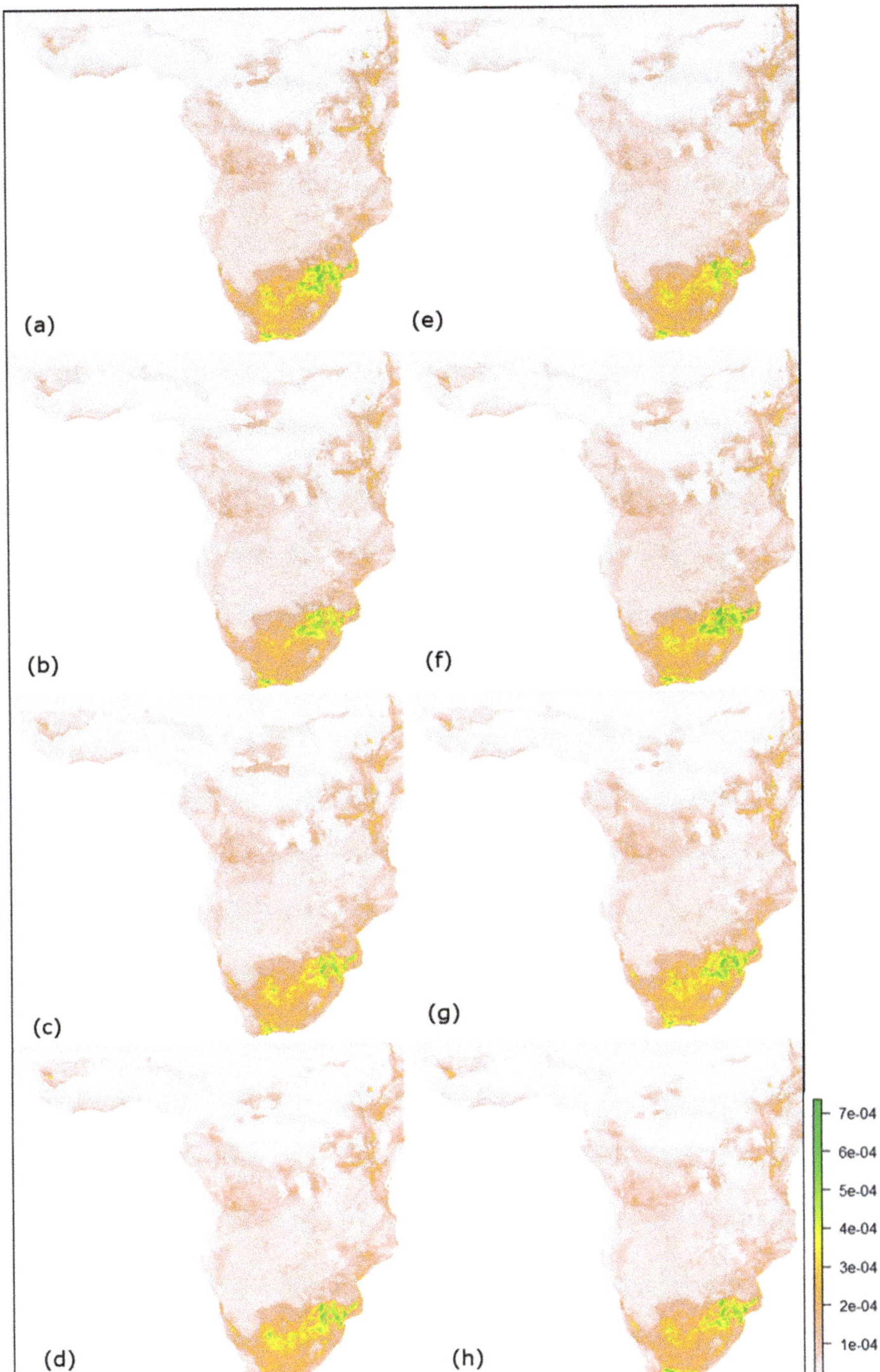

Figure 13. Predicted future suitability maps for January-May in Africa: (**a–d**) and (**e–h**) represent the SSPs 2.6, 4.5, 7.0 and 8.5 for 2030 and 2050, respectively.

5. Discussions

The study presented here analysed the habitat suitability for the Jacobin cuckoo in different seasons, with particular reference to India, using the species' occurrences (1991–2020) in the ML-based Maxent model with environmental variables. The occurrences were obtained from the GBIF database, an observatory composed of data from public institutions, e.g., museums, and citizen observations. The Maxent model's predictive performance achieves higher AUC values, which denotes that this model is excellent and accurate. The results obtained using the Maxent method for predicting the potential suitability of the Jacobin cuckoo are different in all three seasons of India, i.e., June-September (the southwest monsoon), October-December (the northeast monsoon) and January-May (winter and summer). The important environmental variables affecting its habitat suitability are the precipitation of the driest month, precipitation seasonality, precipitation of the warmest quarter, mean temperature of the wettest quarter and wind. The model predictions showed that the species suitability followed the same pattern of both Indian monsoon seasons, i.e., southwest and northeast. Therefore, based on the results, the bird's migration can be linked with monsoons in the assessed regions—India and Africa. Nevertheless, in order to examine India's southwest monsoon season, the datasets were divided into two subsets—1991–2005 and 2006–2020, and then, the Maxent model with the environmental data was executed. From the results, it was surprisingly interesting to see that the monsoon patterns started declining in 1991 in a few regions of the northern and eastern parts of India during the June-September period, which might be because of anthropogenic activities, deforestation, etc. However, in Africa, the climatic conditions are always suitable for this bird's residency starting from October and lasting until April. When the rainy and wet season in Africa ends, the birds start migrating to different parts of the world where they get more favourable climate conditions. The future suitability of the Pied cuckoo bird was modelled here with a full set of climatic conditions under four scenarios (SSPs): 2.6, 4.5, 7.0 and 8.5 for 2030 (averaged for 2021–2040) and 2050 (averaged for 2041–2060), using the results of the current suitability and its projected bioclimatic variables. As per the future predictions carried out in this study, the potentially suitable climatic distribution will shrink in the future (2050 < 2030 < current) under different climate change scenarios, indicating that there could be a change in the monsoon season in India, as well as in Africa, which will result in less suitability for the Jacobin cuckoo. Such a direct link of this bird with the monsoon season helps to critically analyse the likely climatic change activities and which environmental variables play an influential role for its suitability and to support its migratory movements.

6. Conclusions

This study concluded that the ecological systems will be altered with respect to the climate changes, and the favourable habitats of species will shift under different climate change scenarios. The present study demonstrated an example of the modelling or the prediction of these shifts by using citizen observations, which provided the required set of data or the observations to apply robust ML models. Thus, the use of citizen science methods was essential for enabling such an analysis. Future suitability modelling using CMIP6 future datasets revealed that the precipitation and wettest climates might decline while warm and dry climates may rise.

The wettest season and precipitation are major elements in Jacobin cuckoos' distribution, and various collaborative programs are required to maintain the suitability of various migratory birds like the Jacobin cuckoo in such a changing and unexpected potential warming of the Earth. However, such predicted changes are based only on climatic factors and are not necessarily related to the distribution of human-occupied land use like urban settlements and dispersal ability.

Author Contributions: Conceptualisation and methodology, Priyanka Singh and Sameer Saran; investigation, validation, formal analysis, writing—original draft preparation, data curation and visualisation, Priyanka Singh; supervision, project administration and resources, Sameer Saran and writing—review and editing, Sultan Kocaman. All authors have read and agreed to the published version of the manuscript.

Funding: This research has received no external funding.

Institutional Review Board Statement: Not Applicable.

Informed Consent Statement: Not Applicable.

Data Availability Statement: Not Applicable.

Conflicts of Interest: The authors declare no conflict of interest.

References

1. Araújo, M.B.; Guisan, A. Five (or so) challenges for species distribution modelling. *J. Biogeogr.* **2006**, *33*, 1677–1688. [CrossRef]
2. Matthiopoulos, J.; Fieberg, J.; Aarts, G.A.; Beyer, H.L.; Morales, J.M.; Haydon, D.T. Establishing the link between habitat selection and animal population dynamics. *Ecol. Monogr.* **2015**, *85*, 413–436. [CrossRef]
3. Hirzel, A.H.; Le Lay, G.; Helfer, V.; Randin, C.F.; Guisan, A. Evaluating the ability of habitat suitability models to predict species presences. *Ecol. Model.* **2006**, *199*, 142–152. [CrossRef]
4. Phillips, S.J.; Anderson, R.P.; Dudík, M.; Schapire, R.E.; Blair, M.E. Opening the black box: An open-source release of Maxent. *Ecography* **2017**, *40*, 887–893. [CrossRef]
5. Yan, H.; Feng, L.; Zhao, Y.; Wu, D.; Zhu, C. Prediction of the spatial distribution of Alternanthera philoxeroides in China based on ArcGIS and MaxEnt. *Glob. Ecol. Conserv.* **2020**, *21*, e00856. [CrossRef]
6. Rondinini, C.; Stuart, S.; Boitani, L. Habitat Suitability Models and the Shortfall in Conservation Planning for African Vertebrates. *Conserv. Biol.* **2005**, *19*, 1488–1497. [CrossRef]
7. Tikkanen, O.-P.; Heinonen, T.; Kouki, J.; Matero, J. Habitat suitability models of saproxylic red-listed boreal forest species in long-term matrix management: Cost-effective measures for multi-species conservation. *Biol. Conserv.* **2007**, *140*, 359–372. [CrossRef]
8. Remya, K.; Ramachandran, A.; Jayakumar, S. Predicting the current and future suitable habitat distribution of Myristica dactyloides Gaertn. using MaxEnt model in the Eastern Ghats, India. *Ecol. Eng.* **2015**, *82*, 184–188. [CrossRef]
9. Millar, C.S.; Blouin-Demers, G. Habitat suitability modelling for species at risk is sensitive to algorithm and scale: A case study of Blanding's turtle, Emydoidea blandingii, in Ontario, Canada. *J. Nat. Conserv.* **2012**, *20*, 18–29. [CrossRef]
10. Latif, Q.S.; Saab, V.A.; Dudley, J.G.; Markus, A.; Mellen-McLean, K. Development and evaluation of habitat suitability models for nesting white-headed woodpecker (Dryobates albolarvatus) in burned forest. *PLoS ONE* **2020**, *15*, e0233043. [CrossRef] [PubMed]
11. Tittensor, D.P.; Baco, A.R.; Brewin, P.E.; Clark, M.R.; Consalvey, M.; Hall-Spencer, J.; Rowden, A.; Schlacher, T.; Stocks, K.I.; Rogers, A.D. Predicting global habitat suitability for stony corals on seamounts. *J. Biogeogr.* **2009**, *36*, 1111–1128. [CrossRef]
12. Sharma, S.; Arunachalam, K.; Bhavsar, D.; Kala, R. Modeling habitat suitability of Perilla frutescens with MaxEnt in Uttarakhand—A conservation approach. *J. Appl. Res. Med. Aromat. Plants* **2018**, *10*, 99–105. [CrossRef]
13. Ahmed, S.E.; McInerny, G.; O'Hara, K.; Harper, R.; Salido, L.; Emmott, S.; Joppa, L.N. Scientists and software–surveying the species distribution modelling community. *Divers. Distrib.* **2015**, *21*, 258–267. [CrossRef]
14. Elith, J.; Graham, C.H.; Anderson, R.P.; Dudík, M.; Ferrier, S.; Guisan, A.; Hijmans, R.J.; Huettmann, F.; Leathwick, J.R.; Lehmann, A.; et al. Novel methods improve prediction of species' distributions from occurrence data. *Ecography* **2006**, *29*, 129–151. [CrossRef]
15. Hernandez, P.A.; Graham, C.H.; Master, L.L.; Albert, D.L. The effect of sample size and species characteristics on performance of different species distribution modeling methods. *Ecography* **2006**, *29*, 773–785. [CrossRef]
16. Kadmon, R.; Farber, O.; Danin, A. A Systematic Analysis of Factors Affecting the Performance of Climatic Envelope Models. *Ecol. Appl.* **2003**, *13*, 853–867. [CrossRef]
17. Skidmore, A.K.; Gauld, A.; Walker, P. Classification of kangaroo habitat distribution using three GIS models. *Int. J. Geogr. Inf. Syst.* **1996**, *10*, 441–454. [CrossRef]
18. Stockwell, D.R.; Peterson, A.T. Effects of sample size on accuracy of species distribution models. *Ecol. Model.* **2002**, *148*, 1–13. [CrossRef]
19. Wisz, M.S.; Hijmans, R.J.; Peterson, A.T.; Graham, C.H.; Guisan, A.; NCEAS Predicting Species Distributions Working Group. Effects of sample size on the performance of species distribution models. *Divers. Distrib.* **2008**, *14*, 763–773. [CrossRef]
20. Chapman, A.D.; Grafton, O. *Guide to Best Practices for Generalising Sensitive/Primary Species Occurrence-Data, Version 1.0*; Global Biodiversity Information Facility: Copenhagen, Denmark, 2008; Volume 27. [CrossRef]
21. Graham, C.H.; Ferrier, S.; Huettman, F.; Moritz, C.; Peterson, A.T. New developments in museum-based informatics and applications in biodiversity analysis. *Trends Ecol. Evol.* **2004**, *19*, 497–503. [CrossRef] [PubMed]
22. Wieczorek, J.; Guo, Q.; Hijmans, R. The point-radius method for georeferencing locality descriptions and calculating associated uncertainty. *Int. J. Geogr. Inf. Sci.* **2004**, *18*, 745–767. [CrossRef]

23. Feeley, K.J.; Silman, M.R. Modelling the responses of Andean and Amazonian plant species to climate change: The effects of georeferencing errors and the importance of data filtering. *J. Biogeogr.* **2010**, *37*, 733–740. [CrossRef]
24. Guo, Q.; Liu, Y.; Wieczorek, J. Georeferencing locality descriptions and computing associated uncertainty using a probabilistic approach. *Int. J. Geogr. Inf. Sci.* **2008**, *22*, 1067–1090. [CrossRef]
25. Graham, C.H.; Elith, J.; Hijmans, R.J.; Guisan, A.; Peterson, A.T.; Loiselle, B.; The Nceas Predicting Species Distributions Working Group. The influence of spatial errors in species occurrence data used in distribution models. *J. Appl. Ecol.* **2007**, *45*, 239–247. [CrossRef]
26. Dickinson, J.L.; Shirk, J.; Bonter, D.N.; Bonney, R.; Crain, R.L.; Martin, J.; Phillips, T.B.; Purcell, K. The current state of citizen science as a tool for ecological research and public engagement. *Front. Ecol. Environ.* **2012**, *10*, 291–297. [CrossRef]
27. Elith, J.; Leathwick, J.R. Species Distribution Models: Ecological Explanation and Prediction Across Space and Time. *Annu. Rev. Ecol. Evol. Syst.* **2009**, *40*, 677–697. [CrossRef]
28. Franklin, J. *Mapping Species Distributions: Spatial Inference and Prediction*; Cambridge University Press: Cambridge, UK, 2010.
29. Pocock, M.J.O.; Tweddle, J.C.; Savage, J.; Robinson, L.; Roy, H.E. The diversity and evolution of ecological and environmental citizen science. *PLoS ONE* **2017**, *12*, e0172579. [CrossRef] [PubMed]
30. Tulloch, A.I.; Possingham, H.P.; Joseph, L.N.; Szabo, J.; Martin, T.G. Realising the full potential of citizen science monitoring programs. *Biol. Conserv.* **2013**, *165*, 128–138. [CrossRef]
31. Singh, P.; Saran, S. Maximum Entropy Modeling Using Citizen Science: Use Case on Jacobin Cuckoo as an Indicator of Indian Monsoon. *ISPRS Ann. Photogramm. Remote. Sens. Spat. Inf. Sci.* **2020**, *5*, 23–30. [CrossRef]
32. Whitelaw, G.; Vaughan, H.; Craig, B.; Atkinson, D. Establishing the Canadian Community Monitoring Network. *Environ. Monit. Assess.* **2003**, *88*, 409–418. [CrossRef]
33. Cooper, C.B.; Dickinson, J.; Phillips, T.; Bonney, R. Citizen Science as a Tool for Conservation in Residential Ecosystems. *Ecol. Soc.* **2007**, *12*, 11. [CrossRef]
34. Howe, J. The rise of crowdsourcing. *Wired Mag.* **2006**, *14*, 1–4.
35. Irwin, A. Constructing the Scientific Citizen: Science and Democracy in the Biosciences. *Reconfiguring Nature* **2017**, *10*, 281–310. [CrossRef]
36. Burke, J.A.; Estrin, D.; Hansen, M.; Parker, A.; Ramanathan, N.; Reddy, S.; Srivastava, M.B. Participatory Sensing. UCLA: Center for Embedded Network Sensing. 2006. Available online: https://escholarship.org/uc/item/19h777qd (accessed on 28 May 2021).
37. Haklay, M. Citizen Science and Volunteered Geographic Information: Overview and Typology of Participation. In *Crowdsourcing Geographic Knowledge*; Springer Science and Business Media LLC: Dordrecht, The Netherlands, 2013; pp. 105–122.
38. Wiggins, A.; Crowston, K. From Conservation to Crowdsourcing: A Typology of Citizen Science. In *Proceedings of the 44th Hawaii International Conference on System Sciences*; Institute of Electrical and Electronics Engineers (IEEE): Piscataway, NJ, USA, 2011; pp. 1–10.
39. Wilderman, C.C. SHERMANS CREEK: Portrait of a watershed. In *Technical Status Report*; Dickinson College: Carlisle, UK, 2004.
40. Brown, P.; Morello-Frosch, R.; Zavestoski, S. *Contested Illnesses: Citizens, Science, and Health Social Movements*; University of California Press: Berekelay, CA, USA, 2011.
41. Schade, S.; Tsinaraki, C.; Roglia, E. Scientific data from and for the citizen. *First Monday* **2017**, *22*. [CrossRef]
42. Campbell, J.; Bowser, A.; Fraisl, D.; Meloche, M. Citizen science and data integration for understanding Marine Litter. In *Data for Good Exchange*; IIASA: New York, NY, USA, 2019.
43. Hsu, Y.C.; Cross, J.; Dille, P.; Tasota, M.; Dias, B.; Sargent, R.; Nourbakhsh, I. Smell Pittsburgh: Engaging Community Citizen Science for Air Quality. *ACM Trans. Interact. Intell. Syst.* **2020**, *10*, 1–49. [CrossRef]
44. Kocaman, S.; Gokceoglu, C. A CitSci app for landslide data collection. *Landslides* **2018**, *16*, 611–615. [CrossRef]
45. Kocaman, S.; Gokceoglu, C. On the Use of Citsci and Vgi in Natural Hazard Assessment. *ISPRS Int. Arch. Photogramm. Remote. Sens. Spat. Inf. Sci.* **2018**, *XLII-5*, 69–73. [CrossRef]
46. Woodall, G.M.; Hoover, M.D.; Williams, R.; Benedict, K.; Harper, M.; Soo, J.-C.; Jarabek, A.M.; Stewart, M.J.; Brown, J.S.; Hulla, J.E.; et al. Interpreting Mobile and Handheld Air Sensor Readings in Relation to Air Quality Standards and Health Effect Reference Values: Tackling the Challenges. *Atmosphere* **2017**, *8*, 182. [CrossRef]
47. Can, R.; Kocaman, S.; Gokceoglu, C. Development of a CitSci and artificial intelligence supported GIS platform for landslide data collection. *ISPRS Int. Arch. Photogramm. Remote. Sens. Spat. Inf. Sci.* **2020**, *XLIII-B5-2*, 43–50. [CrossRef]
48. Can, R.; Kocaman, S.; Gokceoglu, C. A Convolutional Neural Network Architecture for Auto-Detection of Landslide Photographs to Assess Citizen Science and Volunteered Geographic Information Data Quality. *ISPRS Int. J. Geo-Inf.* **2019**, *8*, 300. [CrossRef]
49. Yalçin, I.; Kocaman, S.; Gokceoglu, C. A CitSci Approach for Rapid Earthquake Intensity Mapping: A Case Study from Istanbul (Turkey). *ISPRS Int. J. Geo-Inf.* **2020**, *9*, 266. [CrossRef]
50. Kelemen-Finan, J.; Scheuch, M.; Winter, S. Contributions from citizen science to science education: An examination of a biodiversity citizen science project with schools in Central Europe. *Int. J. Sci. Educ.* **2018**, *40*, 2078–2098. [CrossRef]
51. Kobori, H.; Dickinson, J.L.; Washitani, I.; Sakurai, R.; Amano, T.; Komatsu, N.; Kitamura, W.; Takagawa, S.; Koyama, K.; Ogawara, T.; et al. Citizen science: A new approach to advance ecology, education, and conservation. *Ecol. Res.* **2016**, *31*, 1–19. [CrossRef]
52. Bonney, R.; Cooper, C.B.; Dickinson, J.; Kelling, S.; Phillips, T.; Rosenberg, K.V.; Shirk, J. Citizen Science: A Developing Tool for Expanding Science Knowledge and Scientific Literacy. *Bioscience* **2009**, *59*, 977–984. [CrossRef]

53. Koomen, M.H.; Rodriguez, E.; Hoffman, A.; Petersen, C.; Oberhauser, K. Authentic science with citizen science and student-driven science fair projects. *Sci. Educ.* **2018**, *102*, 593–644. [CrossRef]
54. Bloom, E.H.; Crowder, D.W. Promoting Data Collection in Pollinator Citizen Science Projects. *Citiz. Sci. Theory Pr.* **2020**, *5*, 5. [CrossRef]
55. Brzeski, K.E.; Gunther, M.S.; Black, J.M. Evaluating river otter demography using noninvasive genetic methods. *J. Wildl. Manag.* **2013**, *77*, 1523–1531. [CrossRef]
56. Kocaman, S.; Gokceoglu, C. CitSci as a New Approach for Landslide Researches. In *Proceedings of the Lecture Notes*; Springer: Cham, Switzerland, 2019; pp. 161–183.
57. Clery, D. Galaxy Zoo Volunteers Share Pain and Glory of Research. *Science* **2011**, *333*, 173–175. [CrossRef] [PubMed]
58. Reed, J.; Raddick, M.J.; Lardner, A.; Carney, K. An Exploratory Factor Analysis of Motivations for Participating in Zooniverse, a Collection of Virtual Citizen Science Projects. In Proceedings of the 2013, 46th Hawaii International Conference on System Sciences, Wailea, Maui, HI, USA, 7–10 January 2013; pp. 610–619.
59. Follett, R.; Strezov, V. An Analysis of Citizen Science Based Research: Usage and Publication Patterns. *PLoS ONE* **2015**, *10*, e0143687. [CrossRef] [PubMed]
60. Parrish, J.K.; Burgess, H.; Weltzin, J.F.; Fortson, L.; Wiggins, A.; Simmons, B. Exposing the Science in Citizen Science: Fitness to Purpose and Intentional Design. *Integr. Comp. Biol.* **2018**, *58*, 150–160. [CrossRef]
61. Broeder, L.D.; Devilee, J.; Van Oers, H.; Schuit, J.; Wagemakers, A. Citizen Science for public health. *Health Promot. Int.* **2016**, *33*, 505–514. [CrossRef]
62. Donnelly, A.; Crowe, O.; Regan, E.; Begley, S.; Caffarra, A. The role of citizen science in monitoring biodiversity in Ireland. *Int. J. Biometeorol.* **2013**, *58*, 1237–1249. [CrossRef]
63. Damoulas, T.; Henry, S.; Farnsworth, A.; Lanzone, M.; Gomes, C. Bayesian Classification of Flight Calls with a Novel Dynamic Time Warping Kernel. In *Proceedings of the 2010 Ninth International Conference on Machine Learning and Applications*; Institute of Electrical and Electronics Engineers (IEEE): Piscataway, NJ, USA, 2010; pp. 424–429.
64. Stoeckle, M. Taxonomy, DNA, and the Bar Code of Life. *Bioscience* **2003**, *53*, 796–797. [CrossRef]
65. Turner, W.; Spector, S.; Gardiner, N.; Fladeland, M.; Sterling, E.; Steininger, M. Remote sensing for biodiversity science and conservation. *Trends Ecol. Evol.* **2003**, *18*, 306–314. [CrossRef]
66. Hemmi, A.; Graham, I. Hacker science versus closed science: Building environmental monitoring infrastructure. *Inf. Commun. Soc.* **2013**, *17*, 830–842. [CrossRef]
67. Sullivan, P.J.; Acheson, J.; Angermeier, P.L.; Faast, T.; Flemma, J.; Jones, C.M.; Knudsen, E.E.; Minello, T.J.; Secor, D.H.; Wunderlich, R.; et al. Defining and implementing best available science for fisheries and environmental science, policy, and management. *Fisheries* **2006**, *31*, 460.
68. Singh, P.; Saran, S.; Kumar, D.; Padalia, H.; Srivastava, A.; Kumar, A.S. Species Mapping Using Citizen Science Approach Through IBIN Portal: Use Case in Foothills of Himalaya. *J. Indian Soc. Remote. Sens.* **2018**, *46*, 1725–1737. [CrossRef]
69. Novacek, M.J. Engaging the public in biodiversity issues. *Proc. Natl. Acad. Sci. USA* **2008**, *105*, 11571–11578. [CrossRef]
70. Silvertown, J.; Cook, L.; Cameron, R.; Dodd, M.; McConway, K.; Worthington, J.; Skelton, P.; Anton, C.; Bossdorf, O.; Baur, B.; et al. Citizen Science Reveals Unexpected Continental-Scale Evolutionary Change in a Model Organism. *PLoS ONE* **2011**, *6*, e18927. [CrossRef] [PubMed]
71. Conrad, C.C.; Hilchey, K.G. A review of citizen science and community-based environmental monitoring: Issues and opportunities. *Environ. Monit. Assess.* **2010**, *176*, 273–291. [CrossRef] [PubMed]
72. Freeman, S.N.; Noble, D.G.; Newson, S.E.; Baillie, S.R. Modelling population changes using data from different surveys: The Common Birds Census and the Breeding Bird Survey. *Bird Study* **2007**, *54*, 61–72. [CrossRef]
73. Michener, W.K.; Jones, M.B. Ecoinformatics: Supporting ecology as a data-intensive science. *Trends Ecol. Evol.* **2012**, *27*, 85–93. [CrossRef]
74. Yu, J.; Kelling, S.; Gerbracht, J.; Wong, W.-K. Automated data verification in a large-scale citizen science project: A case study. In *Proceedings of the 2012 IEEE 8th International Conference on E-Science*; Institute of Electrical and Electronics Engineers (IEEE): Piscataway, NJ, USA, 2012; pp. 1–8.
75. McClintock, B.T.; Bailey, L.L.; Pollock, K.H.; Simons, T.R. Unmodeled observation error induces bias when inferring patterns and dynamics of species occurrence via aural detections. *Ecology* **2010**, *91*, 2446–2454. [CrossRef]
76. Boakes, E.H.; McGowan, P.J.K.; Fuller, R.; Chang-Qing, D.; Clark, N.E.; O'Connor, K.; Mace, G. Distorted Views of Biodiversity: Spatial and Temporal Bias in Species Occurrence Data. *PLoS Biol.* **2010**, *8*, e1000385. [CrossRef] [PubMed]
77. Edwards, J.L.; Lane, M.A.; Nielsen, E.S. Interoperability of Biodiversity Databases: Biodiversity Information on Every Desktop. *Science* **2000**, *289*, 2312–2314. [CrossRef]
78. Wheeler, Q.D. What if GBIF? *Bioscience* **2004**, *54*, 718. [CrossRef]
79. Telenius, A. Biodiversity information goes public: GBIF at your service. *Nord. J. Bot.* **2011**, *29*, 378–381. [CrossRef]
80. Suarez, A.V.; Tsutsui, N.D. The value of museum collections for research and society. *BioScience* **2004**, *54*, 66–74. [CrossRef]
81. Ponder, W.F.; Carter, G.A.; Flemons, P.; Chapman, R.R. Evaluation of Museum Collection Data for Use in Biodiversity Assessment. *Conserv. Biol.* **2001**, *15*, 648–657. [CrossRef]
82. Pennisi, E. Taxonomic Revival. *Science* **2000**, *289*, 2306–2308. [CrossRef] [PubMed]

83. Shaffer, H.; Fisher, R.N.; Davidson, C. The role of natural history collections in documenting species declines. *Trends Ecol. Evol.* **1998**, *13*, 27–30. [CrossRef]
84. Lozier, J.D.; Aniello, P.; Hickerson, M.J. Predicting the distribution of Sasquatch in western North America: Anything goes with ecological niche modelling. *J. Biogeogr.* **2009**, *36*, 1623–1627. [CrossRef]
85. Thorn, J.S.; Nijman, V.; Smith, D.; Nekaris, K.A.I. Ecological niche modelling as a technique for assessing threats and setting conservation priorities for Asian slow lorises (Primates: Nycticebus). *Divers. Distrib.* **2009**, *15*, 289–298. [CrossRef]
86. Herrando-Moraira, S.; Vitales, D.; Nualart, N.; Gómez-Bellver, C.; Ibáñez, N.; Massó, S.; Cachón-Ferrero, P.; González-Gutiérrez, P.A.; Guillot, D.; Herrera, I.; et al. Global distribution patterns and niche modelling of the invasive Kalanchoe × houghtonii (Crassulaceae). *Sci. Rep.* **2020**, *10*, 1–18. [CrossRef] [PubMed]
87. Préau, C.; Nadeau, I.; Sellier, Y.; Isselin-Nondedeu, F.; Bertrand, R.; Collas, M.; Capinha, C.; Grandjean, F. Niche modelling to guide conservation actions in France for the endangered crayfish Austropotamobius pallipes in relation to the invasive Pacifastacus leniusculus. *Freshw. Biol.* **2019**, *65*, 304–315. [CrossRef]
88. Alhajeri, B.H.; Fourcade, Y. High correlation between species-level environmental data estimates extracted from IUCN expert range maps and from GBIF occurrence data. *J. Biogeogr.* **2019**, *46*, 1329–1341. [CrossRef]
89. Sung, S.; Kwon, Y.-S.; Lee, D.K.; Cho, Y. Predicting the Potential Distribution of an Invasive Species, Solenopsis invicta Buren (Hymenoptera: Formicidae), under Climate Change using Species Distribution Models. *Entomol. Res.* **2018**, *48*, 505–513. [CrossRef]
90. Huettmann, F.; Artukhin, Y.; Gilg, O.; Humphries, G. Predictions of 27 Arctic pelagic seabird distributions using public environmental variables, assessed with colony data: A first digital IPY and GBIF open access synthesis platform. *Mar. Biodivers.* **2011**, *41*, 141–179. [CrossRef]
91. Gomes, V.H.F.; Mayle, F.E.; Gosling, W.D.; Vieira, I.C.G.; Salomão, R.P.; Ter Steege, H. Modelling the distribution of Amazonian tree species in response to long-term climate change during the Mid-Late Holocene. *J. Biogeogr.* **2020**, *47*, 1530–1540. [CrossRef]
92. Hastings, R.; Rutterford, L.A.; Freer, J.J.; Collins, R.A.; Simpson, S.D.; Genner, M.J. Climate Change Drives Poleward Increases and Equatorward Declines in Marine Species. *Curr. Biol.* **2020**, *30*, 1572–1577.e2. [CrossRef] [PubMed]
93. Siqueira, S.D.F.; Higuchi, P.; Da Silva, A.C. Contemporary and Future Potential Geographic Distribution of Cedrela fissilis Vell. under Climate Change Scenarios. *Revista Árvore* **2019**, *43*, 43. [CrossRef]
94. Bender, I.M.A.; Kissling, W.D.; Böhning-Gaese, K.; Hensen, I.; Kühn, I.; Nowak, L.; Töpfer, T.; Wiegand, T.; Dehling, D.M.; Schleuning, M. Projected impacts of climate change on functional diversity of frugivorous birds along a tropical elevational gradient. *Sci. Rep.* **2019**, *9*, 1–12. [CrossRef]
95. Ikeda, D.H.; Max, T.L.; Allan, G.J.; Lau, M.K.; Shuster, S.M.; Whitham, T.G. Genetically informed ecological niche models improve climate change predictions. *Glob. Chang. Biol.* **2017**, *23*, 164–176. [CrossRef] [PubMed]
96. Sharma, R.K.; Goyal, A.K.; Sharma, M. Ecology and Evolution of Nest Parasitism in Indian Cuckoo. *Nat. Environ. Pollut. Technol.* **2015**, *14*, 847–853.
97. GBIF.org. GBIF Occurrence. Available online: https://doi.org/10.15468/dl.gd9tk3 (accessed on 23 April 2021).
98. Southwest Monsoon Season. Available online: http://www.imdchennai.gov.in/swweb.htm (accessed on 31 October 2020).
99. A Study of the Northeast Monsoon Rainfall of Tamilnad. Available online: https://metnet.imd.gov.in/mausamdocs/1413_F.pdf (accessed on 31 October 2020).
100. Fick, S.E.; Hijmans, R.J. WorldClim 2: New 1-km spatial resolution climate surfaces for global land areas. *Int. J. Climatol.* **2017**, *37*, 4302–4315. [CrossRef]
101. Global Climate and Weather Data. Available online: https://www.worldclim.org/data/index.html (accessed on 31 October 2020).
102. Séférian, R.; Nabat, P.; Michou, M.; Saint-Martin, D.; Voldoire, A.; Colin, J.; Decharme, B.; Delire, C.; Berthet, S.; Chevallier, M.; et al. Evaluation of CNRM Earth System Model, CNRM-ESM2-1: Role of Earth System Processes in Present-Day and Future Climate. *J. Adv. Model. Earth Syst.* **2019**, *11*, 4182–4227. [CrossRef]
103. Eyring, V.; Bony, S.; Meehl, G.A.; Senior, C.A.; Stevens, B.; Stouffer, R.J.; Taylor, K.E. Overview of the Coupled Model Intercomparison Project Phase 6 (CMIP6) experimental design and organization. *Geosci. Model Dev.* **2016**, *9*, 1937–1958. [CrossRef]
104. Dormann, C.F.; Elith, J.; Bacher, S.; Buchmann, C.; Carl, G.; Carré, G.; Marquéz, J.R.G.; Gruber, B.; Lafourcade, B.; Leitão, P.J.; et al. Collinearity: A review of methods to deal with it and a simulation study evaluating their performance. *Ecography* **2013**, *36*, 27–46. [CrossRef]
105. Naimi, B. Usdm: Uncertainty Analysis for Species Distribution Models. R Package ver. 1.1–18. 2017. Available online: https://cran.r-project.org/web/packages/usdm/index.html (accessed on 12 January 2021).
106. Zuur, A.F.; Ieno, E.N.; Elphick, C.S. A protocol for data exploration to avoid common statistical problems: Data exploration. *Methods Ecol. Evol.* **2010**, *1*, 3–14. [CrossRef]
107. Phillips, S.J. A brief tutorial on Maxent. *AT&T Res.* **2005**, *190*, 231–259.
108. Qin, A.; Liu, B.; Guo, Q.; Bussmann, R.W.; Ma, F.; Jian, Z.; Xu, G.; Pei, S. Maxent modeling for predicting impacts of climate change on the potential distribution of Thuja sutchuenensis Franch., an extremely endangered conifer from southwestern China. *Glob. Ecol. Conserv.* **2017**, *10*, 139–146. [CrossRef]
109. Dingliang, X.; Zhanqing, H. The principle of maximum entropy and its applications in ecology. *Biodivers. Sci.* **2011**, *19*, 295–302. [CrossRef]

110. Yang, X.; Kushwaha, S.; Saran, S.; Xu, J.; Roy, P. Maxent modeling for predicting the potential distribution of medicinal plant, Justicia adhatoda L. in Lesser Himalayan foothills. *Ecol. Eng.* **2013**, *51*, 83–87. [CrossRef]
111. He, J.; Kolovos, A. Bayesian maximum entropy approach and its applications: A review. *Stoch. Environ. Res. Risk Assess.* **2018**, *32*, 859–877. [CrossRef]
112. Kleidon, A.; Malhi, Y.; Cox, P. Maximum entropy production in environmental and ecological systems. *Philos. Trans. R. Soc. B Biol. Sci.* **2010**, *365*, 1297–1302. [CrossRef]
113. Fourcade, Y.; Engler, J.O.; Rödder, D.; Secondi, J. Mapping Species Distributions with MAXENT Using a Geographically Biased Sample of Presence Data: A Performance Assessment of Methods for Correcting Sampling Bias. *PLoS ONE* **2014**, *9*, e97122. [CrossRef]
114. Pearson, R.G.; Raxworthy, C.J.; Nakamura, M.; Peterson, A.T. ORIGINAL ARTICLE: Predicting species distributions from small numbers of occurrence records: A test case using cryptic geckos in Madagascar. *J. Biogeogr.* **2006**, *34*, 102–117. [CrossRef]
115. Kumar, S.; Stohlgren, T.J. Maxent modeling for predicting suitable habitat for threatened and endangered tree *Canacomyrica monticola* in New Caledonia. *J. Ecol. Nat. Environ.* **2009**, *1*, 094–098. [CrossRef]
116. Fielding, A.H.; Bell, J.F. A review of methods for the assessment of prediction errors in conservation presence/absence models. *Environ. Conserv.* **1997**, *24*, 38–49. [CrossRef]
117. Swets, J.A. Measuring the accuracy of diagnostic systems. *Science* **1988**, *240*, 1285–1293. [CrossRef]
118. Guhathakurta, P.; Rajeevan, M.; Sikka, D.R.; Tyagi, A. Observed changes in southwest monsoon rainfall over India during 1901–2011. *Int. J. Climatol.* **2015**, *35*, 1881–1898. [CrossRef]
119. Cheaib, A.; Badeau, V.; Boé, J.; Chuine, I.; Delire, C.; Dufrêne, E.; François, C.; Gritti, E.S.; Legay, M.; Pagé, C.; et al. Climate change impacts on tree ranges: Model intercomparison facilitates understanding and quantification of uncertainty. *Ecol. Lett.* **2012**, *15*, 533–544. [CrossRef]

International Journal of *Geo-Information*

MDPI

Article

CWDAT—An Open-Source Tool for the Visualization and Analysis of Community-Generated Water Quality Data

Annie Gray [1,*], Colin Robertson [2] and Rob Feick [3]

1 Department of Geography and Environmental Management, University of Waterloo, Waterloo, ON N2L 3G1, Canada
2 Department of Geography and Environmental Studies, Wilfrid Laurier University, Waterloo, ON N2L 3C5, Canada; crobertson@wlu.ca
3 School of Planning, University of Waterloo, Waterloo, ON N2L 3G1, Canada; rob.feick@uwaterloo.ca
* Correspondence: ma2gray@uwaterloo.ca

Abstract: Citizen science initiatives span a wide range of topics, designs, and research needs. Despite this heterogeneity, there are several common barriers to the uptake and sustainability of citizen science projects and the information they generate. One key barrier often cited in the citizen science literature is data quality. Open-source tools for the analysis, visualization, and reporting of citizen science data hold promise for addressing the challenge of data quality, while providing other benefits such as technical capacity-building, increased user engagement, and reinforcing data sovereignty. We developed an operational citizen science tool called the Community Water Data Analysis Tool (CWDAT)—a R/Shiny-based web application designed for community-based water quality monitoring. Surveys and facilitated user-engagement were conducted among stakeholders during the development of CWDAT. Targeted recruitment was used to gather feedback on the initial CWDAT prototype's interface, features, and potential to support capacity building in the context of community-based water quality monitoring. Fourteen of thirty-two invited individuals (response rate 44%) contributed feedback via a survey or through facilitated interaction with CWDAT, with eight individuals interacting directly with CWDAT. Overall, CWDAT was received favourably. Participants requested updates and modifications such as water quality thresholds and indices that reflected well-known barriers to citizen science initiatives related to data quality assurance and the generation of actionable information. Our findings support calls to engage end-users directly in citizen science tool design and highlight how design can contribute to users' understanding of data quality. Enhanced citizen participation in water resource stewardship facilitated by tools such as CWDAT may provide greater community engagement and acceptance of water resource management and policy-making.

Keywords: citizen science; data quality; web application; water quality; community-based monitoring

Citation: Gray, A.; Robertson, C.; Feick, R. CWDAT—An Open-Source Tool for the Visualization and Analysis of Community-Generated Water Quality Data. *ISPRS Int. J. Geo-Inf.* **2021**, *10*, 207. https://doi.org/10.3390/ijgi10040207

Academic Editors: Sultan Kocaman and Wolfgang Kainz

Received: 2 February 2021
Accepted: 22 March 2021
Published: 1 April 2021

1. Introduction

Citizen science (CS) encompasses a wide range of topics and investigations, from ornithology to astronomy to meteorology [1]. Despite this heterogeneity, certain barriers are common to many citizen science initiatives [2]. Differences in training, research priorities/interests, and modes of communicating information that often exist between (and within) CS initiatives and the formal scientific community can limit the degree to which CS initiatives influence decision-making processes [3,4]. Other commonly discussed challenges include volunteer retention [5], the generation of actionable information from raw data [6], data sharing and communication, and overall data quality [7]. Reservations regarding the quality/reliability of citizen-collected and citizen-generated data held by much of the wider scientific community are well documented as a barrier not only to the sustainability of CS initiatives, but to the uptake of citizen-generated data in formal scientific circles [5,7–11]. There is a longstanding discourse in the literature regarding data quality barriers in citizen science initiatives. Specific data concerns include comparisons of data from different sources [12], differing metadata standards [13], species identifications [14], and factors such as uncertainty, accuracy, bias, and precision [2,15]. Despite these well-known challenges, Fonte et al. (2015) [16] noted an overall dearth of guidance on CS data quality control and quality assurance (QAQC). The ability of citizens and CS initiatives to independently analyze, interpret, and communicate reliable, actionable results from their own high-quality data has been identified as a key challenge for community-based monitoring [17] and the output of reliable, actionable information has been observed as an important driver for citizen science volunteers [5].

One mechanism by which these barriers can be addressed is the development of open-source data analysis and support tools [18]. The development of such statistical and computational resources can support local data management (including quality assurance/quality control—QAQC), reinforce notions of data sovereignty, and promote capacity building in the field of citizen science [14]. Not only do data analysis tools have the potential to address the challenge of data QAQC, but they also offer to other interrelated benefits to CS initiatives and their participants such as: education/knowledge generation, the mobilization of local expertise, capacity building, greater levels of engagement, and increased data sovereignty [19]. Citizen data collectors need easy-to-use tools and interfaces to help them to summarize and visualize their data, assess the quality of their observations, build their understanding of data quality and scope, and to see value in the data they are creating in light of the bigger scientific or regional questions at play. iNaturalist, for example, has a robust user community mechanism where more senior/experienced naturalists correct and verify observations from newer participants, often providing detailed explanation and learning in the process (available at https://www.inaturalist.org, accessed on 10 March 2021). Other examples in the field of citizen science include Mackenzie DataStream (available at https://mackenziedatastream.ca/, accessed on 10 March 2021); eBird Canada (available at https://ebird.org/canada/home, accessed on 10 March 2021); and the CitSci.org website (https://www.citsci.org/, accessed on 10 March 2021), which allows citizen science initiatives to register their projects and offers numerous supports such as application programming interfaces. Such tools can catalyze local contextual interpretations which is one of the key benefits of citizen participation and can lead to higher quality information. Many open-source tools have been developed specifically for citizen science purposes, including software architectures, databases, and mobile applications [20].

Open-source data analysis tools and interfaces can provide extra levels of accessibility and transparency by allowing users to view and learn about the operations performed on their data and, when appropriate, to independently modify the software code to fit their needs better [20,21]. This is important for building trust among users who collect data and those rely on them for scientific analysis, enhancing technical capacity within communities [22] and increasing participant engagement [1]. Open-source software is typically available free of charge and restrictive licensing requirements which further enhances tool accessibility and can promote the development of communities of practice around open toolsets and methods [23].

The direct involvement of potential end-users in the design and development processes is critical to the success of any analysis or decision support tool. Ongoing engagement can mitigate issues such as retention and user satisfaction by recognizing the interdependencies between technologies and their intended social contexts [24,25]. Repositories such as Github allow tool developers to make their source code publicly available, potentially leading to code "forking" and increased interoperability [21]. Additionally, the ability to independently visualize, analyze, verify, and communicate citizen science data can support the formation of local policies and solutions, which may then be communicated to the wider scientific community and to government as needed—a benefit of citizen science recognized in Muenich et al. (2016) [26] and Weeser et al. (2018) [27]. The independent development of questions, policies, and answers by citizen scientists can leverage local knowledge and strengthen existing relationships between citizen science initiatives and scientists by promoting a two-way exchange of information, ideas, and feedback [28]. This is in contrast to the typically unidirectional flow of information from scientists to citizens [29,30], which can accentuate power inequities through a complete reliance on 'third party' experts for results.

In this paper, we present a novel R/Shiny based web application, the Community Water Quality Data Analysis Tool (CWDAT), that is designed for citizen science initiatives that focus on community-based water quality monitoring (CBWQM). The remainder of this introduction will present and discuss the research context of community-based water quality monitoring, with a focus on the connections between monitoring challenges and the benefits offered by open-source data analysis tools such as CWDAT.

Conservation and protection of the world's freshwater resources is a vital global goal enshrined in Sustainable Development Goal 6 on Clean Water and Sanitation which aims to ensure availability and sustainable management of water and sanitation for all. Hydrometric monitoring networks provide vital information on hydrological processes and characteristics such as surface water discharge, watercourse morphology, and water quality [29]. Although regulated national-scale monitoring networks are often the primary sources of hydrometric data, their spatial and temporal coverage can be inadequate when considering local/regional water quality trends and characteristics, due to operational and capacity constraints. Community-based water quality monitoring initiatives can augment regulated monitoring network data by filling the spatial and temporal data gaps and by prioritizing parameters, times, and locations of local interest or concern [31–35].

Many tools have been developed to support the analysis and presentation of data related to water resources, such as AkvaGIS [36,37] and the USEPA's Water Quality Data Analysis Tool (https://github.com/USEPA/Water-Quality-Data-Analysis-Tool, accessed on 10 March 2021). However, most tools are limited in terms of their accessibility (e.g., cost, system requirements, program requirements) and/or are tied to specific protocols for data collection and/or format in terms of the input data for which they are designed. For example, the USEPA Water Quality Data Analysis Tool is designed to work exclusively with the USEPA's WQP Data Discovery Portal. AkvaGIS, while it accepts data directly from the user, requires field-specific data such as piezometer locations which may not be available/appropriate in the context of community-based water quality monitoring. Overall, tools that aim to monitor, manage, and/or predict natural systems often fail to be adopted and used in the contexts for which they were designed [38,39], a barrier especially relevant to the heterogeneous fields of CS and community-based water quality monitoring. Adapting more general-purpose data-analytic tools for water quality data analysis requires significant technical capacity which may not be possible in many community projects geared toward field data collection. A tool designed through an open-source platform, which can be edited and adapted by end-users and developers alike, thus privileging the respective cultures, contexts, information needs, and preferences of the CBWQM initiatives, holds some promise in addressing such challenges.

To address the needs outlined above, an open-source web application—the Community Water Data Analysis Tool (CWDAT)—was developed as part of a wider project aiming to identify and address barriers to citizen science/CBWQM initiatives and utilization of the data generated by such initiatives (i.e., Global Water Citizenship Project, http://gwc-gwf.ca, accessed on 10 March 2021). A prototype version of CWDAT was designed and presented to members of the Canadian community-based water quality monitoring field through a series of surveys and interactive sessions. Based on the feedback received, a second version of CWDAT was developed. The remainder of this paper will elaborate on the prototype's design, the feedback received from potential end-users, and the consequent developments. Finally, the development of CWDAT is discussed within the overarching context of barriers faced by community-based water quality monitoring initiatives and recommendations for future development and capacity building are provided.

2. Methods

CWDAT is an interactive, open-source web application developed using the R/Shiny framework (R version 4.0.2) [40] and hosted using the open-source version of Shiny Server (https://rstudio.com/products/shiny/shiny-server/, accessed on 10 March 2021). As noted in 2017 by Hewitt and Macleod [41], the R/Shiny platform offers such advantages as: low-no cost, suitability on touch devices, ease of development/extension, and potential for scientific innovation, even when compared to other open-source development platforms such as Python and QGIS. The overall goal of CWDAT is to support and enhance CBWQM initiatives by providing a free, user-friendly and customizable tool for independent data validation, visualization, summary, and analysis. CWDAT is neither designed nor intended to replace pre-existing analyses, nor to compete with working relationships already established between citizens and scientists, but rather to complement such connections and to give communities a medium for independent, preliminary interaction with their raw data. An instance of the tool can be accessed through a browser at the following location: https://spatial.wlu.ca/cwdat/, accessed on 10 March 2021. The source code and files are freely available through GitHub (https://github.com/thespatiallabatLaurier/waterquality, accessed on 10 March 2021). The novelty of CWDAT, relative to other open-source tools in the field of citizen science, lies in its ability to read-in users' data, its standalone nature (no other programs or online portals are required), its ability to statistically compare between data sources, its interactive visualization and reporting capabilities, and its specific focus on the field of community-based water quality monitoring.

The development of CWDAT occurred in three stages (see Figure 1). In the first stage, a prototype version of CWDAT was created based on known barriers to CBWQM in terms of data quality, data visualization, and data communication that were identified from academic and grey literature. Section 2.1 outlines the CWDAT interface, its major features, and the initial design choices. In stage 2, the prototype was presented to members of the CBWQM field through a series of surveys, informal discussions, and interactive tasks. Feedback was solicited on the tool's ease of use, its potential to address barriers faced by CBWQM initiatives, and its potential to generate useful information for CBWQM initiatives (see Section 2.2). Stage 3 centered on incorporating user feedback into a second version of CWDAT (see Sections 3 and 4).

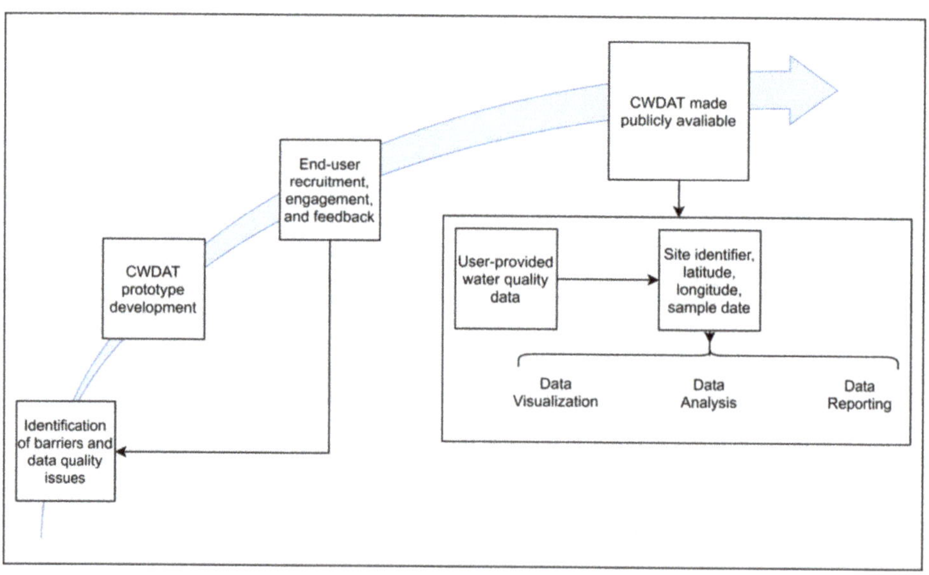

Figure 1. Stages of development for Community Water Data Analysis Tool (CWDAT).

2.1. Prototype Overview and Development

The initial CWDAT prototype included the following sections: Data Upload and Properties; Spatial Visualization; Graphic Visualization; Statistics, and Temporal Coverage Summary. These sections were arranged to facilitate a logical workflow [23] of data visualization and reporting starting with the provision of user-generated water quality data. The Data Upload and Properties component of CWDAT (Figure 2) considered the need for robustness against various naming conventions, dataframe structures, and variables. Additionally, CWDAT is designed to be tolerant of sparse data to recognize that only a subset of variables may be collected at some sites.

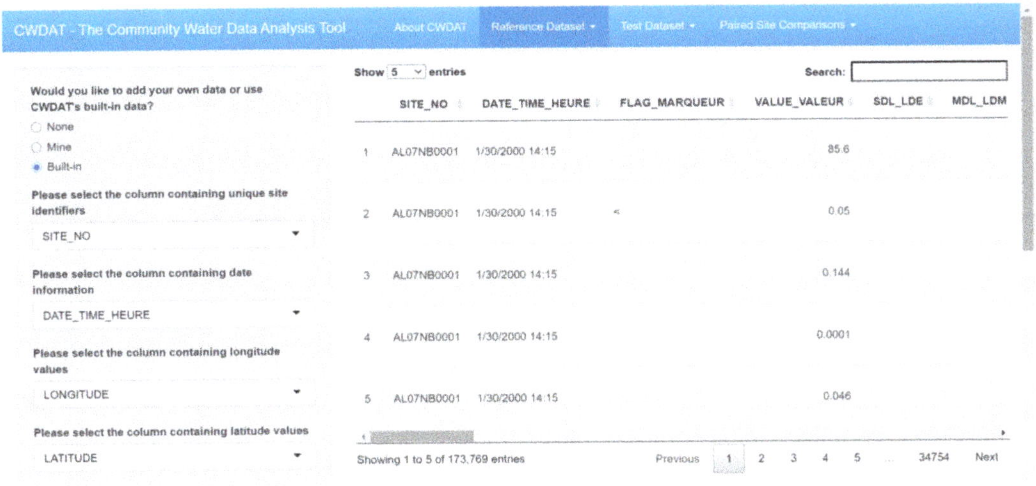

Figure 2. Data upload and properties—users may provide their own water quality data or use CWDAT's built-in data to explore the interface.

CWDAT accepts .csv files or either "long" or "wide" data structures, with the following field requirements: sampling site code, latitude, longitude, and sampling date (additional details and descriptions are provided in Table 1). Data provided in a "long" format require the user to identify which columns contain variable names and variable values. Similarly, data provided in a "wide" format require the user to identify which columns represent specific water quality variables. Once data have been added to CWDAT, the user is asked to identify the requisite columns containing latitude, longitude, date, and site identification code. CWDAT places no limitation on the number of non-requisite columns included in a user-provided .csv file.

The Spatial Visualization page (Figure 3) maps the locations of the monitoring sites. Users may click on a site, select a water quality parameter, and view a scatterplot of the corresponding data. An interactive table with sliders and filters is also provided, for the purpose of outlier identification. Both univariate and bivariate graphing functions are provided by the Graphic Visualization page (Figure 4). Using drop-down menus, users may select the variable(s), months, years, and sites they wish to plot. For reporting purposes, a button is provided for users to download their graphs in .png format. Basic statistical summaries, box plots, and normality testing are offered by the Statistics page (Figure 5) with downloading capabilities. The Temporal Coverage Summary page (Figure 6) was designed to help users identify temporal "gaps" in their data and to report on the volume of data collected. As with previous sections, the user-generated graphs are downloadable.

Table 1. Field requirements for user-provided water quality data.

Required Field	Description	Accepted Data Type(s)
Site identifier	A unique identifier for each discrete location water quality samples were collected or measurements were taken. This can be a name, code, number, or other categorical variable. Multiple observations from a single location should all share the same site identifier.	String, integer, float
Latitude	Latitude coordinate in decimal degrees using the WGS84 system.	Float
Longitude	Longitude coordinate in decimal degrees using the WGS84 system	Float
Date	Date of sample collection. Sample collection time may also be included in this column but is not required.	String, Date, POSIXlt (R)
Indicator/Variable(s)	Water quality indicators (e.g., temperature, pH, etc.). For "long" format data, indicator names will be listed in a single column.	

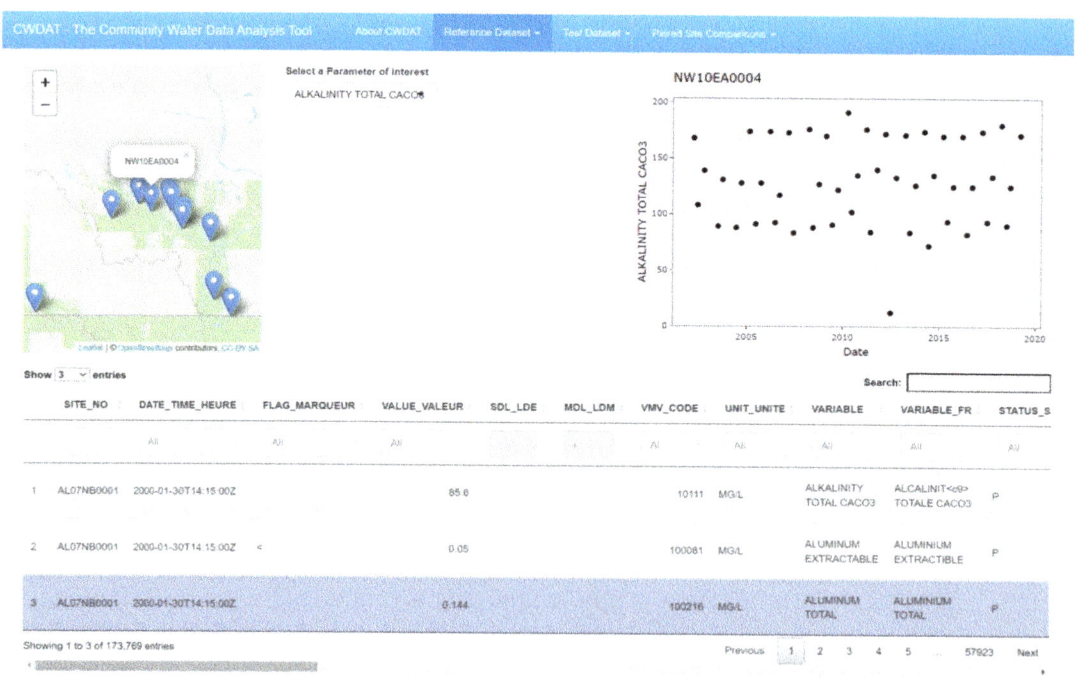

Figure 3. Spatial visualization—users can explore the location, ranges, and values of the data.

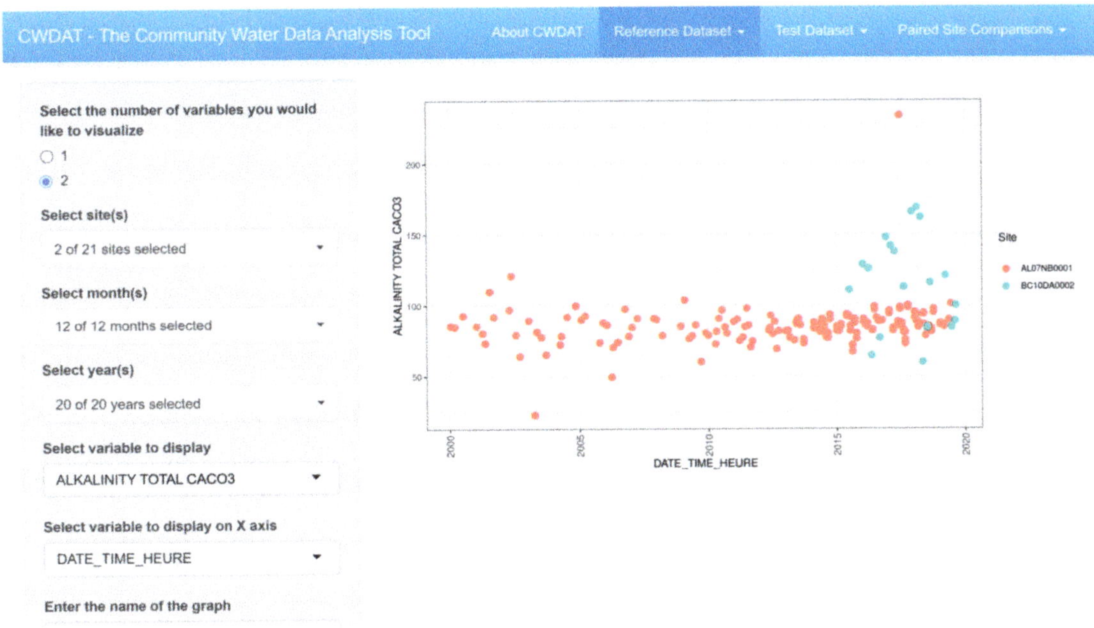

Figure 4. Bivariate graphic visualization—users may specify which site(s), vasriable(s), year(s), and month(s) they wish to visualize.

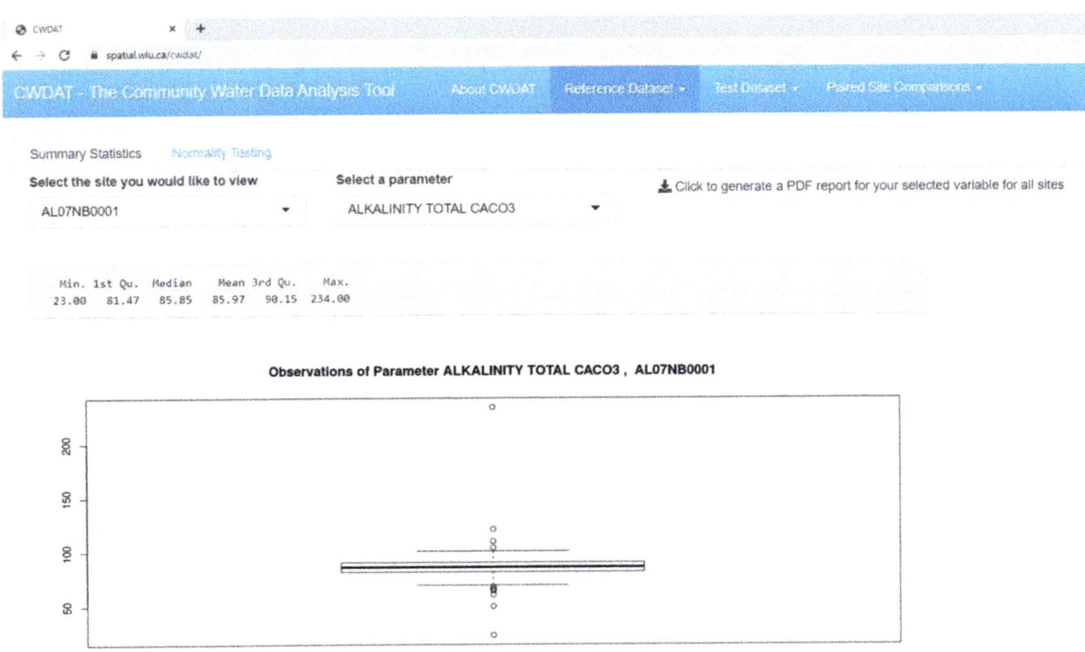

Figure 5. Statistics—users may generate and download PDF reports and individual graphs.

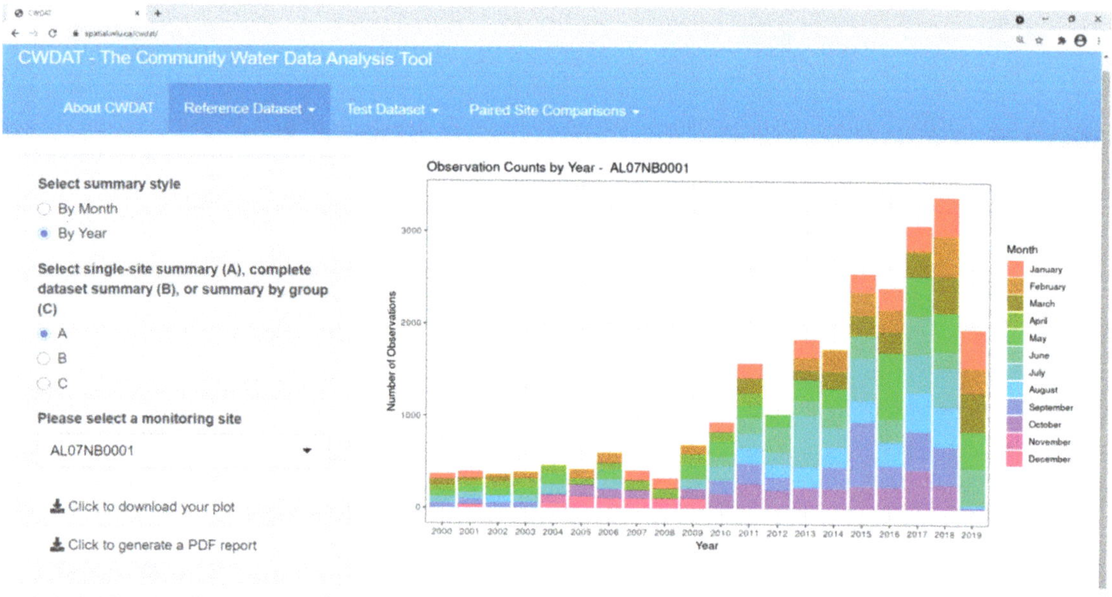

Figure 6. Temporal coverage summary.

In addition to the visualization a preliminary analysis of a single data set, CWDAT offers users the ability to statistically compare the values of one dataset against another. Conceptually, one dataset would serve as a "reference" (for instance data from a regulated monitoring network) and the second dataset would be community-generated. This capability, offered on the Paired Sites Comparison page, is based on the methodology outlined in Kilgour et al. (2017) [42] and allows users to determine if their community-generated "test" data falls within the normal range of the reference data for corresponding sites. Comparing citizen-generated data to an accepted reference is one way to assess the quality or reliability of CBWQM data. This capability was seen to be important for established community based users to check their data prior to submitting them to a larger project database and also for training purposes where a new participant could compare their observations with historical or regional norms.

2.2. CWDAT Community Feedback

The open-source, dynamic nature of the CWDAT application allows for ongoing development and modifications to support the varied needs and preferences of end-users in the community-based water quality monitoring field. To support such development, thirty-two members of the CBWQM field in Canada were asked to share their insight via an independent survey or via facilitated sessions. The potential participants were contacted due to previous collaboration in one of two ways: (1) previous engagement or association with the Global Water Citizenship research project or (2) participation in a roundtable discussion on community-based monitoring jointly convened by the Gordon Foundation, WWF Canada, and Living Lakes Canada in November 2018.

Two options were offered for participation. Option A entailed independent participation using an online survey. The nine questions of Option A. Option B entailed an interactive, facilitated session using both an online survey and interaction with the tool. Option B questions are provided in Supplementary Document S1. In accordance with the survey questions, Option B participants were provided with a step-by-step instructions document (Supplementary Document S2). An informed consent statement was provided to participants upon the commencement of either survey, in accordance with the ethics approval granted by the Wilfrid Laurier University Research Ethics Board. Option A was offered in an attempt to maximize the number of participants, by offering an alternative for those who did not have the time or did not wish to participate in the facilitated session. The independent survey (Option A) focused on the roles, motivations, priorities, and barriers experienced by the participants via multiple selection and short and long answer questions. Option B included all survey questions from Option A, in addition to a set of five interactive tasks using CWDAT. As the inclusion of potential end-users through all stages of software development is critical to user retention, user satisfaction, and uptake [24], facilitated sessions with informal discussion encouraged more meaningful reflection and detailed feedback on CWDAT's potential value. The step-by-step instructions and survey questions are provided as Supplementary Data. Table 2 provides a summary of the Option B tasks, the relevant functions of CWDAT, and related discussion topics.

Table 2. Facilitated session tasks and topics.

Task	Purpose	CWDAT Section(s)	Informal Discussion Topics
1	Upload a .csv file of water quality data	Data Upload and Properties	File structures and sizes; metadata; sampling protocols and users' experiences with data handling and storage
2	Identify and explore outlier values	Spatial Visualization and Statistics	Data QAQC; users' methods and needs; outlier detection
3	Visualize the data's temporal scope	Temporal Coverage Summary	Sampling designs; CBWQM initiative organization and resources
4	Graph a subset of data	Graphic Visualization data	Data presentation; viewer and stakeholder preferences and needs data
5	Determine if a subset of test data is within the normal range of a reference baseline	Paired Site Comparisons	Data validation; QAQC; confidence in results; analysis outcomes

Two .csv files containing sample water quality data were provided to participants. The first file was meant to represent data coming from a CBWQM initiative [43], the second to represent data coming from a regulated water quality monitoring network [44]. Upon completion of each task, participants were prompted to reflect via ordinal rankings, multiple selection, and short/long answer questions. Finally, participants were asked to give their general impression of CWDAT and its potential value to the CBWQM field, and to provide commentary and suggestions for improvement based on their interaction with CWDAT.

3. Results

3.1. Response

Cumulatively for surveys A and B, 22 total hits to the survey links were recorded ($n = 22$). Of these, 14 resulted in survey completions. Recalling the initial recruitment of 32 potential participants, approximately 44% of contacted individuals completed a survey. Of the 14 completions, eight participants requested a facilitated session ($n = 8$) and six completed the independent survey ($n = 6$). Participants' self-declared roles and motivations (multiple select) were primarily scientific research, environmental awareness, and policy and decision-making (Table 3).

Table 3. Participant roles and motivations.

Role	Count	%
Scientist or researcher	9	64
NGO/Not-for-profit	3	21
Outreach	3	21
Data analyst	2	14
Volunteer	2	14
Government or leadership	1	7
Environmental consulting	1	7
Community member	1	7
Motivation	**Count**	**%**
Environmental awareness	7	50
Scientific research	6	43
Planning and decision-making	4	29

3.2. CWDAT Reception

At the end of Option B, users were asked to rank their overall impression of CWDAT based on three criteria: intuitiveness of the interface; relevance to the users' CBWQM data questions; and generation of actionable information, on a scale from 1 (worst)–5 (best). The respective modes were 4, 5, and 5 (n = 8). Most participants emphasized the need for tools such as CWDAT, and many expressed an interest in following the tool's development.

Through informal discussion and interaction with CWDAT, the Option B participants of this study outlined and expanded on numerous barriers they, and their respective CB-WQM/citizen science initiatives, have faced. Some information was solicited in response to participant commentary on the tool and its features. Other information was volunteered by the participants when describing their experience, future hopes of the field, and procedures in their respective community/organization. Highlighted barriers ranged from initiative-specific challenges to perceived and actual characteristics of CBWQM and citizen science fields. Three general categories of barriers were observed from the transcribed feedback as shown in Table 4: metadata standards, data interpretation, and communication/information sharing. Multiple participants affirmed that the CWDAT prototype could be beneficial to the CBWQM field, while stressing the need for ongoing engagement and development.

Table 4. Summary of recurring identified and discussed barriers.

Metadata Standards	Data Interpretation	Communication/Sharing
Controlling for units	No consistent idea of how to use data	Privacy concerns
Inconsistent data labelling variations in instrumentation and laboratory procedures Variations in naming conventions Variations in file format anddata structures	Establishing trends and triggers	Internet capacity
	Long-term analysis capacity	Communication media
		Perceived lack of quality
	Lack of meaningful interpretation, coordination, and common reporting within and between community-based water quality monitoring/citizen scientist initiatives	File sizes

Participant responses to the call for suggestions/next steps for CWDAT included better supplementary information (i.e., explanatory text regarding water quality parameters and plain-language descriptions of the analysis done on the data), enhancement of raw data sharing capabilities, and future engagement with developing initiatives prior to the completion of a publicly available model. Participants' preferred output media included plain-text summaries, graphs, reports, and maps using colour to spatially display water quality parameters, their values, and associated criteria. One participant connected the need for informal, explanatory text to differences between grassroots community members and the wider scientific community, highlighting that the interface must not be too "intimidating". Discussions with other participants placed the same concern in terms of default templates and settings—advanced users may find default settings restrictive, but too many options and settings could overwhelm and deter users less comfortable with technology [24].

3.3. CWDAT Development

In response to the feedback of participants, particularly those who selected Option B and engaged directly with the CWDAT prototype, several changes were made to the CWDAT interface and features. Major additions included a visual theme for the user interface; built-in sample data; the generation of downloadable, editable PDF reports; and plain language descriptions and explanations. Figure 7 shows CWDAT's initial Data Upload and Properties page. Figure 8 shows the same page following participant feedback.

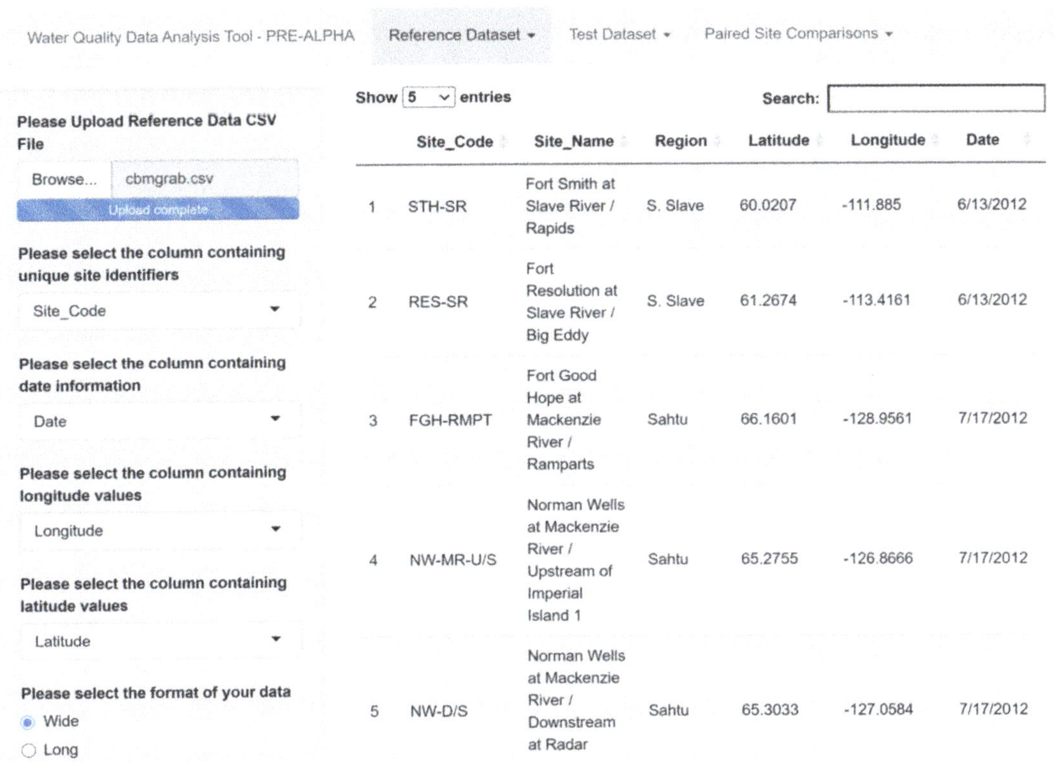

Figure 7. Data upload and properties (initial interface).

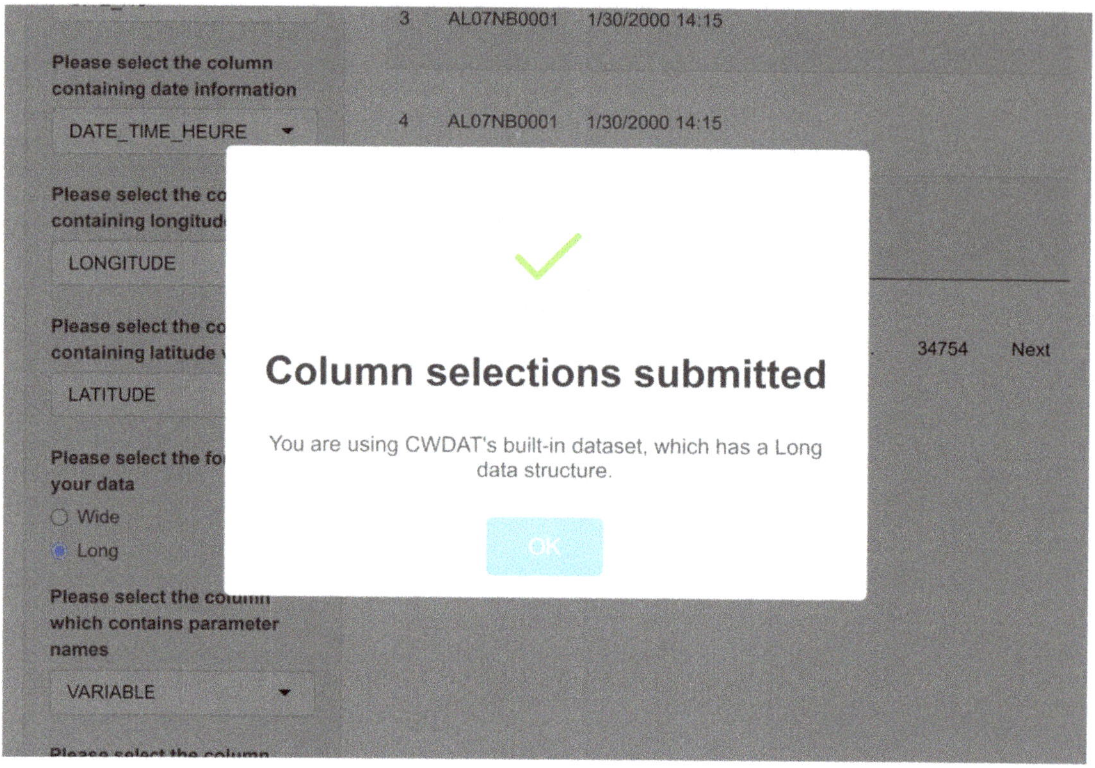

Figure 8. Data upload and properties (following participant feedback).

4. Discussion

4.1. Response and Reception

Although the response rate of 44% was higher than expected, the low overall number of participants, particularly those who interacted with CWDAT via a facilitated session (n = 8) substantially limits any claim of CWDAT's value to the community-based water quality monitoring field. Additionally, a better representation of community members and other grassroots stakeholders would enhance the results and give a truer accounting of CWDAT's potential use within the wider CBWQM field. However, the feedback and discussions described below did provide some insight into the barriers faced by CBWQM initiatives and the ability of CWDAT to address certain needs in the field.

4.2. Prototype Modifications and Implications

The facilitated sessions, while limited to a low number of participants, allowed for in-depth discussions and maximized the insight offered by each participant. Moreover, the provision of a working prototype served as a catalyst for more detailed discussions—both in terms of CWDAT's individual development and its value within the broader context of CBWQM and CS. This was critical as it expanded discussion from more general and abstract concerns and interests focused largely on questions of "what" to address both the "what" and the "how" (interface) [45].

As expected, the workflow from raw data to actionable information, and data quality concerns, are two substantial barriers to sustainable community-based water quality monitoring. The heterogeneous nature of the field, as represented by participants in terms of a dearth of consistency in protocols, reporting, and workflows, is another challenge, a finding consistent with Jollymore et al. (2017) [30] regarding citizen participation in the hydrological sciences.

Field-specific barriers such as water sample data quality must be viewed in the context of initiative-specific barriers and restrictions. For example, the use of laboratory testing, while it can increase the perceived reliability of the data, can create another barrier if consistent laboratory protocols are not used within/across initiatives. If a set of laboratory protocols are established for water quality sample analysis across the field of CBM, it must be considered if all initiatives have the capacity, financial or otherwise, to adhere to such protocols. Participants indicated that the proposed method of statistical paired site comparison is a promising technique which could help to address the discussed barriers. Specifically, the reliance on publicly available datasets can leverage spatial open government data to the benefit of the CBM field, especially as this resource is typically underused outside of the scope of "expert" research projects [46] while remaining accessible.

The provision of the tool's source code, the literature source for the statistics [42], and requested plain language explanatory text within the tool's interface speak to transparency and, where desired, community data sovereignty. Transparency is a guiding pillar of web tool development within the CBM field for enhancing watershed management and planning [47]. Data sovereignty recognizes that some communities (e.g., Citizen groups, Indigenous communities) may want to explore and validate their own monitoring data, yet not share their data with an external citizen science project, government, or industry [48,49]. Further development of data QA/QC functions for the tool, as requested by Survey B participants, included the use of colours to flag extraneous values, reflecting a documented characteristic of Decision Support Systems—the identification of conflicting data [50], which participants connected to the challenge of establishing norms and trends across and within monitoring jurisdictions. The potential for such information (via CWDAT) to help improve the consistency of CBM practices was discussed by some participants in terms of temporal and spatial biases in the data—a line of inquiry consistent with Geldmann et al. (2016) [51], which indicated that modelling the intensity (interpreted in this context as "number") of observations can help to understand spatial and temporal biases at/between monitoring stations.

As discussed at length by one Survey B respondent, while many CBM initiatives have established effective and beneficial working relationships with scientists and formal institutions, the proposed tool has the potential to fill the niche between the grassroots and the highly scientific and technical. By not only allowing users to ask questions of their data but also introducing users to potential questions they had not considered, by virtue of an open-ended design, members of the CBWQM field can be in a better position to understand and leverage their own data (starting small) either independently or in preparation for collaboration. Such discussions aligned well with previously established barriers and best practices in the literature. The use of the open-source R/Shiny framework supports a versatile, open-ended design affirmed by the Option B participants, as opposed to more traditional, "top-down" tool designs. This progression is consistent with findings in Castillo et al. (2016) [52], which suggested future work on Environmental Decision Support Systems will focus on broadening the range of EDSS capabilities and applicability. Although CWDAT should not be considered a full EDSS, the drive toward better features and wider relevance is shared.

The study revealed how obtaining relevant feedback on new software tools in a citizen science context is necessarily time-consuming and application-specific. Thus, identifying generic principles of geospatial capacity building in citizen science initiatives is challenging. However, we found that much of the rich feedback from Survey B respondents, facilitated stronger relationships with the project which are important cornerstones of project sustainability. The technical dimensions of interface design—while important—may be of less overall long-term value than the social dimensions conveyed through the use of the technology as a boundary object between citizen and scientist and/or technologist [53].

5. Conclusions

This paper presented the Community Water Data Analysis Tool, an open-source web application using the R/Shiny platform. CWDAT is intended to support citizen science initiatives in the field of water quality monitoring, especially community-based initiatives. CWDAT's interface allows a user to provide their own water quality data in .csv format and is robust against varying data structures (i.e., long vs. wide), date and time formats, and naming conventions.

A series of facilitated sessions with members of the community-based water quality monitoring field yielded positive feedback for CWDAT, insight into the challenges faced by CBWQM initiatives, and suggestions for future iterations of CWDAT. Feedback on CWDAT was positive and addressed a gap between citizen scientists and the wider scientific community by providing an accessible tool for independent visualization, analysis, and reporting of community-generated water quality data. CWDAT's use of an open-source language (R) with a robust online support community, combined with the provision of CWDAT's source code through Github, allows CBWQM and CS initiatives to modify CWDAT as they see fit. Future iterations of CWDAT will incorporate water quality thresholds and guidelines, the calculation of the Canadian Council of Ministers of the Environment water quality index, and other methods of data presentation and analysis. Overall, feedback from the study participants identified barriers to citizen science initiatives such as data quality and contextual divides between citizens and scientists.

Supplementary Materials: The following are available online at https://www.mdpi.com/article/10.3390/ijgi10040207/s1, Document S1: Survey Questions and Document S2: Instructions.

Author Contributions: Conceptualization, Annie Gray and Colin Robertson; methodology, Annie Gray and Colin Robertson; software, Annie Gray; validation, Annie Gray and Colin Robertson; writing—original draft preparation, Annie Gray, Colin Robertson, and Rob Feick; writing—review and editing, Annie Gray, Colin Robertson, and Rob Feick; visualization, Annie Gray; supervision, Colin Robertson; project administration, Colin Robertson; funding acquisition, Colin Robertson. All authors have read and agreed to the published version of the manuscript.

Funding: This research was funded by the Canada First Research Excellent Fund [Global Water Futures Project].

Institutional Review Board Statement: The study was conducted according to the guidelines of the Declaration of Helsinki and approved by the Research Ethics Board of Wilfrid Laurier University (protocol code 5987, approved 14 February 2019).

Informed Consent Statement: Informed consent was obtained from all subjects involved in the study. Written informed consent has been obtained from the participants to publish this paper.

Data Availability Statement: The participant survey data presented in this study are not available for release due to ethical and privacy considerations governed by our research ethics review. Water quality data presented in this study are publicly accessible via the data portal Mackenzie Data Stream located at https://mackenziedatastream.ca/ and at open.canada.ca.

Acknowledgments: The authors gratefully acknowledge the support of the Gordon Foundation and of members of the community-based water quality monitoring field who took the time to share their insight.

Conflicts of Interest: The authors declare no conflict of interest.

References

1. Haklay, M. Citizen Science and Volunteered Geographic Information: Overview and Typology of Participation. In *Crowdsourcing Geographic Knowledge*; Sui, D., Elwood, S., Goodchild, M., Eds.; Springer: Berlin/Heidelberg, Germany, 2013; pp. 105–122.
2. Kosmala, M.; Wiggins, A.; Swanson, A.; Simmons, B. Assessing data quality in citizen science. *Front. Ecol. Environ.* **2016**, *14*, 551–560. [CrossRef]
3. Jordan, R.C.; Gray, S.A.; Howe, D.V.; Brooks, W.R.; Ehrenfeld, J.G. Knowledge Gain and Behavioral Change in Citizen-Science Programs. *Conserv. Biol.* **2011**, *25*, 1148–1154. [CrossRef]

4. Roy, H.E.; Pocock, M.J.O.; Preston, C.D.; Roy, D.B.; Savage, J.; Tweddle, J.C.; Robinson, L.D. *Understanding Citizen Science & Environmental Monitoring*; Final Report on behalf of UK-EOF; CEH: Lancaster, UK, 2012.
5. Alender, B. Understanding volunteer motivations to participate in citizen science projects: A deeper look at water quality monitoring. *J. Sci. Commun.* **2016**, *15*, A04. [CrossRef]
6. Carlson, T.; Cohen, A. Linking community-based monitoring to water policy: Perceptions of citizen scientists. *J. Environ. Manag.* **2018**, *219*, 168–177. [CrossRef]
7. Bird, T.J.; Bates, A.E.; Lefcheck, J.S.; Hill, N.A.; Thomson, R.J.; Edgar, G.J.; Stuart-Smith, R.D.; Wotherspoon, S.; Krkosek, M.; Stuart-Smith, J.F.; et al. Statistical solutions for error and bias in global citizen science datasets. *Biol. Conserv.* **2014**, *173*, 144–154. [CrossRef]
8. Bonter, D.N.; Cooper, C.B. Data validation in citizen science: A case study from Project FeederWatch. *Front. Ecol. Environ.* **2012**, *10*, 305–307. [CrossRef]
9. Foody, G.M.; See, L.; Fritz, S.; Van Der Velde, M.; Perger, C.; Schill, C.; Boyd, D.S. Assessing the Accuracy of Volunteered Geographic Information arising from Multiple Contributors to an Internet Based Collaborative Project. *Trans. GIS* **2013**, *17*, 847–860. [CrossRef]
10. Goodchild, M.F.; Li, L. Assuring the quality of volunteered geographic information. *Spat. Stat.* **2012**, *1*, 110–120. [CrossRef]
11. Hunter, J.; Alabri, A.; Van Ingen, C. Assessing the quality and trustworthiness of citizen science data. *Concurr. Comput. Pract. Exp.* **2013**, *25*, 454–466. [CrossRef]
12. Huang, G.H.; Xia, J. Barriers to sustainable water-quality management. *J. Environ. Manag.* **2001**, *61*, 1–23. [CrossRef]
13. Connors, J.P.; Lei, S.; Kelly, M. Citizen Science in the Age of Neogeography: Utilizing Volunteered Geographic Information for Environmental Monitoring. *Ann. Assoc. Am. Geogr.* **2012**, *102*, 1267–1289. [CrossRef]
14. MacPhail, V.J.; Colla, S.R. Power of the people: A review of citizen science programs for conservation. *Biol. Conserv.* **2020**, *249*, 108739. [CrossRef]
15. Senaratne, H.; Mobasheri, A.; Ali, A.L.; Capineri, C.; Haklay, M. A review of volunteered geographic information quality assessment methods. *Int. J. Geogr. Inf. Sci.* **2017**, *31*, 139–167. [CrossRef]
16. Fonte, C.C.; Bastin, L.; Foody, G.; Kellenberger, T.; Kerle, N.; Mooney, P.; Olteanu-Raimond, A.-M.; See, L. VGI Quality Control. *ISPRS Ann. Photogramm. Remote Sens. Spat. Inf. Sci.* **2015**, *2*, 317–324. [CrossRef]
17. Conrad, C.C.; Hilchey, K.G. A review of citizen science and community-based environmental monitoring: Issues and opportunities. *Environ. Monit. Assess.* **2011**, *176*, 273–291. [CrossRef]
18. Bonney, R.; Shirk, J.L.; Phillips, T.B.; Wiggins, A.; Ballard, H.L.; Miller-Rushing, A.J.; Parrish, J.K. Next steps for citizen science. *Science* **2014**, *343*, 1436–1437. [CrossRef]
19. Yadav, P.; Darlington, J. Design Guidelines for the User-Centred Collaborative Citizen Science Platforms. *Hum. Comput.* **2016**, *3*. [CrossRef]
20. de Reyna, M.A.; Simoes, J. Empowering citizen science through free and open source GIS. *Open Geospat. Data Softw. Stand.* **2016**, *1*, 1. [CrossRef]
21. Luna, S.; Gold, M.; Albert, A.; Ceccaroni, L.; Claramunt, B.; Danylo, O.; Haklay, M.; Kottmann, R.; Kyba, C.; Piera, J.; et al. Developing Mobile Applications for Environmental and Biodiversity Citizen Science: Considerations and Recommendations. In *Multimedia Tools and Applications for Environmental & Biodiversity Informatics*; Springer International Publishing: Cham, Switzerland, 2018; pp. 9–30.
22. Fernandez-Gimenez, M.E.; Ballard, H.L.; Sturtevant, V.E. Adaptive Management and Social Learning in Collaborative and Community-Based Monitoring: A Study of Five Community-Based Forestry Organizations in the western USA. *Ecol. Soc.* **2008**, *13*, 4. Available online: http://www.ecologyandsociety:vol13/iss2/art4/ (accessed on 10 March 2021). [CrossRef]
23. Brenton, P.; von Gavel, S.; Vogel, E.; Lecoq, M.E. Technology Infrastructure for Citizen Science. In *Citizen Science: Innovation in OpenScience, Society and Policy*, 1st ed.; Hecker, S., Haklay, M., Bowser, A., Makuch, Z., Vogel, J., Bonn, A., Eds.; UCL Press: London, UK, 2018; pp. 63–80.
24. Skarlatidou, A.; Hamilton, A.; Vitos, M.; Haklay, M. What do volunteers want from citizen science technologies? A systematic literature review and best practice guidelines. *J. Sci. Commun.* **2019**, *18*, A02. [CrossRef]
25. Klein, L. What do we actually mean by 'sociotechnical'? On values, boundaries and the problems of language. *Appl. Ergon.* **2014**, *45*, 137–142. [CrossRef]
26. Muenich, R.; Peel, S.; Bowling, L.; Haas, M.; Turco, R.; Frankenberger, J.; Chaubey, I. The Wabash Sampling Blitz: A Study on the Effectiveness of Citizen Science. *Citiz. Sci. Theory Pract.* **2016**, *1*, pe0188507. [CrossRef]
27. Weeser, B.; Kroese, J.S.; Jacobs, S.R.; Njue, N.; Kemboi, Z.; Ran, A.; Breuer, L. Citizen science pioneers in Kenya—A crowdsourced approach for hydrological monitoring. *Sci. Total Environ.* **2018**, *631–632*, 1590–1599. [CrossRef] [PubMed]
28. Capdevila, A.S.L.; Kokimova, A.; Ray, S.S.; Avellán, T.; Kim, J.; Kirschke, S. Success factors for citizen science projects in water quality monitoring. *Sci. Total Environ.* **2020**, *728*. [CrossRef] [PubMed]
29. Hall, G.B.; Chipeniuk, R.; Feick, R.D.; Leahy, M.G.; Deparday, V. Community-based production of geographic information using open source software and Web 2.0. *Int. J. Geogr. Inf. Sci.* **2010**, *24*, 761–781. [CrossRef]
30. Jollymore, A.; Haines, M.J.; Satterfield, T.; Johnson, M.S. Citizen science for water quality monitoring: Data implications of citizen perspectives. *J. Environ. Manag.* **2017**, *200*, 456–467. [CrossRef]
31. Keum, J.; Kaluarachchi, J.J. Development of a decision-making methodology to design a water quality monitoring network. *Environ. Monit. Assess.* **2015**, *187*, 1–14. [CrossRef] [PubMed]

32. Hadj-Hammou, J.; Loiselle, S.; Ophof, D.; Thornhill, I. Getting the full picture: Assessing the complementarity of citizen science and agency monitoring data. *PLoS ONE* **2017**, *12*, e0188507. [CrossRef]
33. Penny, D.; Williams, G.; Gillespie, J.; Khem, R. 'Here be dragons': Integrating scientific data and place-based observation for environmental management. *Appl. Geogr.* **2016**, *73*, 38–46. [CrossRef]
34. Walker, D.; Forsythe, N.; Parkin, G.; Gowing, J. Filling the observational void: Scientific value and quantitative validation of hydrometeorological data from a community-based monitoring programme. *J. Hydrol.* **2016**, *538*, 713–725. [CrossRef]
35. Werts, J.D.; Mikhailova, E.A.; Post, C.J.; Sharp, J.L. An Integrated WebGIS Framework for Volunteered Geographic Information and Social Media in Soil and Water Conservation. *Environ. Manag.* **2012**, *49*, 816–832. [CrossRef]
36. Criollo, R.; Velasco, V.; Nardi, A.; de Vries, L.M.; Riera, C.; Scheiber, L.; Jurado, A.; Brouyère, S.; Pujades, E.; Rossetto, R.; et al. AkvaGIS: An open source tool for water quantity and quality management. *Comput. Geosci.* **2019**, *127*, 123–132. [CrossRef]
37. Perdikaki, M.; Manjarrez, R.C.; Pouliaris, C.; Rossetto, R.; Kallioras, A. Free and open-source GIS-integrated hydrogeological analysis tool: An application for coastal aquifer systems. *Environ. Earth Sci.* **2020**, *79*, 1–16. [CrossRef]
38. Matthies, M.; Giupponi, C.; Ostendorf, B. Environmental decision support systems: Current issues, methods and tools. *Environ. Model. Softw.* **2007**, *22*, 123–127. [CrossRef]
39. Rodela, R.; Pérez-Soba, M.; Bregt, A.; Verweij, P. Spatial decision support systems: Exploring differences in pilot-testing with students vs. professionals. *Comput. Environ. Urban Syst.* **2018**, *72*, 204–211. [CrossRef]
40. R Core Team. *R: A Language and Environment for Statistical Computing*; R Foundation for Statistical Computing: Vienna, Austria, 2020. Available online: https://www.R-project.org/ (accessed on 10 March 2021).
41. Hewitt, R.; Macleod, C. What Do Users Really Need? Participatory Development of Decision Support Tools for Environmental Management Based on Outcomes. *Environments* **2017**, *4*, 88. [CrossRef]
42. Kilgour, B.W.; Somers, K.M.; Barrett, T.J.; Munkittrick, K.R.; Francisy, A.P. Testing Against Normal with Environmental Data. *Integr. Environ. Assess. Manag.* **2017**, *13*, 188–197. [CrossRef] [PubMed]
43. Government of Northwest Territories. *NWT-Wide Community-Based Water Quality Monitoring, Environment and Natural Resources*; Government of the Northwest Territories: Yellowknife, NT, Canada, 2019. [CrossRef]
44. Environment and Climate Change Canada. *Lower Mackenzie River Basin Long-Term Water Quality Monitoring Data—Canada's North*; Record ID 0177c195-13a8-4078-aa85-80b17e9e2cfe; Environment and Climate Change Canada: Gatineau, QC, Canada, 2016.
45. Sharp, H.; Rogers, Y.; Preece, J. *Interaction Design: Beyond Human-Computer Interaction*, 5th ed.; John Wiley: Indianapolis, IN, USA, 2019.
46. Gebetsroither-Geringer, E.; Stollnberger, R.; Peters-Anders, J. Interactive Spatial Web-Applications as New Means of Support for Urban Decision-Making Processes. *ISPRS Ann. Photogramm. Remote Sens. Spat. Inf. Sci.* **2018**, *4*, 59–66. [CrossRef]
47. Sun, A.Y.; Miranda, R.M.; Xu, X. Development of multi-metamodels to support surface water quality management and decision making. *Environ. Earth Sci.* **2015**, *73*, 423–434. [CrossRef]
48. Hummel, P.; Braun, M.; Augsberg, S.; Dabrock, P. Sovereignty and data sharing. *ITU J. ICT Discov.* **2018**, *25*.
49. Kukutai, T.; Taylor, J. *Indigenous Data Sovereignty: Toward an Agenda*; Anu Press: Canberra, Australia, 2016.
50. French, S.; Turoff, M. Decision Support Systems. *Commun. ACM* **2007**, *50*, 39–40. [CrossRef]
51. Geldmann, J.; Heilmann-Clausen, J.; Holm, T.E.; Levinsky, I.; Markussen, B.; Olsen, K.; Rahbek, C.; Tøttrup, A.P. What determines spatial bias in citizen science? Exploring four recording schemes with different proficiency requirements. *Divers. Distrib.* **2016**, *22*. [CrossRef]
52. Castillo, A.; Porro, J.; Garrido-Baserba, M.; Rosso, D.; Renzi, D.; Fatone, F.; Omez, F.V.G.; Comas, J.; Poch, M. Validation of a decision support tool for wastewater treatment selection. *J. Environ. Manag.* **2016**, *184*, 409–418. [CrossRef]
53. Harvey, F.; Chrisman, N. Boundary Objects and the Social Construction of GIS Technology. *Environ. Plan. A Econ. Space* **1998**, *30*, 1683–1694. [CrossRef]

International Journal of
Geo-Information

MDPI

Article

A CitSci Approach for Rapid Earthquake Intensity Mapping: A Case Study from Istanbul (Turkey)

Ilyas Yalcin [1,2], Sultan Kocaman [1] and Candan Gokceoglu [3,*]

1 Department of Geomatics Engineering, Hacettepe University, Beytepe Ankara 06800, Turkey; ilyas.yalcin@hacettepe.edu.tr (I.Y.); sultankocaman@hacettepe.edu.tr (S.K.)
2 Gemlik Asim Kocabiyik Vocational School, Bursa Uludag University, Gemlik Bursa 16600, Turkey
3 Department of Geological Engineering, Hacettepe University, Beytepe Ankara 06800, Turkey
* Correspondence: cgokce@hacettepe.edu.tr

Received: 15 March 2020; Accepted: 17 April 2020; Published: 20 April 2020

Abstract: Nowadays several scientific disciplines utilize Citizen Science (CitSci) as a research approach. Natural hazard research and disaster management also benefit from CitSci since people can provide geodata and the relevant attributes using their mobile devices easily and rapidly during or after an event. An earthquake, depending on its intensity, is among the highly destructive natural hazards. Coordination efforts after a severe earthquake event are vital to minimize its harmful effects and timely in-situ data are crucial for this purpose. The aim of this study is to perform a CitSci pilot study to demonstrate the usability of data obtained by volunteers (citizens) for creating earthquake iso-intensity maps in a short time. The data were collected after a 5.8 Mw Istanbul earthquake which occurred on 26 September 2019. Through the mobile app "I felt the quake", citizen observations regarding the earthquake intensity were collected from various locations. The intensity values in the app represent a revised form of the Mercalli intensity scale. The iso-intensity map was generated using a spatial kriging algorithm and compared with the one produced by The Disaster and Emergency Management Presidency (AFAD), Turkey, empirically. The results show that collecting the intensity information via trained users is a plausible method for producing such maps.

Keywords: citizen science; CitSci; earthquake; intensity mapping; disaster mitigation; spatial kriging

1. Introduction

Sustainable development is the major aim for all countries, and the advancements in geospatial and mobile technologies provide the essential support to improve humans' lives and to protect the planet. According to the "We Are Social and Hootsuite's Global Digital 2019 Report", there are 5.1 billion mobile users worldwide [1]. Moreover, the number of users has a growth rate of 2% compared to the previous year, while the majority of mobile users were encountered in Asia. Considering that most of the devices have an Internet connection, the masses can contribute to science for solving global problems using their mobile phones if they are guided correctly. Citizen Science (CitSci) is a research approach that aims to contribute to scientific processes with the help of ordinary people (i.e., non-professional scientists) and can be utilized by many disciplines. CitSci can also be defined as a type of science developed and adopted by citizens that can help in case of dangerous situations or needs [2].

The increase in the number of natural hazards can also be tied to the population rise, climate change and increased disaster awareness (i.e., event recording and monitoring). Improved disaster management and mitigation can be obtained by better coordination during or after disaster events, which helps to reduce the losses of lives and economic losses as well. The adverse effects of disasters can be monitored after the events so that the necessary measures can be taken. A recent review by [3] demonstrates the potential of CitSci for efficient disaster management.

The disaster-related information can be collected by citizens in the form of volunteered geographic information (VGI) or CitSci [4]. Although the data collection methods can be various, social media platforms especially have often been used for this purpose. In addition, specialized CitSci apps or repositories have been coming into view increasingly. Among the social media platforms, Twitter from Twitter Inc., San Francisco, CA USA, is the most frequently appearing in the literature. In 2017, the tweets sent by the citizens after the earthquake on Lesvos Island in Greece were classified and macroseismic intensity maps of the region were created [5]. In another study, Twitter and a mobile app were used to collect high resolution data after urban flooding [6]. A convolutional neural network (CNN) algorithm was developed in order to categorize data related to flood collected from citizens. In this context, it is concluded that CitSci can be regarded as a data source used in different disciplines.

Earthquake-related studies require extensive field data both for research and disaster mitigation purposes. An earthquake may have destructive effects based on the magnitude and focal depth. It may cause damages in the infrastructure, such as building collapses, road destruction, etc., and ground deformations such as landslides, rockfalls, lateral spreads, liquefaction, surface rupture, etc. Such deformations and failures may result in more damage and more loss of lives than ground shaking. In order to understand these phenomena, which are naturally complex, timely and dense, spatial data are required. Although both CitSci and VGI methods have been employed for this purpose, CitSci can contribute to the studies at advanced levels, such as basic or even high-level interpretation by training the motivated citizen scientists, and not only the basic data collection or validation. However, preparation of training materials and giving training are immensely required for this purpose [4].

The aim of this study is to demonstrate the usability of CitSci approach for producing earthquake intensity maps with a pilot study from Istanbul Earthquake (Mw = 5.8) occurred on 26 September 2019 at 13:59 UTC. The Anatolian Plate is surrounded by North Anatolian Fault Zone, Eastern Anatolian Fault Zone, and Aegean horst-graben system and all these are seismically active. During the last two decades, several large earthquakes such as 1999 Golcuk (Mw = 7.5), 1999 Duzce (Mw = 7.2), 2003 Bingol (Mw = 6.4), 2011 Van (Mw = 7.1), 2020 Manisa (Akhisar) (Mw = 5.4), 2020 Elazig (Mw = 6.8) occurred. Following the Istanbul Mw 5.8 earthquake that is the subject of this study, approximately 150 aftershocks occurred with intensities ranging from 1.0 Mw to 4.1 Mw [7]. Although the proximity to the faults plays an important role in the intensity level felt after an earthquake, it is not the only factor. The ground conditions of the earthquake-affected area, the construction quality and the number floors of the building have strong influence on the intensity level. There is almost no analytical solution for determining the earthquake intensity in different regions and field observations are compulsory for producing reliable intensity maps.

In this study, a mobile app (named "I felt the quake") was developed for this purpose and integrated in a spatial database management system for data storage and management (Supplementary Materials). The citizen scientists participated to the study were informed about the aims of the study, the importance of their contributions and the correct use of the app. Using the collected data and spatial analysis methods (i.e., ordinary kriging), an iso-intenstiy map was created and compared with the intensity map published by The Disaster and Emergency Management Presidency (AFAD). The results show that the developed approach is useful for producing accurate and reliable iso-intensity maps, which is an important base for disaster mitigation efforts.

2. Background

VGI is a commonly used term in geography, which emphasizes that humans can contribute to gather geographic information thanks to their senses and superior interpretation intelligence [8]. Many studies have been conducted in order to get support from people in different scientific studies.

Brovelli et al. [9] have developed potholes and architectural barriers application related to urban monitoring and mapping of tourism points of interest, and investigated the participation of the public in the applications of Geographical Information System (GIS). These applications were presented to users through the Open Data Kit Collect (ODK Collect) application [10], which is an Android application created with open source codes to collect data from the user through survey forms. The data were collected not only from the Android platform but also from the web with the Enketo [11] application which can work with ODK. With potholes application in the study, the broken roads in the Northern Italy region were detected and, thus, the maintenance of the potholes was accelerated. In another application on architectural barriers, it was aimed to determine the physical structures that would create undesired situations for disabled citizens in the city. In the last application developed by the same research group, points of interest for tourism were provided. Considering all of studies carried out by Brovelli et al. [9], it was clearly seen that although potholes application had more campaign time, less users showed interest in this application than other applications. It was concluded that more advertising, mapping party and gamification methods should be employed in order to spread the use of applications. In addition, it was emphasized that the familiarization of the people with the technology is of high importance for the spread of the applications. From this point of view, it was aimed to make these applications become widespread by the use of paid students under the age of thirty in the application of architectural barriers and the use of the mapping party in the tourism application.

Boyd et al. [12] conducted a study integrated with remote sensing and volunteered user data, stating that recent, spatial and reliable data were needed to end slavery activities. The study area is "Brick Belt" region which comprises North India, Nepal, Pakistan and Bangladesh. The purpose of choosing this region was the abundance of brick kilns. It was known that most of the workforce of the kilns was made up of socially excluded people. Therefore, the aim of the study was to determine the number of the kilns. High resolution satellite images of the region were obtained and the kilns on these images were marked by volunteers. In order to collect data by volunteers, a CitSci platform called Zooniverse was used. This platform has 1.6 million users worldwide as of 2020, and facilitates the collection of data required for various scientific studies. The platform serves as a bridge that connects volunteers and researchers [13]. In the study of Boyd et al. [12], the users were directed to a website called "Slavery from Space" and they made it possible to detect the kilns through images. The study was carried out with 120 volunteered users. Approximately 55,383 kilns were detected. The Rajasthan region was selected for the control study. In addition, gas emissions from these kilns and their effects on the ecosystem were discussed.

Koskinen et al. [14] conducted a study based on a participatory GIS (PGIS) approach using open source software to draw mapping of forest plantations at the regional level in Tanzania. In order to collect data, Collect Earth [15] software of Open Foris platform [16] was used. Collect Earth is an open source software developed by the Food and Agriculture Organization of the United Nations (FAO) to monitor land use and change over high spatial and temporal resolution images [17]. It helps data collection on land use and allows the worldwide use of the application since it is open source. In this study, 22 participants were requested to complete survey data entries about forest types in the study area with the help of Collect Earth software and the reference data was created. The study was concluded by interpreting the effect of user-oriented data on the classification.

Hicks et al. [18] analyzed 106 CitSci projects within the scope of disaster risk reduction and aimed to look at citizen science from a wider window and to establish principles for the correct development of this science. In this study, six principles were introduced to guide the implementation of CitSci studies in order to reduce disaster risks. Similarly, 10 principles were introduced by the European Citizen Science Association (ECSA) [19] and translated into 30 languages. In the study, a new definition for citizen science was provided by replacing the word "information" with the word "science", since it was thought that the data collected by the citizen scientists may be in the social and cultural context and they may contribute to the applications on the focus subject. In addition, an online map which provides country based demonstrations of citizen science projects was created within the website developed in the study [20]. The website can be defined as a collective and extensive in which anybody can follow the all studies focusing on reducing disaster risks, necessary briefing can be provided, and studies can be made widespread.

Liang et al. [21] aimed to increase their contribution to earthquake research by raising the awareness of Taiwanese people on earthquakes through the applications they developed. In this way, with the awareness of the public, frequent earthquakes in Taiwan can be survived with least damage. The Earthquake Science Information (TESIS) [22] web application developed by the group provides earthquake reports from Taiwan official institutions. This platform both helps the public to have the credible information after an earthquake and the scientists to identify the fault lines and other hazards that could lead to other hazards. They developed two applications to enable CitSci. The first one shows earthquake intensity by determining the weight from the data obtained from surveys. This application was designed based on "Did You Feel It" (DYFI) [23] developed by the United States Geological Survey (USGS). The DYFI application is a data collection platform that allows people to provide earthquake-related information through online surveys, which can then be used to create earthquake intensity maps. In this application, it was seen that there were differences between user data and institutional data. Anticipating that the differences result from the length of the survey given to users, they decided to shorten it. Secondly, sensors were installed on the computers of more than 200 educators in Taiwan, and data on earthquake waves were collected online. With the publication of the collected data as an educational argument, it may be guiding scientific studies. In addition, unlike these studies, users were trained and an Ushahidi-based application was developed to enter the situations that may occur after the earthquake such as cracks. Ushahidi [24] is a paid online application that helps collect data like ODK. This application can facilitate the management and reporting of data from mobile platforms. In this study, more than a hundred users were trained and a sample reporting application was implemented.

Kong et al. [25] developed a mobile app called MyShake [26] which could warn about earthquake shocks in advance. In the app, the accelerometer signals on mobile devices were analyzed by an artificial neural network (ANN) algorithm to determine if they point out an earthquake event or not. The signals were recorded when mobile devices were at a fixed position (e.g., hands-free on a table) and compared with the signals at the time of an earthquake. The ANN method utilized the changes in the amplitude and the frequency of the signals. In the study, it was emphasized that citizens can perform this task with their mobile devices, which can be particularly useful in countries without an earthquake early warning system.

A mobile and web-based CitSci application, called LaMA, was developed to detect landslide incidents [27,28]. In the application, photographs and the location of a landslide observed by the users were collected in a database. The data can be collected by users via web browser and a CNN architecture was developed for verification of landslide photographs as well [29]. In order to train the CNN algorithm, images were obtained from various data repositories and websites and it is possible to classify the landslide photographs with high accuracy.

AFAD of Turkey developed an earthquake mobile application called eAfad [30]. The purpose of the application is to inform citizens about earthquakes in Turkey and the surrounding areas [31]. The data are being taken from the Earthquake Data Center. Information on earthquakes with a magnitude of 4.0 Mw or larger are disseminated by the app by AFAD. The app also has features for visually impaired people. The first results of the application were presented by Eravci et al. [32]. On the other hand, a Web-GIS framework based on open source technologies was proposed by [33,34] with a mobile app for data collection with the theme of earthquake data collection, but neither practical implementation for open use nor a case study existed.

3. Data Collection and Analysis Methodology

3.1. System Design and Implementation

The mobile earthquake app ("I felt the quake") developed in this study works on an Android platform and has been used to collect data regarding the location, date, time, the intensity level as well as the name of the data provider who felt the quake (Table 1). The name field is, however, filled out optionally. The data have been recorded in a spatial database. The date and time fields in the app represent the event occurrence time, whereas the data upload (transaction) time is also sent to the database automatically. The geographical location, where the person was present at the time of earthquake, must also be entered by the user by selecting on a map (Figure 1). After pressing the save button, the name, surname, date, time, user-interpreted intensity value, transaction date and time (taken from the mobile phone system), and the location (geographical coordinate values) are sent to the server and constitute a tuple in the spatial database.

Table 1. Data fields and types collected in the developed app.

Column	Data Type
Id	Integer
Name	Character
Surname	Character
Intensity	Character
Latitude	Double precision
Longitude	Double precision
Date	Character
Time	Character
Transaction date and time (*not shown on the interface*)	Timestamp

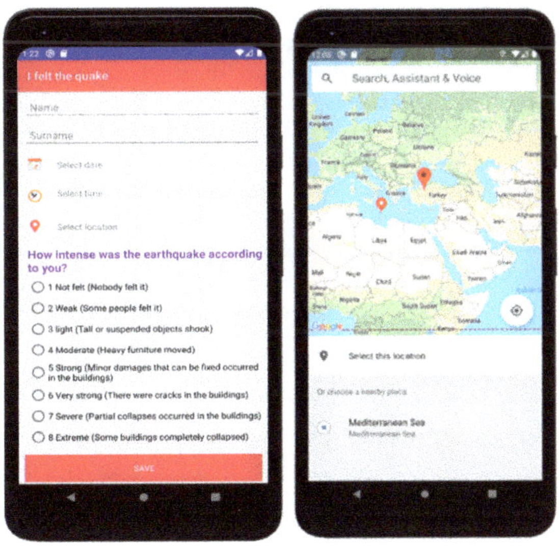

(**a**) (**b**)

Figure 1. (**a**) Interface of the 'I felt the quake' application; (**b**) a screenshot from 'Select location' interface.

The intensity values shown in Figure 1a and explained further in Table 2 are revised from the Modified Mercalli Intensity Scale (MM) developed in 1931 by Wood and Neumann [35]. The MM scale is also in use by USGS [36]. In order to design the interface effectively and use only the meaningful values for the participants, the scale values were reduced to 8 as shown in Table 2. In the MM, the values 1 and 2 are too weak to be felt by a person, and the values 11 and 12 are too strong and destructive, so that the use of the app would be unnecessary. Therefore, these four values were not included in the app to simplify the interface.

Table 2. The relationship between Modified Mercalli Intensity Scale and values used in the developed app.

Scale Value in the App	Description of the Intensity Scale in the Developed App	Corresponding Modified Mercalli Intensity Scale
1 Not felt	Nobody felt it	3
2 Weak	Some people felt it	4
3 Light	Tall or suspended objects shook	5
4 Moderate	Heavy furniture moved	6
5 Strong	Minor damages that can be fixed occurred in the buildings	7
6 Very strong	There were cracks in the buildings	8
7 Severe	Partial collapses occurred in the buildings	9
8 Extreme	Some buildings completely collapsed	10

As a spatial database management system (DBMS), the open source PostgreSQL with PostGIS [37] extension, which enables spatial functionality in the DBMS, is utilized here. The selection was based on open source environment and enabling spatial queries via indices. The app was developed in an Android Studio IDE (Integrated Development Environment). It can be installed on mobile devices that support Android 4.0.3 and further versions from the Google Play Store website [38]. While designing the application interface, it was aimed to keep it simple and comprehensible for non-expert users. Further attention was paid to keep the application size (storage requirement) relatively small. The geolocation information entry by the citizen scientists was enabled using the Google-powered PlacePicker API (Application Programming Interface) in order to fully provide the application with backward data entry while not weakening the interface simplicity. Since the case study focuses on Istanbul and surrounding provinces, the app was prepared in Turkish as default with English language support. The overall system architecture design is presented in Figure 2. In the system architecture in Figure 2, the working direction of the system is defined by arrows. Information about the earthquake intensity felt by the user is entered manually to the app. The connection between the Android app and the server is provided via the Internet. A web service is used to provide the connection between the Android app and PostgreSQL database. Amazon Web Services (AWS) from Amazon.com Inc., Seattle, U.S.A. was used as the server system to ensure continuous operation of the system. Input data provided from the user is recorded in the database management system installed on the server. Then, the toast message is returned to the user indicating that the provided data has been saved.

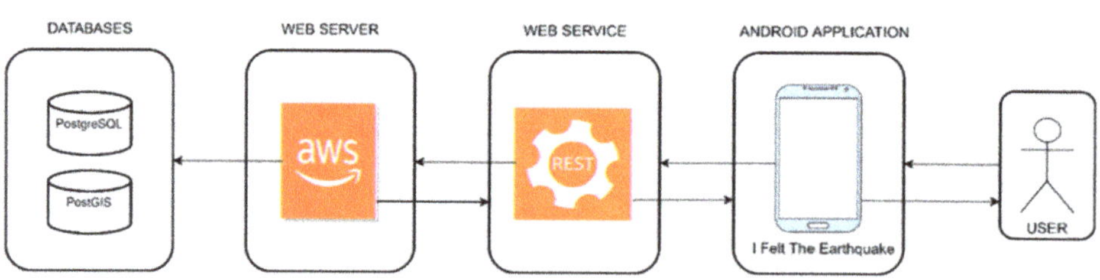

Figure 2. System architecture design of the application.

The application was activated on Google Play Store on 12/10/2019. All participants were informed about the main goals of the study and instructed on the use of the app personally, since external validation of such data is almost impossible due to the nature of the problem (i.e., different ground conditions and being on different floors may cause variations in the intensity level felt by the user). Although more data came from various regions across Turkey that are not only from the trained users, the data was filtered for the trained users via name and surname. On the other hand, only the data of the Marmara Region were employed in the analysis and the results. Similarly, the Taiwan Scientific Earthquake Reporting System (TSER) provides training to their users in order to ensure the data quality [21]. The difficulties in determining the volunteers' profiles (e.g., background, knowledge and skills, etc.) or ensuring the soundness of the data were reported as known issues in CitSci studies as stated by [39].

3.2. Geostatistical Analysis for Iso-Intensity Map Production

Geostatistical analysis was applied to the earthquake data collected with the app in order to produce a continuous iso-intensity map. Geostatistical approaches are often used to identify and analyze spatial changes in natural phenomena, and enable statistical determination of the spatial relationships among sample data in a region. Such approaches cover the mathematical and statistical principles generally used by experts in geology and mining [40]. In the field of geostatistics, Krige's work in the field of mining in the Witwatersrand region of South Africa is accepted as pioneering [41]. Matheron emphasized the importance of regional variable for geostatistics [42].

Kriging, one of the spatial interpolation methods, is widely used in geostatistical analysis studies. There are many types of kriging interpolation methods, such as ordinary kriging, simple kriging, and universal kriging [43]. Kriging interpolation is a method that estimates the values of unknown data by statistically employing the values of sample data in an area. This method uses semivariance values between point pairs. Semivariance reflects the degree of uniqueness between point pairs with increasing distance from covariance [40]. The graphic on which semivariance represented is called variogram. Three active terms define a variogram, which are nugget (C0), range (C1), and sill (C0 + C1). The nugget is used to identify the discontinuity encountered in the inability to detect similarity between points close to each other. Range is used to define the distance required to reach the variogram threshold. Beyond this distance, location-based dependence ends. Sill is the maximum value that the variogram reaches. Variogram will take values around the sill value after it reaches to the sill value [44].

Using the statistical calculations in Kriging method, the variance for each unknown point is calculated, which indicates the reliability of the interpolation. It also enables weight calculations for unknown points from known points through semi-variogram. In order to obtain neutral results, it is restricted to Lagrangian multiplier (λ). Therefore, the sum of the weights obtained is expected to be equal to one [44]. From this point of view, by looking at the weight of each unknown point, it is possible to establish a distance–proximity relationship to the known points. In other words, while the highly weighted points are close, the low-weighted points remain distant.

In this study, location-based earthquake data were modeled using the ordinary kriging interpolation method. Ordinary kriging has similar aspects to simple kriging. In the ordinary kriging method, the local mean is not known within the search area, but taken as constant [40].

4. Results Obtained

This study was based on the data collected between 12/10/2019–13/11/2019 after the Istanbul earthquake (Mw = 5.8). During this period, a total of 156 records were obtained from the app. Out of those, 99 of them were provided by the trained users. The iso-intensity map of the study was produced using the spatial kriging method and compared with the one produced by AFAD Earthquake Department responsible for the Marmara Region [7]. The iso-intensity map produced by AFAD is shown in Figure 3. The map was georeferenced in the study according to The European Petroleum Survey Group (EPSG) 4326 projection in order to match with the projection system of the location data collected from the users using a number of 2D ground control points extracted from existing maps. The data points collected in the study are denoted on the AFAD map in Figure 4. It should be noted that all 12 MM intensity values exist on the legend of the map produced by AFAD. The differences between user data and intensity map produced by AFAD can be seen clearly, as expected due to the nature of the problem.

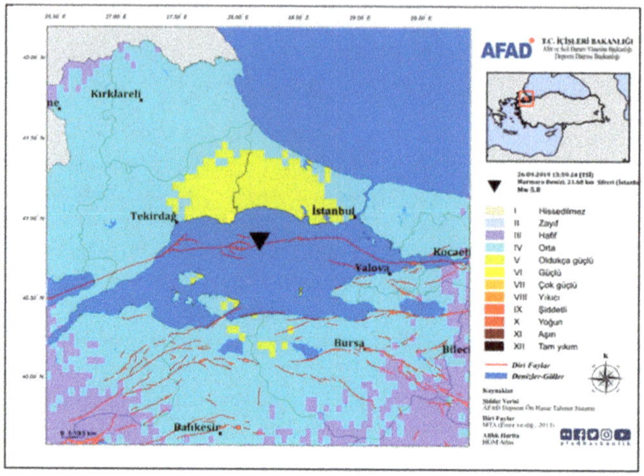

Figure 3. AFAD Earthquake Intensity Map produced after the Istanbul Earthquake (Mw = 5.8) [7].

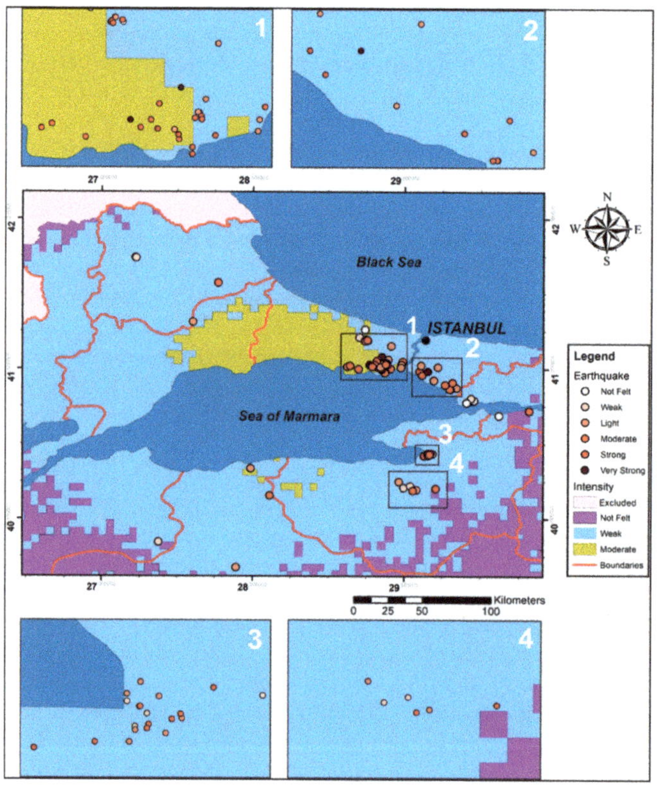

Figure 4. Data points collected in the study and AFAD Iso-intensity map; additional zoomed views for four sub-parts.

A statistical analysis between of the study data provided by the 99 volunteers and the corresponding value on AFAD intensity map is given in Table 3. As can be seen in the table, the scale 2—Weak

(Some people felt it) largely complies with the AFAD map (15 out of 16 data points are in accordance with the map). Many points with scales of 3—Light and 4—Moderate are located on the Weak areas according to the AFAD map, as shown in Figure 4. A total of seven data points came with Strong and Very strong scales (Table 3), although these scales did not exist in the AFAD map. However, the volunteers who provided these values stated that these buildings were marked as severely damaged by the relevant municipalities and the decisions for demolition were taken. Examples photos from these buildings were also provided by the volunteers as proof. In Figure 5, two sample photos from one building which provided the scale value as Strong are shown. In Figure 6, a photo from another building with Very Strong scale is provided.

The resulting iso-intensity map obtained after spatial interpolation with ordinary kriging method can be seen in Figure 7. The minimum and the maximum intensity values obtained from the kriging method in the map area are 4.7 and 6.6, respectively.

Table 3. The number of intensity values obtained on the AFAD map and the data collected in the study.

Frequency	App Scale Number	Number of Compatible Scale Values (cells) in AFAD Intensity Map
6	1 (Not felt)	0
16	2 (Weak)	15
29	3 (Light)	1
41	4 (Moderate)	7
4	5 (Strong)	No data
3	6 (Very strong)	No data

Figure 5. Column damages of a building to be demolished. Photos were provided by the study participant upon request of the authors. The intensity value provided by the participant was "Strong".

Figure 6. Image sent by a user after the earthquake. The intensity value provided by the participant was "Very Strong".

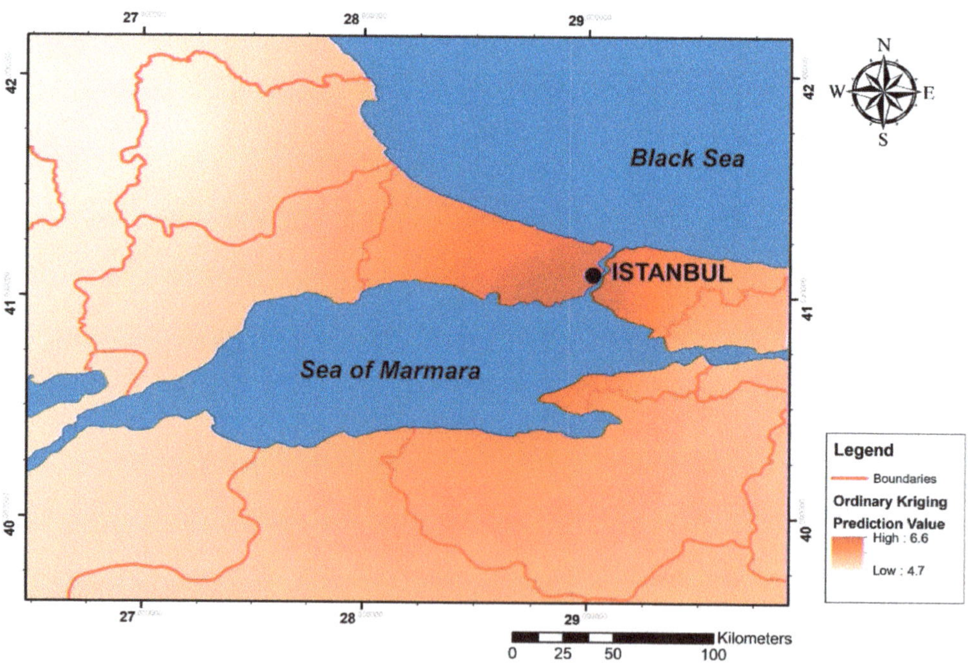

Figure 7. Interpolation map produced from the data collected in the study using the ordinary kriging method.

Figure 8 shows the drawing of the variogram obtained from the application data. The horizontal axis represents the distance (in degrees) that each point pair, which is used to calculate the intensity variance, has. The variances in the user-provided intensity values range between 0.1–2.4 (i.e., between 0.05–1.2 for semi-variance). As expected, there are high similarities between the intensity values at close distances. On the other hand, both small and high variations are observed at large distances due to the radial behavior of the intensity values. The intensity values point pairs located at far distances from the earthquake epicenter are similar and small. On the other hand, the point pairs having a point close to the epicenter and the other point very far exhibit larger variations.

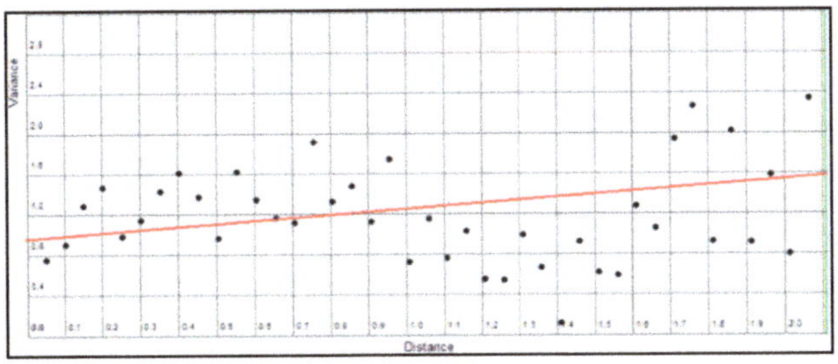

Figure 8. Variogram produced from the data collected in the study.

5. Discussion

5.1. Production of Intensity Maps

In this study, differences were observed between the map produced by AFAD and the information received from the citizen scientists. Production of reliable earthquake iso-intensity maps is crucial for disaster mitigation efforts and the increasing use of CitSci methods and platforms will support the development in this field. Therefore, citizens need to be informed about the importance of their high-quality contributions.

Data quality is an important aspect of CitSci studies. The term quality in this study involves correctness of the location data and the other information content, and the spatial distribution of the collected data. Inadequacy in the number and distribution of the data points would lead to weak conclusions, although the proposed kriging method can handle this issue to some extent. On the other hand, provision of training to the volunteers regarding the purpose of the project, the consequent stages and the outcomes is crucial for the success of the study.

5.2. The Use of the CitSci Method and the Issuess (Lessons Learned)

The widely used mobile devices can bring people who have similar experiences after a disaster event together to share their ideas. Surveys show that there is an increasing dialogue between people having similar ideas with the use of mobile technologies. Even in underdeveloped societies, the use of mobile technologies may facilitate scientific research with CitSci, and people can be connected at advanced levels in comparison to sensors. Instead of using costly sensors, each human can act as a sensor to bring a new dimension to scientific studies. In addition, they can help to reduce the costs depending on the scale of the studies. Furthermore, although people can collect data on a specific subject just like a sensor, they can also produce information on diverse subjects. Thus, instead of establishing different infrastructure for multiple studies, the required information can be obtained from the CitSci network, which can be the case for earthquakes that trigger further events (e.g., landslides).

Increasing the number of CitSci studies can help to advance the technology and the public awareness of disasters. By training the volunteers in the scope of different CitSci studies and producing data in these studies can provide better insight on different problems of the society and scientific solutions. Thus, not only the information society that produces data, but also the elevated information society, which has a high perception and understanding on many issues, can be created. Awareness of people with the increase of knowledge can protect the world against many issues, especially the environmental ones. In addition, when conducting CitSci studies, it may not be required to classify people according to their level of education, because the trainings given before the studies and the data obtained can be passed through some filters to produce information from people at all levels. This allows people from various educational backgrounds to participate in CitSci projects.

When the proposed approach, i.e., using specially designed applications for data collection, is compared with social media-based disaster data collection efforts, the main advantages are collection of structured data with correct geolocation information via trained users that allows rapid processing and design flexibility. A study on using crowd sourcing information collected from Twitter has shown that several issues were encountered with the existence of geotags, spams, information volume, linguistic differences, and insufficient geographical coverage [45]. Another study on using Twitter data for flood assessment has also shown that geographical location of tweets can be different from the event location, the geographical distribution can be an issue, and finding the relevant information inside a large numbers of tweets is challenging as well [46]. On the other hand, the disadvantages of using a specifically developed app could be listed as the maintenance of the app, the need of multi-platform (e.g., operating system) support, and convincing the users to install and use an additional app on their mobile phones.

6. Conclusions

In this study, the iso-intensity map of the Istanbul 5.8-Mw earthquake, which occurred on 26 September 2019, was produced based on CitSci data and the ordinary kriging method for spatial interpolation. A total of 99 data points collected over the Marmara Region of Turkey were employed for this purpose. Prior to the data collection, the volunteers were informed about the main goals, the research problem and the expected outcomes of the study. The intensity scale used here was a revised form of Mercalli scale and the data collection was performed using an Android mobile app developed in the study. The data upload was performed online to the server which has a spatial DBMS installed.

The results show that the proposed intensity scale is suitable for producing iso-intensity maps rapidly because the iso-intensity maps are extremely important for disaster management efforts after a large earthquake. However, the assessment of the intensity of a quake at a given location used to be a slow process, as it was usually performed by means of personalized surveys [47]. For this reason, some researchers (i.e., [47–50]) developed some empirical approaches to assess intensity based on some earthquake parameters such as peak ground acceleration (PGA), peak ground velocity (PGV), peak ground displacement (PGD), the magnitude scale, and the epicentral distance, etc. However, the methodology presented here is different and based on volunteer observations. The methodology is quite reliable considering the facts that the citizen scientists were informed about the importance of their high-quality contributions. Due to the nature of the problem, it can be said that CitSci is the only reliable source of data (apart from field work by experts) to produce the iso-intensity maps since the intensity levels depend on the local ground conditions, construction date and quality, and the number of floors where the person was located at the time of the earthquake event. The differences between the intensity map published by AFAD and the citizen collected data confirms the conclusion. Yet, an estimate of differences between areas demonstrated by the intensity map is useful.

Since the citizen scientists are the key to the data quality, provision of the necessary training and technological tools (e.g., specially designed apps) would increase the reliability of such studies utilizing CitSci methods. In addition, supplementary procedures, machine learning and data analysis methods can support the data validation in large scale CitSci projects. Further spatial and logical analysis could also be employed as automatic quality control procedures, such as using a function of the measured earthquake intensity and the geographical location of the provided data as an indicator of the possible values felt by people together. When large amounts of contributions are provided by users, it is also expected to have a normal distribution of errors and, thus, the outliers can be eliminated by the ordinary kriging method and a smooth iso-intensity map can be obtained.

It can also be said that the number of such scientific studies may increase in the future. As a result, CitSci will make more promises in the future by incorporating human efforts with the technological advancements.

Supplementary Materials: The mobile app "Sarsintiyi Hissettim—I felt the quake" is available online at Google Play. Available: https://play.google.com/store/apps/details?id=com.ilyas.asus.postgresqlsample2&hl=en_US.

Author Contributions: Conceptualization and Validation: Candan Gokceoglu; Methodology, Ilyas Yalcin and Sultan Kocaman; Software and Data Curation, Ilyas Yalcin; Formal Analysis and Supervision, Sultan Kocaman; Writing-Original Draft Preparation, Ilyas Yalcin; Writing-Review & Editing, Sultan Kocaman and Candan Gokceoglu. All authors have read and agreed to the published version of the manuscript.

Funding: This research received no external funding.

Acknowledgments: The authors gratefully acknowledge AFAD for provision of the detailed earthquake report and the volunteers who provided their honest opinions about the earthquake event.

Conflicts of Interest: The authors declare no conflict of interest.

References

1. We Are Social. Available online: https://wearesocial.com/blog/2019/01/digital-2019-global-internet-use-accelerates (accessed on 25 November 2019).
2. Irwin, A. *Citizen Science: A Study of People, Expertise and Sustainable Development*; Routledge: Abington, UK, 2002; ISBN 9780203202395.
3. Kocaman, S.; Anbaroglu, B.; Gokceoglu, C.; Altan, O. A Review on Citizen Science (CitSci) Applications FOR Disaster Management. *Int. Arch. Photogramm. Remote Sens. Spat. Inf. Sci.* **2018**, *XLII-3/W4*, 301–306. [CrossRef]
4. Kocaman, S.; Gokceoglu, C. On the use of CitSci and VGI in Natural Hazard Assessment. In Proceedings of the International Archives of the Photogrammetry, Remote Sensing & Spatial Information Sciences, Volume XLII-5, ISPRS TC V Mid-term Symposium "Geospatial Technology—Pixel to People", Dehradun, India, 20–23 November 2018.
5. Arapostathis, S.G.; Lekkas, E.; Kalabokidis, K.; Xanthopoulos, G.; Drakatos, G.; Spirou, N.; Kalogeras, I. Developing Seismic Intensity Maps From Twitter Data; The Case Study Of Lesvos, Greece 2017 Earthquake: Assessments, Improvements and Enrichments on the Methodology. *Int. Arch. Photogramm. Remote Sens. Spat. Inf. Sci* **2018**, *XLII-3/W4*, 59–66. [CrossRef]
6. Wang, R.Q.; Mao, H.; Wang, Y.; Rae, C.; Shaw, W. Hyper-resolution monitoring of urban flooding with social media and crowdsourcing data. *Comput. Geosci.* **2018**, *111*, 139–147. [CrossRef]
7. AFAD. Marmara Denizi Silivri Açıkları (İstanbul) Mw 5.8 Depremine İlişkin Ön Değerlendirme Raporu; 2019. Technical Report T.C. İçişleri Bakanliği Afet ve Acil Durum Yönetimi Başkanlığı, September. Available online: https://deprem.afad.gov.tr/downloadDocument?id=1822 (accessed on 18 April 2020).
8. Goodchild, M.F. Citizens as sensors: The world of volunteered geography. *GeoJournal* **2007**, *69*, 211–221. [CrossRef]
9. Brovelli, M.A.; Minghini, M.; Zamboni, G. Public participation in GIS via mobile applications. *ISPRS J. Photogramm. Remote Sens.* **2016**, *114*, 306–315. [CrossRef]
10. ODK Collect. Available online: https://docs.opendatakit.org/collect-intro/ (accessed on 6 November 2019).
11. Enketo. Available online: https://enketo.org/ (accessed on 6 November 2019).
12. Boyd, D.S.; Jackson, B.; Wardlaw, J.; Foody, G.M.; Marsh, S.; Bales, K. Slavery from Space: Demonstrating the role for satellite remote sensing to inform evidence-based action related to UN SDG number 8. *ISPRS J. Photogramm. Remote Sens.* **2018**, *142*, 380–388. [CrossRef]
13. Zooniverse. Available online: https://www.zooniverse.org/ (accessed on 6 November 2019).
14. Koskinen, J.; Leinonen, U.; Vollrath, A.; Ortmann, A.; Lindquist, E.; d'Annunzio, R.; Pekkarinen, A.; Käyhkö, N. Participatory mapping of forest plantations with Open Foris and Google Earth Engine. *ISPRS J. Photogramm. Remote Sens.* **2019**, *148*, 63–74. [CrossRef]
15. Collect Earth. Available online: http://www.openforis.org/tools/collect-earth.html (accessed on 19 November 2019).
16. Open Foris. Available online: http://www.openforis.org/ (accessed on 19 November 2019).
17. Bey, A.; Sánchez-Paus Díaz, A.; Maniatis, D.; Marchi, G.; Mollicone, D.; Ricci, S.; Bastin, J.-F.; Moore, R.; Federici, S.; Rezende, M. Collect earth: Land use and land cover assessment through augmented visual interpretation. *Remote Sens.* **2016**, *8*, 807. [CrossRef]
18. Hicks, A.; Barclay, J.; Chilvers, J.; Armijos, M.T.; Oven, K.; Simmons, P.; Haklay, M. Global Mapping of Citizen Science Projects for Disaster Risk Reduction. *Front. Earth Sci.* **2019**, *7*. [CrossRef]

19. European Citizen Science Association (ECSA). Available online: https://ecsa.citizen-science.net/documents (accessed on 20 November 2019).
20. Citizensciencedrr (CSDRR). Available online: https://citizensciencedrr.com/project-map/ (accessed on 20 November 2019).
21. Liang, W.-T.; Lee, J.-C.; Chen, K.H.; Hsiao, N.-C. Citizen earthquake science in Taiwan: From science to hazard mitigation. *J. Disaster Res.* **2017**, *12*, 1174–1181. [CrossRef]
22. Taiwan Earthquake Science Information (TESIS). Available online: http://tesis.earth.sinica.edu.tw/new/ (accessed on 23 November 2019).
23. Did You Feel It (DYFI). Available online: https://earthquake.usgs.gov/data/dyfi/ (accessed on 23 November 2019).
24. Ushahidi. Available online: https://www.ushahidi.com/ (accessed on 23 November 2019).
25. Kong, Q.; Allen, R.M.; Schreier, L.; Kwon, Y.-W. MyShake: A smartphone seismic network for earthquake early warning and beyond. *Sci. Adv.* **2016**, *2*, e1501055. [CrossRef]
26. MyShake. Available online: https://myshake.berkeley.edu/ (accessed on 24 November 2019).
27. Kocaman, S.; Gokceoglu, C. A CitSci app for landslide data collection. *Landslides* **2019**, *16*, 611–615. [CrossRef]
28. Can, R.; Kocaman, S.; Gokceoglu, C. A convolutional neural network architecture for auto-detection of landslide photographs to assess citizen science and volunteered geographic information data quality. *ISPRS Int. J. Geo-Inf.* **2019**, *8*, 300. [CrossRef]
29. GeoCitSci. Available online: http://www.geocitsci.com/ (accessed on 24 November 2019).
30. eAfad. Available online: https://play.google.com/store/apps/details?id=com.basarsoft.afaddeprem&hl=tr (accessed on 9 February 2020).
31. AFAD-Earthquake Mobile Application. Available online: https://www.afad.gov.tr/deprem-mobil-uygulamasi (accessed on 9 February 2020).
32. Eravci, M.B.; Yanik, K.G.; Yenilmez, D.; ve Fahjan, Y. Akıllı Telefonlar Aracılığı ile Deprem Sonrası Şiddet Tahmini. In Proceedings of the 2. Türkiye Deprem Mühendisliği ve Sismoloji Konferansı, Antakya, Turkey, 25–27 September 2013.
33. Yalcin, I. Açık Kaynaklı Web Tabanlı Coğrafi Bilgi Sistemi Geliştirilmesi. Master's Thesis, Hacettepe University, Graduate School of Science and Engineering, Ankara, Turkey, May 2018; 112p. Available online: http://openaccess.hacettepe.edu.tr:8080/xmlui/bitstream/handle/11655/4893/10194429.pdf?sequence=1&isAllowed=y (accessed on 10 March 2020).
34. Yalcin, I.; Kocaman, S. Açık Kaynaklı Web Tabanlı Coğrafi Bilgi Sistemi Geliştirilmesi. In Proceedings of the VII. Uzaktan Algılama ve CBS Sempozyumu Uzal-CBS 2018, Eskisehir, Turkey, 18–21 September 2018.
35. Wood, H.O.; Neumann, F. Modified Mercalli intensity scale of 1931. *Bull. Seismol. Soc. Am.* **1931**, *21*, 277–283.
36. The Modified Mercalli Intensity Scale. Available online: https://www.usgs.gov/natural-hazards/earthquake-hazards/science/modified-mercalli-intensity-scale?qt-science_center_objects=0#qt-science_center_objects (accessed on 29 December 2019).
37. PostGIS. Available online: http://www.postgis.net/ (accessed on 24 November 2019).
38. Sarsıntıyı Hissettim. Available online: https://play.google.com/store/apps/details?id=com.ilyas.asus.postgresqlsample2&gl=TR (accessed on 24 November 2019).
39. Gura, T. Citizen science: Amateur experts. *Nature* **2013**, *496*, 259–261. [CrossRef]
40. Oyana, T.J.; Margai, F. *Spatial Analysis: Statistics, Visualization, and Computational Methods*; CRC Press: Boca Raton, FL, USA, 2015; p. 305. ISBN 978-1498707633.
41. Krige, D.G. A statistical approach to some basic mine valuation problems on the Witwatersrand. *J. S. Afr. Inst. Min. Metall.* **1951**, *52*, 119–139.
42. Matheron, G. Principles of geostatistics. *Econ. Geol.* **1963**, *58*, 1246–1266. [CrossRef]
43. Goovaerts, P. *Geostatistics for Natural Resources Evaluation*; Oxford University Press: New York, NY, USA, 1997; 416p, ISBN 0-9-511538-4.
44. O'Sullivan, D.; Unwin, D. *Geographic Information Analysis*; John Wiley & Sons: New Jersey, NY, USA, 2014; p. 406. ISBN 9780470288573.
45. Carley, K.M.; Malik, M.; Landwehr, P.M.; Pfeffer, J.; Kowalchuck, M. Crowd sourcing disaster management: The complex nature of Twitter usage in Padang Indonesia. *Saf. Sci.* **2016**, *90*, 48–61. [CrossRef]
46. Jongman, B.; Wagemaker, J.; Romero, B.R.; De Perez, E.C. Early flood detection for rapid humanitarian response: Harnessing near real-time satellite and Twitter signals. *ISPRS Int. J. Geo-Inf.* **2015**, *4*, 2246–2266. [CrossRef]

47. Alvarez, D.A.; Hurtado, J.E.; Bedoya-Ruíz, D.A. Prediction of modified Mercalli intensity from PGA, PGV, moment magnitude, and epicentral distance using several nonlinear statistical algorithms. *J. Seism.* **2012**, *16*, 489–511. [CrossRef]
48. Wald, D.J.; Quitoriano, V.; Heaton, T.H.; Kanamori, H. Relationships between peak ground acceleration, peak ground velocity, and modified Mercalli intensity in California. *Earthq. Spectra* **1999**, *15*, 557–564. [CrossRef]
49. Atkinson, G.M.; Kaka, S.I. Relationships between felt intensity and instrumental ground motion in the central United States and California. *Bull. Seism. Soc. Am.* **2007**, *97*, 497–510. [CrossRef]
50. Atkinson, G.M.; Sonley, E. Empirical relationships between modified Mercalli intensity and response spectra. *Bull. Seism. Soc. Am.* **2000**, *90*, 537–544. [CrossRef]

International Journal of
Geo-Information

MDPI

Article

Crowdsourcing without Data Bias: Building a Quality Assurance System for Air Pollution Symptom Mapping

Marta Samulowska [1,*], Szymon Chmielewski [2], Edwin Raczko [1], Michał Lupa [3], Dorota Myszkowska [4] and Bogdan Zagajewski [1]

1 Department of Geoinformatics, Cartography and Remote Sensing, Faculty of Geography and Regional Studies, University of Warsaw, Krakowskie Przedmieście 30, 00-927 Warsaw, Poland; edwin.raczko@uw.edu.pl (E.R.); bogdan@uw.edu.pl (B.Z.)

2 Department of Grassland and Landscape Studies, Faculty of Agrobioengineering, University of Life Sciences, 15 Akademicka St., 20-950 Lublin, Poland; szymon.chmielewski@up.lublin.pl

3 Department of Geoinformatics and Applied Computer Science, Faculty of Geology, Geophysics and Environmental Protection, AGH University of Science and Technology, Mickiewicza Av. 30, 30-059 Cracow, Poland; mlupa@agh.edu.pl

4 Department of Clinical and Environmental Allergology, Jagiellonian University Medical College, Botaniczna 3, 31-531 Kraków, Poland; dorota.myszkowska@uj.edu.pl

* Correspondence: m.samulowska@uw.edu.pl; Tel.: +48-225-520-654

Abstract: Crowdsourcing is one of the spatial data sources, but due to its unstructured form, the quality of noisy crowd judgments is a challenge. In this study, we address the problem of detecting and removing crowdsourced data bias as a prerequisite for better-quality open-data output. This study aims to find the most robust data quality assurance system (QAs). To achieve this goal, we design logic-based QAs variants and test them on the air quality crowdsourcing database. By extending the paradigm of urban air pollution monitoring from particulate matter concentration levels to air-quality-related health symptom load, the study also builds a new perspective for citizen science (CS) air quality monitoring. The method includes the geospatial web (GeoWeb) platform as well as a QAs based on conditional statements. A four-month crowdsourcing campaign resulted in 1823 outdoor reports, with a rejection rate of up to 28%, depending on the applied. The focus of this study was not on digital sensors' validation but on eliminating logically inconsistent surveys and technologically incorrect objects. As the QAs effectiveness may depend on the location and society structure, that opens up new cross-border opportunities for replication of the research in other geographical conditions.

Keywords: crowdsourced data quality; GeoWeb; citizen science; outdoor air pollution; symptom mapping

Citation: Samulowska, M.; Chmielewski, S.; Raczko, E.; Lupa, M.; Myszkowska, D.; Zagajewski, B. Crowdsourcing without Data Bias: Building a Quality Assurance System for Air Pollution Symptom Mapping. *ISPRS Int. J. Geo-Inf.* **2021**, *10*, 46. https://doi.org/10.3390/ijgi10020046

Received: 2 December 2020
Accepted: 20 January 2021
Published: 22 January 2021

Publisher's Note: MDPI stays neutral with regard to jurisdictional claims in published maps and institutional affiliations.

1. Introduction

Urban air pollution is well known to cause negative health impacts. Therefore, the monitoring of pollutant concentration levels plays a key role in understanding air quality and its effects on the subjective well-being (SWB) of citizens [1]. The SWB reflects the philosophical notion of a good life, as a proxy for assessing life satisfaction, momentary experiences, and stress. Kim-Prieto et al. [2] also took into account contemporary health hazards, among which air pollution is a key factor [3,4]. The development of smart and sustainable cities can only be accomplished through inclusive growth, using smart people, technologies, and policies [5]. From the perspective of smart people [6]—those who use smart devices to make their everyday living easier and health safer—we found it necessary to develop geospatial web (GeoWeb) solutions to measure the adverse impact of cities on their inhabitants. Ensuring measurement credibility becomes a key scientific challenge in this context. To this end, we carried out research on the example of air pollution health symptoms, an emerging trend particularly related to odor [7] and green pollutant [8] crowdsensing [9,10]. As crowdsensing (or, more generally, crowdsourcing) methods for health symptom mapping are subject to data bias [11], we developed and tested a quality assurance mechanism (QAm) framework (Section 2.2), which can be transferred to similar health-symptom-based studies.

Sparse or irregular monitoring station networks as well as limited access to the reference air pollution data underlie the need for citizen science (CS) activities in the field of air pollution monitoring. Personalized information about exposure to air pollutants, monitoring during acute events or at specific locations, partnerships with local governments, and educational and community-driven purposes are the key benefits of bottom-up environmental monitoring. CS enables the collection of data on much larger spatial and temporal scales and at much finer resolution than would otherwise be possible. The issue of urban air pollution crowdsourcing motivated the implementation of several CS programs, such as those led by Mapping for Change, a community interest company in London (e.g., Pepys Air Quality Project, Science in the City Project, Love Lambeth Air (https://mappingforchange.org.uk/projects/)), as well other international-scale activities found at claircity.eu and citi-sense.eu [12]. All the CS activities mentioned above were hosted on GeoWeb [13–15] (as discussed in Section 1.2). In this approach, citizens are required to act as sensors [12,16].

The CS HackAir project (https://www.hackair.eu) as well PollenApp were the projects that undertook the first attempts to crowdsource urban air pollution data, where air pollution was understood as the compound effect of anthropogenic- and biophysical-sourced particulate matter (PM) [8]. Pan-European CS projects, such as Distributed Network for Odor Sensing Empowerment and Sustainability (D-NOSES) [7], have shown that particular aspects of urban air pollution can be measured through the sense of smell of trained citizen scientists. Despite the subjective measurement nature of the human sense of smell, human-sensed CS (sCS) meets the high-quality method expectations (D-NOSES meets germ. Verein Deutscher Ingenieure, eng. the Association of German Engineers (VDI) 3940 standard). Ambient air pollution sCS, as well as the symptoms it causes, provides a promising source of spatial information. However, the unstructured nature of crowdsourced data [17–19] requires data quality assurance (QA) protocols [20,21], as well as trust and reputation modeling (TRM) [22] procedure development.

To the best of our knowledge, using QA for the purpose of air pollution symptom mapping (APSM) has not yet been investigated. Our goal was to address the challenge of crowdsourced APSM with the use of the GeoWeb platform (Section 2.3) and to solve the data bias problem by using a quality assurance system (QAs; Section 2.2) based on logic rules. By assessing the rejection rate of reports affected by data bias, we provide evidence of reliable air pollution monitoring expressed as the severity of human health symptoms caused by combined factors of anthropogenic and biophysical ambient air pollutants [23]. Extending the paradigm of air pollution referenced by WHO in terms of six main air pollutant [24] concentration levels and their public health impacts [25] to air pollution symptoms (APS), we indicate new possibilities for citizen-driven research and social inclusion in environmental and health-related issues, as experienced by sustainable cities. We also contribute to the development of the sCS data quality methodology.

In our study, we consider health symptoms caused by air pollution as one of the indicators of the current ecological footprint of humanity on the environment. Therefore, we focus on urban air pollution as a case study with the starting point of health symptoms caused by human exposure to air pollutants. This concept highlights the relationship between urban habitats and the SWB of citizens. The issue of air pollution requires spatial information provided thoroughly by a modern spatially variable society [26,27]. This underlines the need to implement a local partnership between monitoring agencies, researchers, and the local community, who all breathe the same air.

1.1. Extending the Paradigm of Urban Air Pollution

In the field of environmental research air quality, information about the quality (i.e., clean or polluted) of air is reported as an air quality index (AQI) [28]. The AQI tracks six major air pollutants: inhalable particulate matter (PM_{10}), fine particulate matter ($PM_{2.5}$), ozone (O_3), sulfur dioxide (SO_2), nitrogen dioxide (NO_2), and carbon monoxide (CO) [24]. The spectrum of pollutant sources includes those related to the development of human civilization (anthropogenic pollutants) [23], as well those from natural sources, which questions the belief that everything that is natural is healthy [29]. Ambient air pollution concentrations above the approved limits [30,31] can cause certain health symptoms. Conversely, health symptoms can reflect air pollution. However, health symptoms resulting from inhalation of polluted air are also stimulated by natural-sourced biophysical PM such as pollen, mold spores [23,32], and volcanic emissions [33], causing human health problems such as respiratory allergies, including allergic asthma, which is regarded as an important disease [23,34,35]. Bastl et al. [8,32] described pollen as one of the green pollutants, which are significant components of the atmosphere and are relevant to air quality information for pollen allergy sufferers. This distinction is important for the comprehensive understanding of APS. Air pollution is specified as the concentration of pollutants measured in physical values (e.g., micrograms per cubic meter), whereas air quality refers to the AQI, as well as to classifications, opinions, and feelings, including the experiences of citizens in terms of air- and air-quality-related SWB [1,4]. This concept extends our understanding of air pollution from pollutant concentration levels to personal health symptoms caused by pollutant inhalation. The quantity and severity of symptoms can explain the air quality; however, a consensus about the terminology involving urban air quality has not yet been reached, and researchers typically distinguish air pollution through pollen exposure [36]. There is no symptom classification for air quality yet. Regardless, both factors shape air quality. Future research is required to understand and quantify the interaction of co-exposure to both types of air pollutants and its impact on the severity of human health symptoms [37].

Symptom mapping is a prerequisite for the spatial explanation of both dependencies. The first attempts at citizen symptom mapping related to green pollutants were made by Bastl et al. [32] and Werchan et al. [38]. Their research proved that citizen symptom load can be mapped efficiently using crowdsourced data; however, the sources of the symptoms cannot be clearly determined. The symptom load index is not directly correlated with annual pollen load and has a strong correlation to allergen content [32], with a linear (often daily) correlation [32,39]. Finding that relationship is beyond the scope of this paper; however, crowdsourced symptom data have shown potential as an indicator of the effects of urban air pollution on citizen well-being. The unstructured nature of crowdsourced data requires rigorous QA mechanisms (QAm). In this study, our aim is to identify a QA system for GeoWeb-based APSM to stream high-quality geospatial data. So far, this data stream does not exist. By sharing trusted and open data on air pollution symptoms, our findings can be used for aerobiological and health-risk-forecasting research.

1.2. Contribution of Citizen Science to Improvements in Air Pollution Mapping

According to Haklay [14], geographical citizen science overlaps volunteered geographic information (VGI), especially in the geographical context of citizen-driven research. GeoWeb plays an essential role in this field. However, it is crucial that CS and VGI not be seen as equal, as the main purpose of VGI is to produce geographical information, while citizen science aims to produce new scientific knowledge [40,41]. Citizens engaged in scientific research projects become citizen scientists [42], who, depending upon their personal interests, motivation, education level, and experience in previous projects, engage with different levels of participation and expect to see the results of their research contribution. They contribute to the project by collecting and analyzing data but may also be involved in defining research questions or even interpreting results [14,43,44]. Considering the scope of citizen participation, Haklay [14] defined four levels of CS: crowdsourcing (first level), distributed intelligence (second level), participatory science (third level), and extreme citizen science (fourth level). Citizen involvement in environmental projects on air pollution is usually based on collecting and analyzing sensor data in the form of online maps. In this way, knowledge is produced. The fundamental questions about the harmful health effects of air pollutants have been asked, so these activities are typified as CS level 1 and CS level 2. Of course, higher levels (depending on the engagement of members) are not excluded. In the case of odor crowdsourcing, which requires training as well as expecting measurement insights back from members, a collection method can be devised (i.e., level 3). Our study was based on the first level of CS, where citizens are engaged in the process of crowdsourcing APS data, producing a new scientific knowledge of APSM, together with researchers. CS provides a solution to research problems [44], while also educating citizens [45].

So far, smartphones have not been considered appropriate equipment for measuring urban air pollution. This is due to the fact that the built-in sensors of smartphones, by default, do not allow users to measure air pollutant concentrations. Therefore, bottom-up activities considering air pollution have usually relied on external, low-cost sensors (initially only capable of PM measurement, these sensors can now also sense all major pollutants, including volatile organic compounds). Loreto et al. [46] emphasized that modern participatory sensing, which is one of three sub-categories of citizen cyberscience [47], has witnessed significant progress related to the fast development and social networking tools of information and communication technologies (ICT), which "allow effective data and opinion collection and real-time information sharing processes". In that context, Guo et al. [48] and Capponi et al. [49] introduced mobile crowdsensing (MCS), which focuses on sensing and collecting data with mobile devices and aggregating data in the cloud. However, there are pollutants that are still exclusive for Internet of Things sensor dust. A great challenge of contemporary CS measurement is odor sensing, which affects both indoor as well as outdoor air quality. Human-sensed air pollution monitoring seems to be an emerging trend.

In this research, we specify the citizens as sensors and participatory sensing concepts, where the senses, subjective impressions, and perception of humans are the only sensors used in the project; therefore, we propose this as sCS. Moreover, by developing a QA mechanism for sCS, this study contributes to bottom-up air pollution monitoring and open-data credibility.

1.3. Importance of Data Quality in Crowdsourced Air Pollution

Data quality issues include errors and biases. Factors affecting the data collected through citizen perceptions result in data biases. CS requires the collaborative contributions of multiple contributors [50], but the assumption of multiple contributors is insufficient to provide high-quality data. Therefore, data quality protocols are an essential part of crowdsourcing-driven research. Although participatory research faces methodological challenges such as biases in data collection [51,52], CS has been proven to be a source of trusted geospatial data [20,53–56], including for health risk mapping [57–59] and risks caused by poor air quality [32,60,61]. The data quality determines its usefulness [62,63]. Thus, the unstructured nature of crowdsourced data requires rigorous data QA protocols [17,64–67].

The starting point for QA in CS is education and the provisioning of technical information and resources [21,68] in order to increase citizen knowledge about the issues of air pollution and to improve their environmental awareness and motivation to provide air-quality-monitoring supporting activities [60,69,70]. Of the range of crowdsourced data quality measures discussed in the academic literature by Haklay [50] and Foody et al. [67], among others, attribute accuracy and completeness are essential. Furthermore, these aspects of geographic data quality have also been recognized by international standards of spatial data quality. The ISO 19157 [71], which handles the diverse perspective of data quality, defines a set of standardized data quality measures, including completeness of data, positional accuracy, and temporal accuracy, which are all grouped as so-called data quality elements (DQEs) [72]. Each DQE is, then, further evaluated, and the result of the evaluation is documented and reported [67]. The principles of the aforementioned ISO 19157 [71] served as the basis for the proposed APSM data quality assurance framework.

The goal of this study was to answer the question of data quality assurance mechanism implementation in GeoWeb-based APSM. For this purpose, we propose a dedicated APSM framework for the QA mechanisms of start-check, sequence, cross-validation, repeating, and time-loop check in order to reduce data bias, such as user response inconsistency, location inaccuracy, and duplicate time–space-related reports (see Section 2). By combining several logic-based QA mechanisms (QAm), we tested the robustness of the QAm to find the strongest QAm set and build a ranked data quality assurance system (QAs). Our research question was, Which QAs best reduces data bias in APSM? The sources of data bias include contradictory entries in the geodatabase attribute table recorded as answers supplied to the specially created APSM survey. However, we did not solve the problem caused by the non-air-pollution-related factors that affect human symptom severity, which act synergistically with air pollution to contribute to spatial database robustness on health-related symptoms [8,73–76].

2. Materials and Methods

For our participatory APSM project, we followed the CS development framework of Bonney et al. [77], starting from the research question and project team formulation through to CS action execution and the dissemination of project findings (Sections 2.1 and 2.3). However, we focused on addressing data bias (Section 2.2) to improve the symptom-based air-pollution-mapping data quality.

2.1. Building a Field Data Collection Strategy

The method was adopted in the city of Lublin and Lubelskie Voivodship, located in eastern Poland. The data collection campaign lasted for one academic semester, from February to May 2018. Starting the campaign in the first quarter of the year is crucial, as anthropogenic pollutants and pollen occur simultaneously at the beginning of the year, especially gaseous pollutants that can act as adjuvants, exacerbating pollen allergenic potency and immunoreactivity [78]. A group of 56 students from different faculties of the University of Life Sciences in Lublin were involved in the project as volunteers and, according to Harlin et al. [79], turned into citizen scientists, of which, 30 spatial management students were the core citizen group of the project. A total of 18 non-academic citizens joined the research and collected data together with the student group. The project was continually open to everybody through the community channel for data and app sharing, which was implemented in GeoWeb. By sharing their conclusions and opinions during the social campaign, the citizens had a direct impact on the optimization of methods used.

To strengthen the data quality potential already before the field data collection campaign began, the scientific student organization of the Spatial Management Faculty of the University of Life Sciences in Lublin organized workshops for the citizen scientists, who were learned about the subject of study and the research project assumptions. Furthermore, they were trained on handling the mobile and web apps (details about apps provided in Section 2.3). The workshops, training, informing, and research project promotion among citizens lasted for the first month.

According to the study requirements, the citizen scientists were asked to collect data only during their daily outdoor activities, preferably once per day. If they observed air-pollution-related symptoms, they were asked to report them as soon as possible. If they caught a cold or were sick, they were expected to stop collecting data until they recovered. The project assumptions and rules for collecting data were included in the mobile app, as a user guide introducing the study. Other educational materials were made available in the narrative web apps.

Our study referred to the patterns of citizen activity characteristic for CS specifics [80]. As such, we implemented a method for citizen engagement improvement, which was based on the monthly activity ranking of users, presenting a number of submitted APS reports. Users were assigned award titles according to the following number of reports sent per month: 1–5, Beginner; 6–10, Pretty Involved; 11–20, Super-Engaged; and >20, Excellent Citizen Scientist. A user-activity-tracking module, included in the operations dashboard app (details in Section 2.3), was public and allowed for competition between the citizens.

2.2. Data Quality Assurance Methods for APSM

Following other studies of human health symptom severity, such as the Symptom Load Index research in pollen allergy sufferers [81], we used a questionnaire sheet. However, we extended the group of potential citizens to all those who suffered from air-quality-related health symptoms, whereas to improve the method of data quality assessment, the user-screening question was implemented in the questionnaire, which helped identify the more sensitive citizens to air pollution. The question asked for any pollen allergens the user is allergic to (birch, alder, hazel, grass, pine, plantain, mugwort, nettle, ragweed, others; it is also possible to answer no allergens; Figure 1a). The survey included close-ended questions about health symptoms related to air quality according to the classification defined in the research on the epidemiology of allergic diseases in Poland (PL: Epidemiologia Chorob Alergicznych w Polsce (ECAP)) [82], accompanying the SWB question, which developed the QA method; a personal profile with the user-screening question; and finally time range and geolocation information. The questions are as follows:

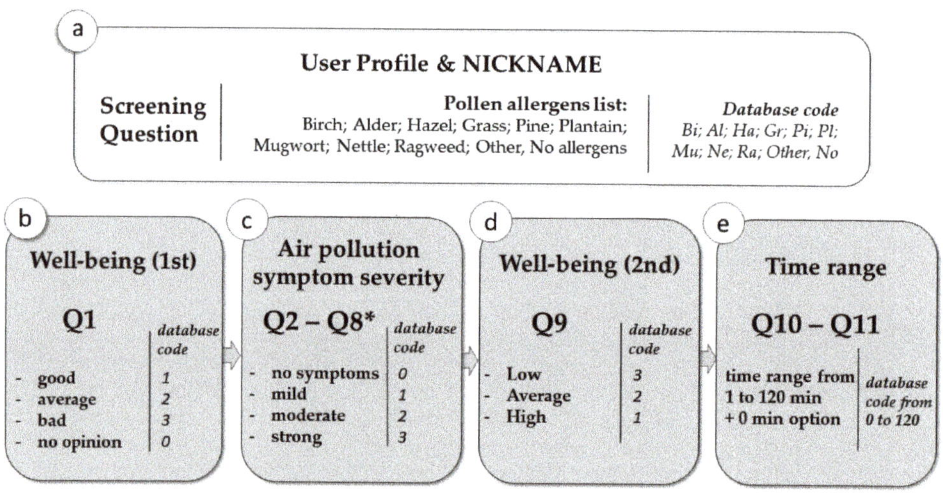

Figure 1. The flowchart of the air pollution symptom mapping (APSM) questionnaire: (**a**) the user-screening question about pollen allergy as well as an individual nickname and survey four-digit code assigned to the users' profile, (**b**) the introducing question about subjective well-being (asked at the very beginning of the survey), (**c**) the set of seven close-ended questions about individual symptoms (question no 8 was answered, "Seldom," "Quite often," "Very often," or "I do not rub eyes"), (**d**) closing questions about subjective well-being, and (**e**) time and geographical location recording.

User-screening question: Which pollen allergens are you allergic to?

Q1. How do you feel today?

Q2. Sneezing. If you are currently experiencing this symptom, please choose the level of severity.

Q3. Nose itching. If you are currently experiencing this symptom, please choose the level of severity.

Q4. Runny nose. If you are currently experiencing this symptom, please choose the level of severity.

Q5. Watering eyes. If you are currently experiencing this symptom, please choose the level of severity.

Q6. Scratchy throat. If you are currently experiencing this symptom, please choose the level of severity.
Q7. Breathing problems. If you are currently experiencing this symptom, please choose the level of severity.
Q8. Do you rub your eyes?
Q9. Could you assess the level of your current self-comfort?
Q10. How long have you been in this location?
Q11. For how long have you felt your symptoms?

The questionnaire design was inspired by sociological research. We followed the rule that filling out a web form shouldn't take more than 10 min [83]. The designed web questionnaire was tested in web form so that it did not exceed the assumed 10 min. Following Malchotra [84], the questionnaire contained mainly dichotomous and multiple-choice questions. The APS questions were grouped together [85]. The data stored in the database were displayed as a text data type in the mobile app, which is simple and intuitive for the user. The data were also coded in the database in the short integer data type (except for the user-screening question, which was coded in text data type) and were used for data analysis and statistics, as shown in the flowchart of the APSM questionnaire (Figure 1). The proposed method includes data forms, which are the basis of the developed conditional statements.

2.2.1. QA Methods Applied during the Data Collection Process

We initially adopted three QA methods in the mobile survey app in order to improve data quality during the data collection process. These methods were based on the quality measures of ISO 19157 [71]: positional and temporal accuracy, data completeness, and consistency. The first QA method eliminated identical reports sent from the same location within a certain time interval (5 min) from the database in case the same report was duplicated. The report duplication was confirmed with the same nickname of the citizen who made the report. Then reports lacking geolocation were excluded from the database by the second QA method. The third method controlled the Global Navigation Satellite System (GNSS) positioning accuracy of the reported APS observations, under the assumption that reports with a horizontal accuracy error greater than 100 m are outliers, which were eliminated in the data collection stage. If the surveys were accepted under the three QA methods described above, they were finally checked with the completeness quality measure. Surveys that were not completed, in terms of obligatory questions, were automatically blocked against submission through the mechanism configured in the app so that the database was free from incompleteness.

2.2.2. Logic-Based QA Mechanisms Implemented after the Data Collection Process

The proposed logic-based QA framework for APSM has a four-tier structure, starting with a set of logically related questions, the basis of the quality assurance (QA) mechanisms, which in various configurations form five mechanisms and then QA systems (Figure 2).

The logic formula for each conditional statement was built to filter and eliminate data bias. The conditional statements were the primary components for QA mechanisms. To determine the potential logic rules, the following general criteria based on the database-coded answers from the questionnaire were assumed: Q2–Q7, which are the questions reporting citizen health symptoms, should have been implemented in minimum six conditional statements, while Q1 and Q9 (SWB questions) should have been implemented in minimum two conditional statements each. Then the conditional statements identified data inconsistencies, based on false hypothesis, such that false results were returned (Table 1). The QA mechanisms of start-check, sequence, cross-validation, repeating, and time-loop check were proposed, with one or two levels of robustness (Table 2). These QA mechanisms were implemented in the database after the data collection process was finished.

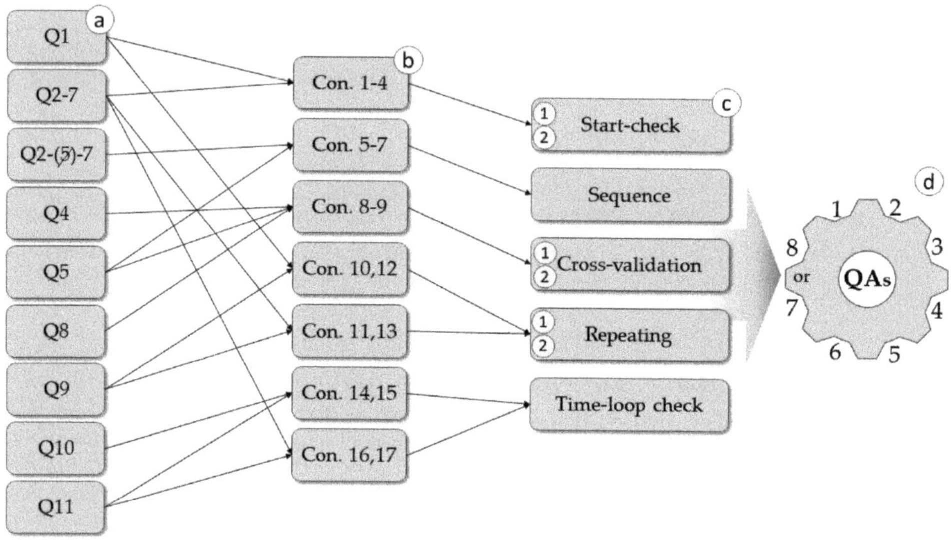

Figure 2. The four-tier structure of the data quality assurance (QA) framework: (**a**) set of logically related questions with number-coded answers, (**b**) conditional statements being the primary logical-based formula, (**c**) combination of several conditional statements creating a QA mechanism tier, and (**d**) combining mechanisms creating an advanced QA system.

Table 1. Conditional (Con.) statements, the basis of the quality assurance (QA) mechanisms.

Conditional Statement Number	Logic Formula
Con.1	$if[(Q1 = 0)\ or\ ((Q1 = 1)\ and\ (any(Q2:Q7) = 3))]\ then\ false,$
Con.2	$if[(Q1 = 0)\ or\ ((Q1 = 3)\ and\ (all(Q2:Q7) \neq 3))]\ then\ false,$
Con.3	$if[(Q1 = 0)\ or\ ((Q1 = 1)\ and\ (any(Q2:Q7) > 1))]\ then\ false,$
Con.4	$if[(Q1 = 0)\ or\ ((Q1 = 3)\ and\ (all(Q2:Q7) \neq 3))]\ then\ false,$
Con.5	$if[all((Q2:Q4)\ or\ (Q6:Q7)) = 1)\ and\ (Q5 = 3)]\ then\ false,$
Con.6	$if[all((Q2:Q4)\ or\ (Q6:Q7)) = 3)\ and\ (Q5 = 1)]\ then\ false,$
Con.7	$if[all((Q2:Q4)\ or\ (Q6:Q7)) = 0)\ and\ (Q5 = 2)]\ then\ false,$
Con.8	$if[((Q5 = 0)\ and\ (Q8 = 3))\ or\ ((Q4 = 0)\ and\ (Q8 = 3))]\ then\ false,$
Con.9	$if[((Q5 = 0)\ and\ (Q8 \neq 0))\ or\ ((Q4 = 0)\ and\ (Q8 \neq 0))]\ then\ false,$
Con.10	$if[((Q9 = 1)\ and\ (Q1 = 3))\ or\ ((Q9 = 3)\ and\ (Q1 = 1))]\ then\ false,$
Con.11	$if[((Q9 = 1)\ and\ (any(Q2:Q7) = 3))\ or\ ((Q9 = 3)\ and\ (all(Q2:Q7) \neq 3))]\ then\ false,$
Con.12	$if[((Q9 = 1)\ and\ (Q1 \neq 1))or\ ((Q9 = 2)\ and\ (Q1 \neq 2))\ or\ ((Q9 = 3)\ and\ (Q1 \neq 3))]\ then\ false,$
Con.13	$if[((Q9 = 1)\ and\ ((any(Q2:Q7) = 3)\ or\ (any(Q2:Q7) = 2)))or$ $or\ ((Q9 = 3)\ and\ (all(Q2:Q7) \neq 3))]\ then\ false,$
Con.14	$if(Q11 \geq Q10)\ then\ false,$
Con.15	$if[(Q11 = 1)\ or\ (Q11 > 120)]\ then\ false,$
Con.16	$if[(all(Q2:Q7) = 0)\ and\ (Q11 \neq 0)]\ then\ false,$
Con.17	$if[(all(Q2:Q7) \neq 0)\ and\ (Q11 = 0)]\ then\ false.$

To date, such a method has not been implemented in an sCS air-pollution-related symptom-mapping project. As a data quality assurance framework for APSM, we propose a data quality assurance system (QAs) that is a combination of each QA mechanism, depending on the QA mechanism robustness variant (Table 3). The choice of the QA system variant depends on the project character: each QA mechanism works independently and, so, can be implemented in the APSM project separately or in any combination, if needed.

Table 2. QA mechanisms implemented in the citizen air-pollution-related symptoms questionnaire, with less (1) and more (2) robust variants.

QA Mechanism	QA Mechanism Code	QA Mechanism Components: Conditional Statement Combination
Start-check (SC)	SC1 SC2	Con.1 or Con.2 Con.3 or Con.4
Sequence (Sq)	Sq	Con.5 or Con.6 or Con.7
Cross-validation (CV)	CV1 CV2	Con.8 Con.9
Repeating (Rp)	Rp1 Rp2	Con.10 or Con.11 Con.12 or Con.13
Time-loop check (TC)	TC	Con.14 or Con.15 or Con.16 or Con.17

Table 3. QA system variants applied for air pollution symptom mapping (APSM).

QA System (QAs) Variant	QA System Components: QA Mechanism Combination
QAs_1	SC1 or Sq or CV1 or Rp1 or TC
QAs_2	SC1 or Sq or CV1 or Rp2 or TC
QAs_3	SC1 or Sq or CV2 or Rp1 or TC
QAs_4	SC2 or Sq or CV1 or Rp1 or TC
QAs_5	SC2 or Sq or CV1 or Rp2 or TC
QAs_6	SC2 or Sq or CV2 or Rp1 or TC
QAs_7	SC1 or Sq or CV2 or Rp2 or TC
QAs_8	SC2 or Sq or CV2 or Rp2 or TC

The core principle for the research is the logic-based data quality assurance procedure. The start-check, cross-validation, and time-loop check mechanisms are based primarily on the medical assumptions included in the logic rule framework. The other QA mechanisms, sequence and repeating, are purely logical and result from social and psychological survey methods for monitoring, the typical respondent-answering process, and data quality control [86].

The start-check mechanism was used to verify the report consistency at the beginning of the survey, excluding reports whose symptom severity answers were not consistent with the general well-being question. The quality assurance method of applying a general question about the issue preceding the detailed questions was used by Bastl et al. [8] and Bousquet et al. [87]; however, these studies did not report success in using this conditional statement. We examined this in two variants of robustness. Components and algorithms of the start-check mechanism are presented in Tables 1 and 2, which were implemented in the database work as follows: Variant 1 is less robust and assumes that the report is consistent when the citizens assess their current comfort as "good" (1), and then the answers to Q2–Q7 should be between "no symptoms" (0) and "moderate symptom severity" (2). If the answer for Q1 is poor self-comfort (3), then at least one question between Q2–Q7 must be answered as "strong symptom severity" (3). Variant 2 is stricter and assumes that the possible answers for Q2–Q7 can only be "no symptoms" (0) or "mild symptom severity" (1) when the current comfort is rated as "good" (1). If the answer for Q1 was "poor comfort" (3), the same conditional statement was used as in variant 1. For both variants, the reports with the answer "I have no opinion" (0) for Q1 were excluded.

The sequence mechanism was applied to exclude user automatism in providing answers, which is often caused by a citizen giving rash answers or not reading the questions. Therefore, each report with all questions answered by responses with the same place in the sequence (e.g., every first answer for each question) were eliminated from the database. The QA mechanism is based on the method of rearranging the order of possible answers [88,89] for one question in the sequence of similarly asked questions. The standard order of the answers for questions Q2–Q7 was 1, 2, 3, 0. Q5 was an exception, with an answer order of 3, 0, 1, 2.

The cross-validation mechanism was used to reject responses by using three essentially related questions. If the answer to the additional question was not consistent with one of the two previously answered related questions, the report was excluded. The APS mentioned in Q8 (rubbing eyes) should be related to symptoms of a runny nose (Q4) or watering eyes (Q5). The mechanism was tested in two variants, where variant 2 is stricter.

The repeating mechanism determines the consistency of the report, according to the other previously answered questions, by asking the same question but in a different way. If the repeated answer is not consistent with the former one [21,88,90,91], the report is excluded from the database. This was used with Q9, which repeated the content about the citizen's self-comfort in Q1. Additionally, the mechanism tested the report's compatibility based on the repeated SWB question and symptom severity ones (Q2–Q7). The mechanism was examined in two variants of the robustness level analogous to the previous mechanisms.

The time-loop check mechanism was used to eliminate reports that did not align to the geolocation of the citizens, according to the length of their stay in a place in comparison to the duration of their symptoms. Previous studies [92–94] have shown that human allergic reactions to pollen range from 10 to 20 min, usually resolving after 1–2 h for an early-phase reaction and 3–4 h with resolution after 6–12 h, sometimes even 24 h, for a late-phase reaction. The late-phase reaction is preceded by an early-phase reaction. The early-phase reaction is characterized by symptoms such as allergic rhinitis (including sneezing, itching, and rhinorrhea) [34,95], while the late phase is connected with nasal congestion and obstruction [34,96]. For anthropogenic-sourced air pollutants PM_{10} and $PM_{2.5}$, the reaction time (according to the pollutant type) ranges between 2 and 10 min for the most sensitive subjects and resolves after 30 min [97,98]. For the purpose of APSM, we adopted the time range for the duration of the human early-phase reaction to air pollutants as between 1 min and 2 h. This mechanism assumes that all reports with a time loop value higher than the duration the user remained in the place are eliminated, as such information indicates a late-phase reaction, which is not connected with the actual geolocation of the citizen.

2.3. GeoWeb Method Supporting Data Quality Assurance

The GeoWeb platform, which was the technological basis of this project, to maintain project openness, consists of a mobile app for field data collection based on the configurable Survey123 for the ArcGIS application (Survey123; Esri Inc., Redlands, CA, USA) and a data mapping module, including dashboards (web mapping applications based on ArcGIS Online (AGOL) components; see Figure 3), available to citizens for free on a smartphone but also via a web browser.

The mobile app (A) includes the APS survey for field data collection. The educational and technical training materials (B) are provided in a narrative web mapping application (Esri story map templates), which are served through Representational State Transfer (REST) services. The collected data are sent to the geodatabase through REST services, and raw data are presented in the dashboard app (C) in real time. Then, the raw data are QA-checked in the database using ArcGIS Desktop integrated with the REST and R statistical software (R: a language and environment for statistical computing; R Foundation for Statistical Computing, Vienna, Austria; www.R-project.org) (D), which is used for analysis of the QA mechanisms and QA system variant robustness, as well as the survey result statistics. Finally, the QA-checked data are returned to the database and presented as APSM results in the open web mapping applications, including dashboards (C). The results were reported in the dashboard app as a percentage of the observations eliminated by the most robust QA mechanism combination, presenting the total robustness of the implemented QA system.

The questions proposed for the survey in Section 2.2 were included in the mobile survey, which was divided into individual pages, in order to help the user to quickly navigate between questions without scrolling down the whole form [99,100]. This helped to avoid mistakes in filling out similar APS questions and focus on the relevant section of the survey. An offline mode in the mobile app was secured to collect actual geolocation, even if without the internet. We used a user authorization option with an alphanumeric nickname and a four-digit code at the beginning of the survey in order to help follow the reports of each citizen, while also providing them with anonymity. To improve the quality of collected data, we implemented closed questions with single- or multiple-choice type and data range functionality, which prevented submitting certain reports that were inconsistent with the rules of the project.

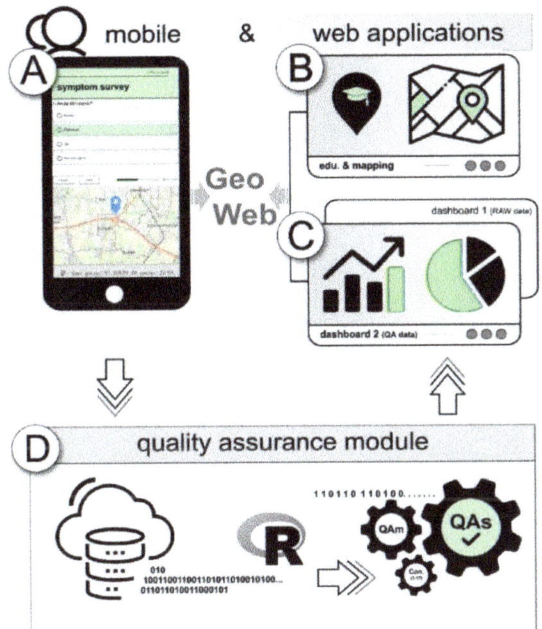

Figure 3. Geospatial web architecture implemented for the APSM project: (**A**) mobile survey app for field data collection, (**B**) web app with educational and training materials, (**C**) dashboard app for monitoring the data collection process and presenting APSM results, and (**D**) quality assurance module.

Within the field data collection campaign, the reports were mapped, in real time, in a point-symbolized data layer (Figure A4, Appendix A). Cooperation between the Survey123 mobile app and the web mapping module was based on the typical WebGIS architecture [101]. The dashboard displayed the user activity-ranking widget, which was based on a list of 10 most active citizens, presented as their nicknames together with their award titles, as well as their number of reports in the last month (Section 2.1). This functionality helped to enhance the engagement of the citizens during the data collection process.

The design concept and interface description of GeoWeb applications, along with the crowdsourced symptom map formed from the Lublin case study, are provided in Appendix A.

3. Results

Data quality assurance is key for useful and effective sCS, and so, the results mostly focus on the implemented QA mechanisms and the robustness analysis of the QA system variants for air-pollution-related symptom mapping. We first focused on the field data collection campaign, which provided us with the input for our analysis. The citizen activity curves indicated the varying and regularly decreasing activity of the citizens. For QA robustness results, our key focus was the eight QA system variants, the combinations of five logic-based QA mechanisms, which rejected from 18.3% to 28.6% of reports with data bias. As a result, we created and proposed a QA framework for APSM, based on the ArcGIS platform, which was configured and customized to the study requirements. The air-pollution-related symptom map and GeoWeb apps are presented in Appendix A.

3.1. Data Collection Campaign Outcomes

During the data collection campaign, 1936 APS reports were sent by citizens to the cloud, which were used as the input to the QA-mechanism-developed database. The citizen scientists involved in the APSM project became members of the Citizen Science Community of University of Warsaw (CS-Community-UW; Warsaw, Poland) (https://arcg.is/WeqfK), which was set up to share all project data, materials, and apps in GeoWeb, available at https://arcg.is/1iDD18, in order to provide continuous access to the results. When citizens collected data, their activity was controlled and updated every month in a user activity module implemented in the dashboard app (Section 3.3). Analysis of the activity curve indicated that citizen activity peaked for 11 days at the beginning of the campaign (39–48 reports per day) and then decreased. After this, activity stabilized at between 4 and 29 reports per day, with two peaks of 32 and 34 reports per day (Figure 4). The students were more active than non-academic citizens. The most active student collected 25 reports in April, whereas the most active non-academic citizen collected 7 reports in March.

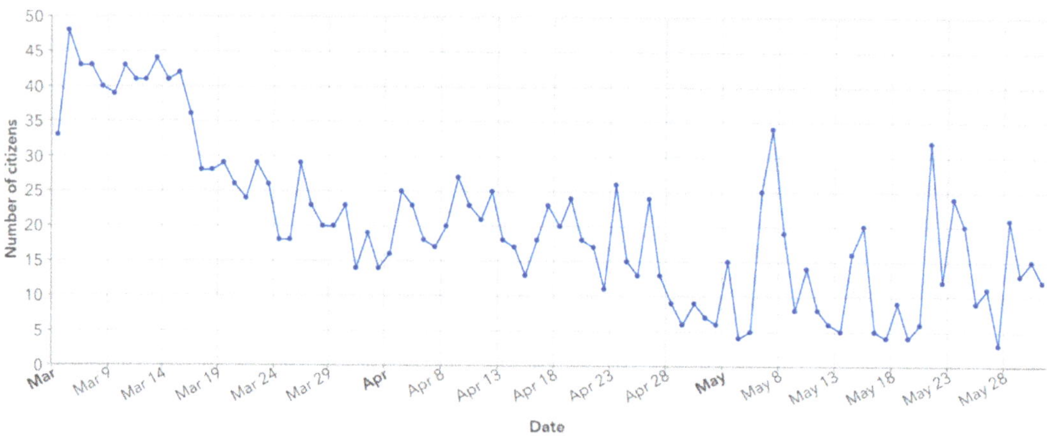

Figure 4. Citizen activity curves during the field data collection campaign.

3.2. Robustness of QA Mechanisms

During the initial stage of data set filtering, 19 outliers with no geolocation were eliminated. Another 68 APS records were excluded as they were duplicate reports at the same geolocation within a short (5 min) interval. The horizontal positioning accuracy varied from 0 to 100 m, where 93% of reports had a horizontal accuracy between 0 and 30 m. We filtered 26 outliers that had a horizontal accuracy error exceeding 100 m, which were rejected. As a result, 1823 reports of the 1936 remained and were used as the object of our logic-based QA implementation study. After the implementation of the QA mechanisms in the database, their robustness was analyzed.

According to Table 4, the three most robust QA mechanisms for our specific case study were: repeating in two variants (more robust, repeating2 (Rp2); less robust, repeating1 (Rp1)) and start-check in the more robust variant (SC2). The repeating2 mechanism excluded 23.1% of reports, repeating1 eliminated 10.6% of reports, and start-check2 eliminated 6.9% of reports. Thus, most data bias resulting from inconsistencies between the first and the repeated question was identified by the repeating mechanism.

Three QA mechanisms—start-check, cross-validation, and repeating—which were implemented in two variants (i.e., less robust and more robust) were analyzed in terms of the report reduction in each variant. For the start-check mechanism, the two variants reduced 71 of the same reports, while start-check2 eliminated 54 more reports than start-check1. Cross-validation1 and cross-validation2 reduced the same 54 reports, while cross-validation2 additionally eliminated 54 observations. For the repeating mechanism, repeating1 and repeating2 were compatible for 194 reports, while repeating2 was 12.5% more robust than repeating1, excluding 228 additional APS reports (Table 5).

Table 4. QA mechanism robustness.

QA Mechanism		Reports				QA Mechanism Robustness Rank
Name	**Code**	**Accepted**	**Rejected**	**Total**	**Rejected (% of Total)**	
Start-check	SC1	1752	71	1823	3.9	5
	SC2	1698	125	1823	6.9	3
Sequence	Sq	1806	17	1823	0.9	8
Cross-validation	CV1	1769	54	1823	3.0	6
	CV2	1716	107	1823	5.9	4
Repeating	Rp1	1629	194	1823	10.6	2
	Rp2	1401	422	1823	23.1	1
Time-loop check	TC	1771	52	1823	2.9	7

Table 5. QA mechanism variant compatibility.

		Start-Check2		Cross-Validation2		Repeating2	
		Accepted	**Rejected**	**Accepted**	**Rejected**	**Accepted**	**Rejected**
Start-check1	Accepted	1698	54				
	Rejected	0	71				
Cross-validation1	Accepted			1716	53		
	Rejected			0	54		
Repeating1	Accepted					1401	228
	Rejected					0	194

The largest data bias was related to the inconsistency between the general SWB question and its repeated query (repeating1: 10.6% and repeating2: 23.1%). This result could have been produced by the inaccuracy of the repeated-question structure or by citizens misunderstanding the question. The start-check2 mechanism rejected 6.9% of reports, which means that the severity symptoms did not align with the general SWB assessment. The sequence mechanism was the least robust (0.9%), indicating that citizens rarely filled in the form automatically, that is, without carefully reading the answers. A high rejection rate was observed using the QA mechanisms start-check2 (6.9%), cross-validation2 (5.9%), and repeating2 (23.1%), which were between 46% and 57% more robust than their alternative variants (start-check1, cross-validation1, and repeating1, respectively). They rejected a higher percentage of reports. Due to its high rejection rate, start-check2 was found to also reject some consistent reports.

Finally, according to the methodology (Section 2), the QA mechanisms were combined into eight QA system variants ($QAs_{1\text{–}8}$; Table 6). Implementing each subsequent QA mechanism changed the database, considering the QA system functions. For this study, the most robust QA systems were QAs_8 and QAs_7 (28.6%) and QAs_2 and QAs_5 (27.3%), which best reduced the number of falsely filled in reports in the survey. They increased the quality of the collected data but rejected a percentage of reports that might have contained valid information. As a result, some valuable data would be lost. The least robust were QAs_1 (18.3%) and QAs_3 (20.0%), which could not identify all the reports with false data, thus decreasing the quality and validity of the research results. As mentioned above, two pairs of QAs variants reduced the same data set and replicated the result database (QAs_2 with QAs_5; QAs_7 with QAs_8), thus allowing their interchangeable use. To reduce replication in the results, the studied QA framework was limited to six QAs variants (with QAs_5 and QAs_7 removed). For another location (i.e., country, continent) and society structure, the robustness ranking of the six QAs variants could be different and the particular QA mechanisms could be more or less effective than in the considered case study in Lublin.

Table 6. QA system variant robustness.

QA System Variant	QA Mechanisms Combination	Reports				QA System Robustness Rank
		Accepted	Rejected	Total	Rejected (% of total)	
QAs_1	SC1 or Sq or CV1 or Rp1 or TC	1490	333	1823	18.3	6
QAs_2	SC1 or Sq or CV1 or Rp2 or TC	1325	498	1823	27.3	2
QAs_3	SC1 or Sq or CV2 or Rp1 or TC	1459	364	1823	20.0	5
QAs_4	SC2 or Sq or CV1 or Rp1 or TC	1445	378	1823	20.7	4
QAs_5	SC2 or Sq or CV1 or Rp2 or TC	1325	498	1823	27.3	2
QAs_6	SC2 or Sq or CV2 or Rp1 or TC	1422	401	1823	22.0	3
QAs_7	SC1 or Sq or CV2 or Rp2 or TC	1302	521	1823	28.6	1
QAs_8	SC2 or Sq or CV2 or Rp2 or TC	1302	521	1823	28.6	1

3.3. APSM Results after QA System Implementation

APSM results calculated on the QA_{s8}-checked data set were added to the GeoWeb dashboard app (Figure 5) such that each user could track the APSM data of the whole crowdsourcing campaign, presenting the impact of air pollution on his/her SWB status. From the QA_{s8}-checked database, the most frequently observed health symptom was a runny nose, which was reported in 46.98% surveys during the whole campaign. The least frequently reported symptoms were breathing problems (only 6.92% surveys). The QA_{s8} had an impact on the percentage of surveys reporting any APS (Table 7). Most surveys reporting breathing problems were rejected, reducing their percentage in the database by 47.34%. The percentage of surveys reporting a runny nose remained almost unchanged, as it increased by 0.86%. The percentage of surveys reporting other symptoms reduced by 14.88% to 35.42%.

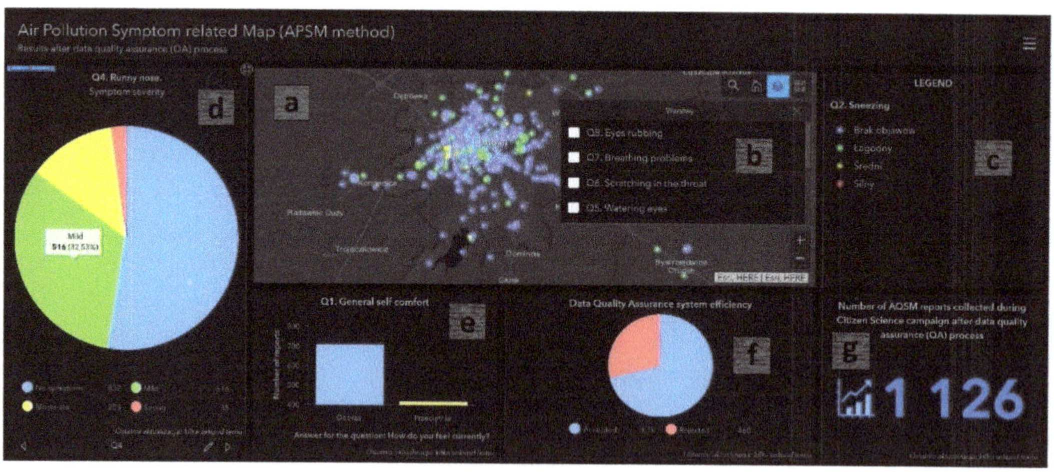

Figure 5. Web mapping dashboard app presenting results of QA-checked data using QAs_8: **(a)** map, **(b)** layer list, **(c)** legend, **(d)** pie chart of citizen symptom severity, **(e)** bar chart of citizen general comfort, **(f)** pie chart of QA system robustness, and **(g)** indicator of QA-checked reports.

Table 7. QA_{s8} impact on the percentage of surveys with reported air-pollution-related symptoms.

	% of Surveys with Reported Symptoms		QA_{s8} Impact on the Results
	Raw Data	QA_{s8}-Checked Data	
Sneezing	29.98	25.52	−0.149
Nose itching	30.27	21.98	−0.274
Runny nose	46.58	46.98	0.009
Watering eyes	26.77	21.52	−0.196
Scratching in the throat	22.73	14.68	−0.542
Breathing problems	13.14	6.92	−0.473

The information about the percentage of surveys reporting individual symptom severity values were presented in the dashboard app (Q2–Q8; Figure 5f), which includes additionally the percentage robustness of the QA_{s8} implemented in the project (Figure 5g).

Moreover, the web application presented the results in terms of the spatial location of eligible symptom-related layers (Q1–Q8; Figure 5a–c), the number of QA-checked reports depending on the spatial extent (Figure 5d), and the general level of citizen comfort (Figure 5e). For more details about the GeoWeb app, see Appendix A.

4. Discussion

In this study, we introduced social innovation into the urban air pollution issue, where citizens act to assess air pollution using their symptoms, thereby extending the paradigm of air pollution. We considered the spatial context; therefore, a map was used to spatially model symptom severity (Figure A5, Appendix A). The web mapping application is public and provides information about air pollution in specific areas of the city such that citizens can learn about which areas could positively or negatively impact their SWB, according to information about the severity of the symptoms observed by the APSM project members within individual months. The air pollution level based on health symptom observations could be compared with other methods, such as satellite observations and in situ measurements. Such in situ ones were described and studied by [81], who compared pollen counts recorded by the European Aeroallergen Network and pollen-induced symptoms reported by pollen allergy sufferers, precisely the Pollen App users' community. The health symptoms registered by the APSM community can be compared with green pollutants' concentration levels recorded by the Polish Aerobiological Monitoring Network [102]. In the context of other air pollutants, Sentinel 5p imagery can be taken as a point of comparison, especially because it is served near real time (NRT) [103], although at a more global scale.

Although the potential for using crowdsourced data to monitor urban air pollution was demonstrated here, the minimum report sample can be considered a limitation of the project. Citizen science is an emerging trend in Poland, and so, specific motivation mechanisms need to be elaborated, based on the citizen motivation recommendations developed in other projects [104,105]. As APSM focuses on citizens who are interested in the effect of air pollution on their health and well-being, our study is wider than other projects that are dedicated only to people diagnosed with health problems. This means that the motivation mechanisms involved differ from those that are appropriate for patients (e.g., free medical consultations).

Crowdsourcing projects rely on a suite of methods to boost data quality and account for data bias [20]. To gain better data quality in CS, three-step mechanisms are generally recommended: taking considerations before, during, and after the data collection process [21]. The APSM method, using logic rules to reject inconsistent database entries, was successfully implemented after the data collection process. Depending on the expected data quality level, different mechanisms were tested.

From a practical point of view, the data quality (i.e., completeness, spatial accuracy, thematic resolution, timeliness, and logic consistency) should be suitable for the project purpose. The quality of CS data is expected to be similar to that of the data collected by professionals. According to Wiggins et al. [21], some general solutions for improving crowdsourced data quality include volunteer training (workshops), a large sample size, data filters, data mining algorithms, a qualifying system, voting for the best, reputation scores, online data and metadata sharing, citizen contribution feedback, reuse of data, and replicate studies; however, the purpose of our study was to create a data quality framework.

This confirms that data quality control mechanisms are an indispensable element in any citizen-driven research, hence also being effective in CS activities such as that considered in this paper. The removal of falsely completed surveys increased the collected data's quality and usefulness. In our research, only 5.8% of data were eliminated due to positioning accuracy, either a lack of geolocation or duplicated reports at the same geolocation. Farman [106] identified 12% of crowdsourced data to have accuracy-related errors. Thus, we conclude that in our sCS, this type of error did not pose a significant problem in terms of data quality; however, subjective data bias definitely does. Still, the problem of human bias in data poses a problem that must always be considered during data analysis. Human bias introduced into data can be mitigated by using clearly stated survey questions, providing additional training, and limiting the scope of the survey. We found that up to 29% of all collected surveys regarding air pollution objectively contained useless or false information. Kosmala et al. [20] reported subjective data bias at levels between 5% and 35%, depending on the simplicity of the tasks assigned to citizens, Hube et al. [107] presented data bias at 15–17% in a crowdsourced data set, and Eickhoff [108] pointed out an accuracy rate reduction of 20% due to cognitive data bias. We recognize that our percentage share of data bias was high, highlighting the absolute necessity of a QA mechanism framework for sCS health-related projects. As QA mechanisms are created through the use of logic rules, they are easy to understand and can be crafted to match particular (expected or observed) error types in collected survey data. We found that QA mechanisms can be used to remove surveys that contain clearly defined errors. Moreover, by choosing a single QA system (Table 3) and combining rejected reports with the username, each user's quality rank can be calculated. A user who passes the QA system could then be rewarded with trust and reputation statuses. Such citizen trust models have been proposed by Alabri and Hunter [109], who developed a social trust metrics framework, and Langley et al. [110], who applied a reputation model and used a reputation score system to determine the threshold for accepting volunteered data. This should be based, for each citizen, on the ratio of reports accepted by the QA system to the total number of their surveys: the higher the ratio of accepted reports, the higher the level of citizen trust.

The APSM data quality mechanisms implemented after the data collection stage—but referenced to a particular user's reports—could be used to develop a user motivation system (which, in the current study, was limited only to user activity). Furthermore, the technological implementation of QA systems as cloud services may enable the ranking of user trust and reputation during the data collection process. Then, not only the quantity but also the quality of user reports can be analyzed and their level of reputation could be assessed and presented during the campaign. In large-scale CS projects such as iNaturalist (iNaturalist.org), the trust and reputation of citizens are based on their activity: "The users' community ensures that data is reliable, but it also gives the opportunity for fellow users to gain knowledge" [111].

In future research, during the data collection process, some new solutions can be implemented, such as GNSS trajectories. Currently, Q10 (Figure 1) requires users to estimate how long they have stayed in a location. Using GNSS or Global System for Mobile Communications (GSM) data to characterize user mobility patterns and analyze user spatial trajectories could increase data quality and make the application smarter. Changes in user trajectories could also result in an individual push notification to maintain or cancel the APS, depending on the change of location. Finally, to gain better data quality, improvements before the data collection stage may also be proposed. APSM was carried out as a Polish case study. As CS has been recognized as an emerging trend in Poland, we found it necessary to promote the sCS concept through the European CS platform (https://eu-citizen.science) and engage citizens in air pollution monitoring by organizing training sessions. What differentiates CS from other VGI activities is the fact that CS can be taken up by any volunteers who have undertaken standardized training. Learning how to observe one's own symptoms in reaction to air pollution, relating them with air quality information and green pollutant concentration levels, and regular symptom recording were considered prerequisite parameters for ensuring the quality of APSM. The D-NOSES project [7] proved that the sense of smell of individuals can be calibrated through training on odor pollution and workshops exploring odor perception in the D-NOSE method.

In the study, we applied a user activity rank model, which showed that citizen activity decreased over time, which is typical for a CS project [112]. The level of citizen engagement and motivation was the highest at the beginning of the crowdsourcing campaign (100 reports per day), dropping after 14 days. The two peaks in the last month of the field data collection campaign could have resulted from the motivational workshops with an educator where citizens rankings were discussed, thus increasing competition between the citizens (35 and 40 reports per day, May 2018). In summary, for a case study of Lublin or any city of similar size and structure, a citizen group larger than 74 people is needed and regular workshops, as well social media campaigns engaging people to participate directly in the CS group forum, are necessary to maintain their activity.

Due to the intuitive access and operation of the presented tools, such methods and tools are suitable for scientists, educators, and evaluators alike. The ability to reduce data bias in real time is not possible without a programmed web-based mechanism functionality. AGOL-configurable capabilities allow for data filtering, but the filters are too basic for the conditional statement combinations that form the QA mechanisms.

Conveniently, our database was set up on REST services, such that the QA mechanisms and their combinations could be implemented and analyzed using desktop software, in direct connection with AGOL apps, which ensured the stability of the REST service-based data source.

5. Conclusions

In summary, the APSM project was the first research in Europe that focused on assessing air pollution based on the health symptoms of citizens and their subjective well-being. This source of air pollution monitoring could be complementary to other methods. The highly subjective form of the data source could be burdened with data bias and specific errors. Therefore, it requires a QA framework for APSM projects, which was proposed and implemented in this study.

Of the five QA mechanisms employed, the most robust were those aimed at removing inconsistent user answers, which were intentionally repeated in the survey (i.e., the repeating QA mechanism). These results suggest that some of the methods employed might lead to a decrease in user engagement, as some users were not consistent with their own answers in the same survey. This finding may be due to a natural phenomenon associated with the human condition or to a survey questionnaire that lacks user engagement. Future surveys employing sCS as a data source might expect many haphazardly completed user surveys. Analyzing the QAm effectiveness results, we assume that some of the QA mechanisms' effectiveness can be increased, and we recommend some modifications to the examined QA framework. The sequence mechanism should be additionally enhanced by a functionality measuring how much time it took to fill out the form. This will help capture the user automatism in filling out the survey. Furthermore, the results provided evidence that APS assessment is much easier for citizens than identification of their SWB. This conclusion is confirmed with the level of the repeating mechanism's robustness, which eliminated the most reports, and was based on the repeated SWB questions. Therefore, we recommend that researchers rather focus on the health symptom questions and not repeat SWB questions in the survey, as it could be too difficult for the citizens to verify. The screening question could be developed in future research as well. Moreover, the QA method for APSM can be extended by the abundance and frequency of the surveys in a close or the same geolocation in the near time. This will be a mechanism that potentially improves the reliability of collected data. If some citizens give very close responses in the same or a close time and location, then the quality of data should have a potentially higher level of trust than those that are not confirmed by any surveys of other citizens.

The focus of our research was not on validation with digital sensors but on elimination of logically inconsistent answers and technologically incorrect objects. However, the APSM method can capture the moment when air pollution changes. The observed health symptom severity can be validated with air pollution concentrations measured by air-quality-monitoring stations. Having information about whether citizens are diagnosed as pollen allergy sufferers, and by collating this information with the current concentration of pollen species, the chance for confirmation of the impact of air pollution on citizen SWB is higher. The completed database has potential for further research to test the thesis, if citizens more sensitive to air pollution provide data of better quality than those who do not report any pollen allergy or other relevant preexisting conditions. Thus, people who are more sensitive to air pollution can be potentially more interested in providing high-quality data than those who have no air-pollution-related health problems. However, understanding the mechanisms underlying citizen scientists' motivations requires further research.

Citizens, together with scientists, built a reliable model of the impact of air pollution on the well-being of citizens in their city. As a result of our research, we can confirm that not only QA mechanisms but also citizen activity are necessary for CS contribution to geospatial data quality improvement.

Due to the proposed QA framework, the data obtained with regard to the measured air pollution could be output as a spatial model of city well-being.

Author Contributions: Marta Samulowska: project idea, QA method elaboration and development, GeoWeb development; Szymon Chmielewski: project idea, smart cities, crowdsourcing campaign running and evaluation; Edwin Raczko: QA method development, R statistics software implementation; Michał Lupa: allergy symptom mapping idea, QA method verification; Dorota Myszkowska: air quality health symptoms, aerobiology and allergology issues; Bogdan Zagajewski: project idea evaluation, QA method verification. All authors prepared the manuscript. All authors have read and agreed to the published version of the manuscript.

Funding: The publishing costs were covered by the sources of the Polish Ministry of Science and Higher Education (project no. 500-D119-12-1190000).

Institutional Review Board Statement: Not applicable.

Informed Consent Statement: Not applicable.

Data Availability Statement: Not applicable.

Acknowledgments: The authors would like to thank the Esri Poland team. We are also grateful to the editors and anonymous reviewers for their constructive comments and suggestions that helped to improve this manuscript.

Conflicts of Interest: The authors declare no conflict of interest.

Abbreviation

Acronym	Defined in Section	Meaning (Section)
CS	citizen science	Citizen-driven research that citizens (non-experts) participate in by cooperating with researchers (Introduction, Section 1.2).
QAm	quality assurance mechanism	Conditional-statement-based mechanism for data bias controlling. Five data quality assurance mechanisms are proposed in this study (Introduction, Section 1.3; Section Materials and Methods, Section 2.2).
QAs	quality assurance system	Combinations of data quality assurance mechanisms. In the study, we studied and analyzed eight QAs variants, depending on their robustness levels, QAs_1–QAs_8 (Materials and Methods, Section 2.2).
GeoWeb	geospatial web	Geographically related tools and web services for individuals and groups (Abstract, Introduction, Materials and Methods Section 2.3).

SC (SC1, SC2)	start-check mechanism in variant 1 and variant 2	The start-check mechanism is used to verify the report quality at the beginning of the survey and controls the report quality during the analysis of each symptom severity answer.The mechanism studied and proposed in two variants of robustness (variant 1: less robust; variant 2: more robust) (Materials and Methods, Section 2.2).
Rp (Rp1, Rp2)	repeating mechanism in variant 1 and variant 2	The repeating mechanism determines the quality of the report, according to other previously asked questions, by asking the same question but in a different way.The mechanism studied and proposed in two variants of robustness (variant 1: less robust; variant 2: more robust) (Materials and Methods, Section 2.2).
Sq	sequence mechanism	The sequence mechanism was applied to exclude user automatism in providing answers (Materials and Methods, Section 2.2).
CV (CV1, CV2)	cross-validation mechanism in variant 1 and variant 2	The cross-validation mechanism was used to reject responses using three essentially related questions.The mechanism studied and proposed in two variants of robustness (variant 1: less robust; variant 2: more robust) (Materials and Methods, Section 2.2).
TC	time-loop check mechanism	The time-loop check mechanism was used to eliminate reports that did not align with the geolocation of the citizens, according to the length of their stay in the place, in comparison to the duration of their symptoms (Materials and Methods, Section 2.2).
SWB	subjective well-being	Reflects the philosophical notion of people's good life, a proxy of their life satisfaction, momentary experiences, and stress (Introduction).
sCS	human-sensed CS	Citizen science measurement relying on one of the human senses (Introduction).
APSM	air pollution symptom mapping	Air pollution monitoring, expressed on the map as the severity of human health symptoms caused by combined factors of anthropogenic and biophysical ambient air pollutants (Introduction).
APS	air pollution symptoms	Human health symptoms related to air pollution, caused by combined factors of anthropogenic and biophysical ambient air pollutants (Introduction).
AP	air pollution	Air pollution refers to six major air pollutants: inhalable particulate matter (PM_{10}), fine particulate matter ($PM_{2.5}$), ozone (O_3), sulfur dioxide (SO_2), nitrogen dioxide (NO_2), and carbon monoxide (CO) (Introduction).
AQ	air quality	Air quality refers to the AQI as well as to classifications, opinions, and feelings (including citizens' experiences) of air- and air-quality-related SWB. However, a consensus about urban air quality terminology has not been reached, and researchers distinguish air pollution through pollen exposure [49] (Introduction).
AQI	air quality index	The AQI tracks six major air pollutants—inhalable particulate matter (PM_{10}), fine particulate matter ($PM_{2.5}$), ozone (O_3), sulfur dioxide (SO_2), nitrogen dioxide (NO_2), and carbon monoxide (CO)—to describe the air quality with the use of an objective scale (Introduction).
AGOL	ArcGIS Online	WebGIS platform by Esri Inc. (Materials and Methods).
Q1–Q12	question 1–question 12	The 12 questions about air-pollution-related symptoms and factors related to APS, but also additional information about subjective well-being, asked to citizens in the mobile survey (Materials and Methods, Section 2.2).
Con.1–Con.17	conditional statement 1–conditional statement 17	The 17 conditional statements that, in specific combinations, are the basis of the developed data quality assurance mechanisms (QAm) (Materials and Methods, Section 2.2).
Survey123, cascade, time slider	Survey123 for ArcGIS mobile app, Esri Story Map Cascade app template, Esri Time Aware app template	Configurable mobile apps and web app templates based on ArcGIS.
PM	particulate matter	A mixture of particle pollution, both solid and liquid droplets found in the ambient air. PM is characterized by particle size and chemical composition. A PM fraction of 2.5 µm or less ($PM_{2.5}$) is especially important for evaluating health as well as environmental risks (Introduction).

Appendix A APSM-Dedicated GeoWeb Tools Supporting Quality Assurance

The tools for the APSM project were based on GeoWeb. We used ArcGIS platform components, which were available to the technologist as a puzzle structure, which allowed for direct customization of the applications to implement the APSM assumptions and requirements. For the project, we configured the mobile app and a set of web apps was publicly shared for citizens.

Appendix A.1 Mobile App for Crowdsourcing

The mobile app was based on Survey123 for ArcGIS components and is available at the public link https://arcg.is/0HWXrO. The survey consists of six information pages to facilitate its use and clear navigation (Figure A1a). It is available in two languages, Polish and English (Figure A1b). The app starts with an introduction with a short user guide (Figure A1c) in order to explain the research rules and how to use the survey app (page 1), followed by the user's basic information (nickname, four-digit code, and student/non-academic status; Figure A1d), helping the users in the citizen group to control the data collection process (page 2). The next pages (3–5) include 12 APSM questions, which are completed with the user's geolocation and the date of the report (page 6). All obligatory questions are marked with a red star (Figure A1e). The third page focuses only on the general well-being level of the citizens (Figure A1f), which is the basis for the start-check mechanism. Then, the citizens answer questions about their individual symptoms using drop-down lists of answers (Figure A1g). In the summary (page 5), the citizens specify their level of well-being, choosing from a star rating scale, where one star means the lowest and three stars indicate the highest level of well-being (Figure A1h). Then, using the calculator appearance widget, the users report the length they have stayed in their location and the symptoms observed. These values are expressed in minutes, provided for question 11 (Figure A1i). Question 11, regarding the length of the observation of symptoms, is fixed in the app as relevant only when any APS are observed. If Q2–Q7 are answered as "no symptoms," then Q11 does not appear in the survey. On the last page of the app, a map widget is presented to mark the current location and date (Figure A1j). Here, app users are told that all reports with a horizontal positioning accuracy error greater than 100 m are automatically eliminated, as these values are considered GNSS positioning accuracy outliers. When completing the survey, the user can check the current location status at any time (i.e., latitude, longitude, and horizontal accuracy; Figure A1k). The default date is set to the current date. The geolocation defaults to the current GPS location of the user, as well. When the survey is completed, a bottom-right submit tick is made active, and the report is ready to send to the cloud geodatabase.

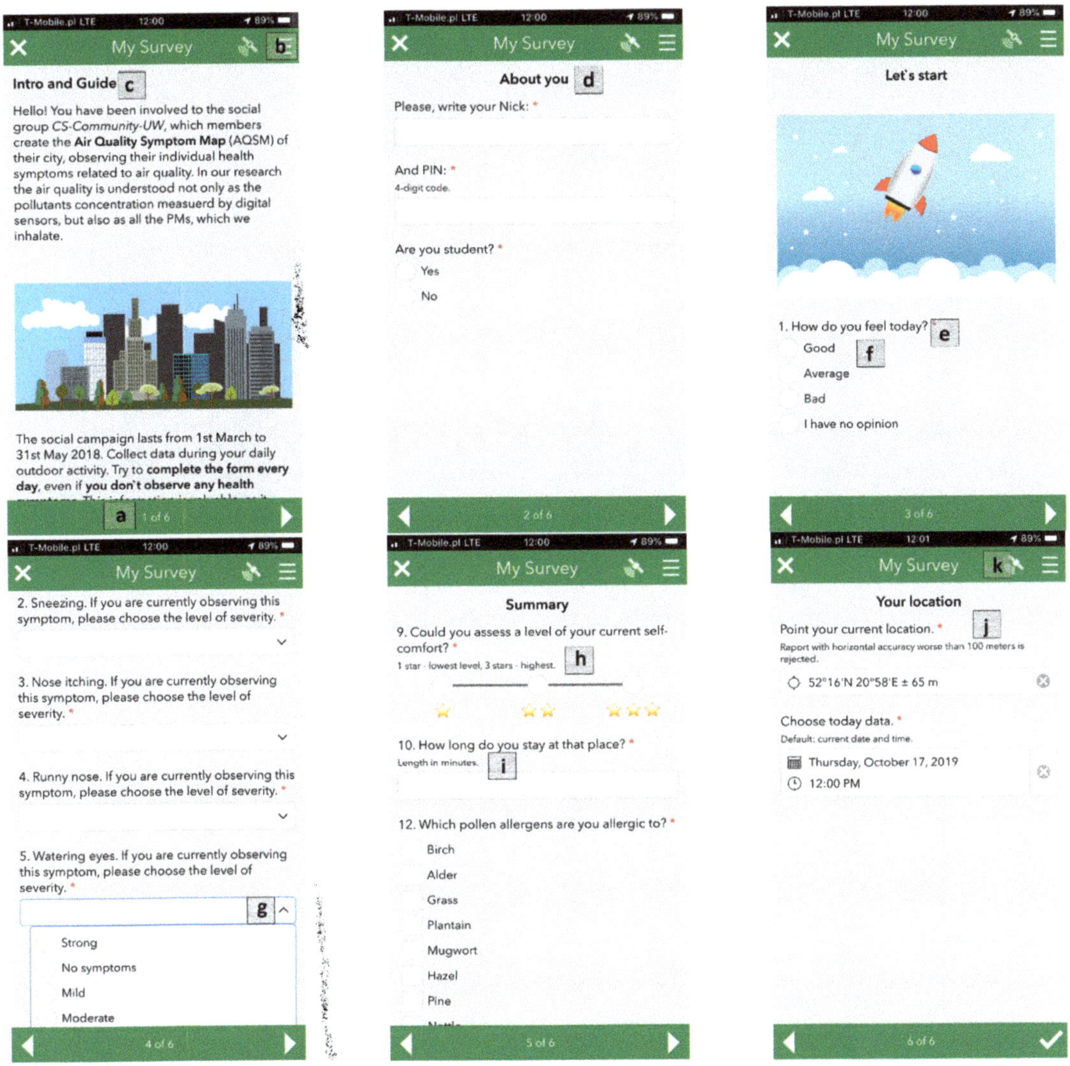

Figure A1. Six-page mobile survey app user interface: (**a**) six-page navigation, (**b**) button to choose a language, (**c**) introduction and user guide, (**d**) user's basic information, (**e**) obligatory question mark, (**f**) general well-being question, (**g**) health symptom questions, (**h**) repeated well-being question, (**i**) length present in the place report, (**j**) location and date, and (**k**) current location status.

Appendix A.2 APSM: Result Sharing through Web Apps

The resulting web app is available at https://arcg.is/1iDD18 as an open application for each person interested in the project results. The site is primarily used to provide result feedback for the citizens engaged in the study. The app was configured based on the Map Series template and consists of five applications, which can be opened by selecting five buttons: 1, Introduction to the project; 2, Field data collecting app; 3, Real-time data (before logic-based QA check); 4, APSM results (after logic-based QA check); and 5, APSM time slider (Figure A2).

Figure A2. Web application for the APSM project: collection of five applications.

The first app—Introduction to the project (Figure A3)—is based on the Esri Story Map Cascade template, which is used for building narrative web mapping apps by combining images, maps, and multimedia context with narrative text (https://storymaps-classic.arcgis.com/en/app-list/cascade/). The application has an educational function for the citizens involved. It provides educational materials about air pollution and the APSM project idea, as well as extended mobile and web app tutorials and technical knowledge.

Figure A3. *Cont.*

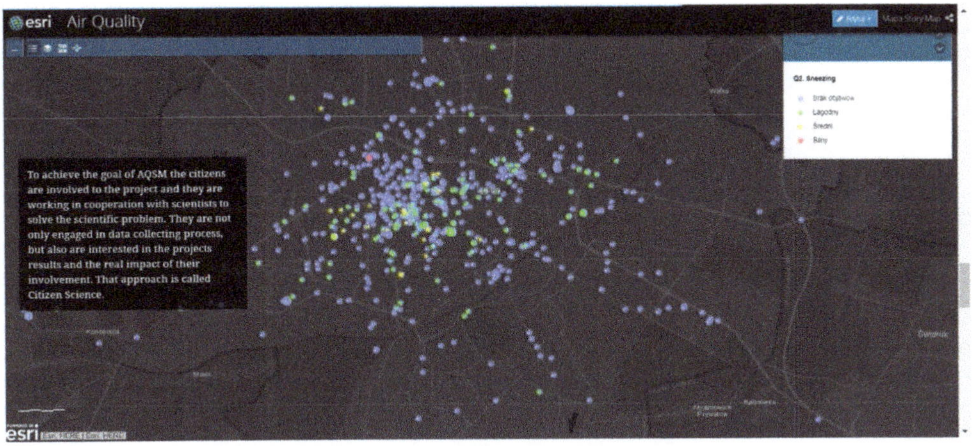

Figure A3. Educational part of the application: cascade story map.

Button number 2 links to a web version of the Survey123-based application for data collection.

Application 3 presents the raw data collected before the QA process and contains the operations-dashboard-based interface, which consists of five modules: a map with the raw-data APSM reports collected during the crowdsourcing campaign (Figure A4a); a legend (Figure A4b); an indicator counting the total number of reports (Figure A4c); a histogram of the citizen activity from the beginning of the crowdsourcing campaign to the current moment (Figure A4d), which changes dynamically, according to the map; and a citizen activity ranking, divided for each month and cumulatively (Figure A4e), as described in Section 2.1.

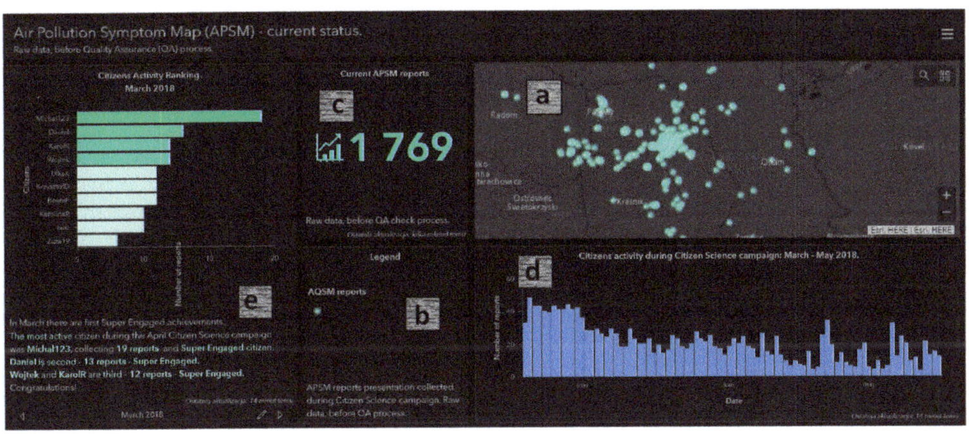

Figure A4. Web mapping application operations dashboard, presenting collected data (before QA check) in real time: (**a**) map, (**b**) legend, (**c**) indicator of current number of the reports, (**d**) citizen activity histogram, and (**e**) citizen activity ranking.

The APSM result application (number 4) is a six-module app that presents the results of the APSM data checked with QAs$_8$, the most robust variant of the QA system. The application includes a map (Figure A5a) with eligible symptom-related layers (Q1–Q8; Figure A5b), a legend for the map (Figure A5c), a pie chart for the severity of each symptom (Q2–Q8) indicated with question-related bookmarks (Figure A5d), a bar chart presenting the level of general citizen comfort (Figure A5e), a pie chart presenting the percentage robustness of the QA system (Figure A5f), and an indicator counting all the QA-checked reports (Figure A5g).

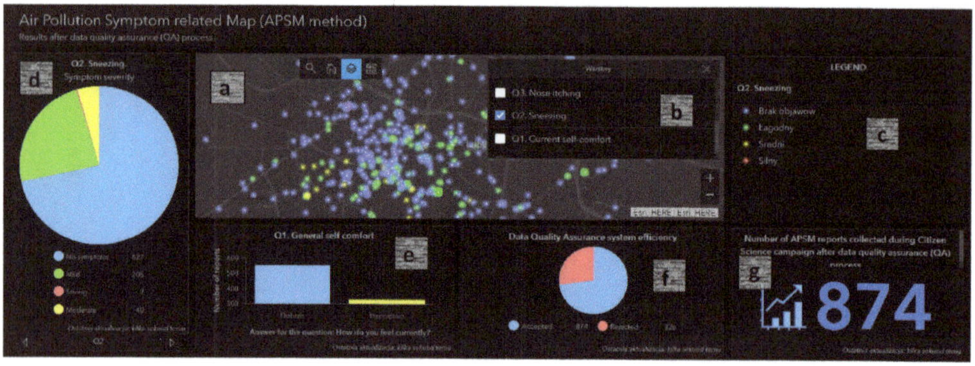

Figure A5. Web mapping application operations dashboard, presenting results of QA_{S8}-checked data: (**a**) map, (**b**) selectable layer list, (**c**) legend, (**d**) pie chart presenting severity of each air pollution related symptom, (**e**) bar chart of citizen general comfort, (**f**) pie chart of QA system robustness, and (**g**) indicator of QA-checked reports changing according to the map extent.

The time slider app (number 5) is based on the Esri Time Aware configurable template. It includes a map of point-symbolized QA-checked reports accompanied by a time slider tool, which displays the increase in collected data over the entire duration of the crowd-sourcing campaign. The time slider can move automatically (with a play button) or can be moved manually to the required date (Figure A6).

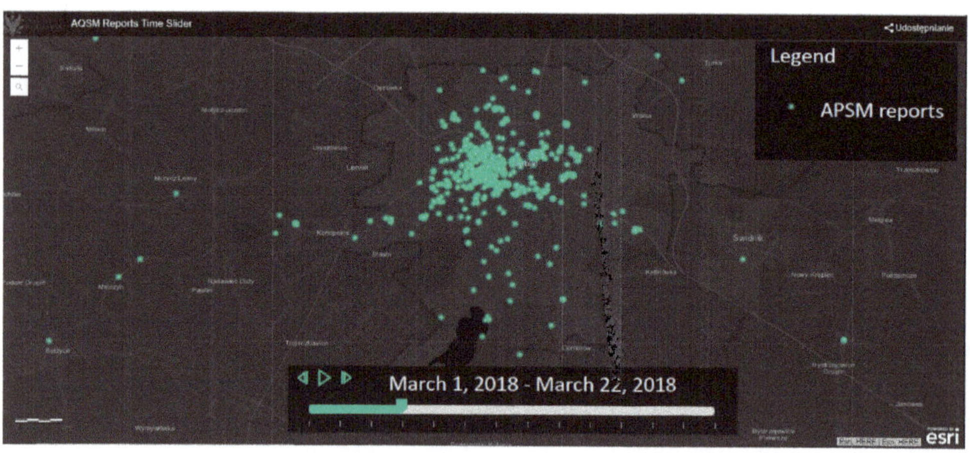

Figure A6. Time slider app following the data collection process over time.

References

1. Laffan, K. Every breath you take, every move you make: Visits to the outdoors and physical activity help to explain the relationship between air pollution and subjective wellbeing. *Ecol. Econ.* **2018**, *147*, 96–113. [CrossRef]
2. Kim-Prieto, C.; Diener, E.; Tamir, M.; Scollon, C.; Diener, M. Integrating the Diverse Definitions of Happiness: A Time-Sequential Framework of Subjective Well-Being. *J. Happiness Stud.* **2005**, *6*, 261–300. [CrossRef]
3. Ferreira, S.; Akay, A.; Brereton, F.; Cuñado, J.; Martinsson, P.; Moro, M.; Ningal, T.F. Life satisfaction and air quality in Europe. *Ecol. Econ.* **2013**, *88*, 1–10. [CrossRef]

4. Signoretta, P.E.; Buffel, V.; Bracke, P. Mental wellbeing, air pollution and the ecological state. *Health Place* **2019**, *57*, 82–91. [CrossRef]
5. Yigitcanlar, T.; Kamruzzaman, M.; Foth, M.; Sabatini-Marques, J.; da Costa, E.; Ioppolo, G. Can cities become smart without being sustainable? A systematic review of the literature. *Sustain. Cities Soc.* **2019**, *45*, 348–365. [CrossRef]
6. Giffinger, R.; Fertner, C.; Kramar, H.; Meijers, E. *City-Ranking of European Medium-Sized Cities*; Centre of Regional Science: Vienna, Austria, 2007; p. 28.
7. Arias, R.; Capelli, L.; Díaz, C. A new methodology based on citizen science to improve environmental odour management. *Chem. Eng. Trans.* **2018**, *68*, 7–12. [CrossRef]
8. Bastl, K.; Kmenta, M.; Geller-Bernstein, C.; Berger, U.; Jäger, S. Can we improve pollen season definitions by using the symptom load index in addition to pollen counts? *Environ. Pollut.* **2015**, *204*, 109–116. [CrossRef]
9. Dutta, J.; Chowdhury, C.; Roy, S.; Middya, A.I.; Gazi, F. Towards Smart City. In Proceedings of the 18th International Conference on Distributed Computing and Networking—ICDCN'17, Hyderabad, India, 4–7 January 2017; ACM Press: New York, NY, USA, 2017; pp. 1–6. [CrossRef]
10. Feng, C.; Tian, Y.; Gong, X.; Que, X.; Wang, W. MCS-RF: Mobile crowdsensing–based air quality estimation with random forest. *Int. J. Distrib. Sens. Netw.* **2018**, *14*. [CrossRef]
11. Zupančič, E.; Žalik, B. Data Trustworthiness Evaluation in Mobile Crowdsensing Systems with Users' Trust Dispositions' Consideration. *Sensors* **2019**, *19*, 1326. [CrossRef] [PubMed]
12. Castell, N.; Kobernus, M.; Liu, H.-Y.; Schneider, P.; Lahoz, W.; Berre, A.J.; Noll, J. Mobile technologies and services for environmental monitoring: The Citi-Sense-MOB approach. *Urban Clim.* **2015**, *14*, 370–382. [CrossRef]
13. Komarkova, J.; Novak, M.; Bilkova, R.; Visek, O.; Valenta, Z. Usability of GeoWeb Sites: Case Study of Czech Regional Authorities Web Sites. In *Business Information Systems*; Springer: Berlin/Heidelberg, Germany, 2007; pp. 411–423. ISBN 9783540720348. [CrossRef]
14. Haklay, M. Citizen Science and Volunteered Geographic Information: Overview and Typology of Participation. In *Crowdsourcing Geographic Knowledge*; Sui, D., Elwood, S., Goodchild, M., Eds.; Springer: Dordrecht, The Netherlands, 2013; pp. 105–122. ISBN 9789400745872. [CrossRef]
15. Jankowski, P.; Czepkiewicz, M.; Zwoliński, Z.; Kaczmarek, T.; Młodkowski, M.; Bąkowska-Waldmann, E.; Mikuła, Ł.; Brudka, C.; Walczak, D. Geoweb Methods for Public Participation in Urban Planning: Selected Cases from Poland. In *Geospatial Challenges in the 21st Century*; Koutsopoulos, K., de Miguel González, R., Donert, K., Eds.; Springer Nature: Cham, Switzerland, 2019; pp. 249–269. [CrossRef]
16. Goodchild, M.F. Citizens as Voluntary Sensors: Spatial Data Infrastructure in the World of Web 2.0. *Int. J. Spatial Data Infrastruct. Res.* **2007**, *2*, 24–32.
17. Moreri, K.K.; Fairbairn, D.; James, P. Volunteered geographic information quality assessment using trust and reputation modelling in land administration systems in developing countries. *Int. J. Geogr. Inf. Sci.* **2018**, *32*, 931–959. [CrossRef]
18. Capineri, C.; Haklay, M.; Huang, H.; Antoniou, V.; Kettunen, J.; Ostermann, F.; Purves, R. (Eds.) *European Handbook of Crowdsourced Geographic Information*; Ubiquity Press: London, UK, 2016; p. 474. ISBN 9781909188792.
19. Kamp, J.; Oppel, S.; Heldbjerg, H.; Nyegaard, T.; Donald, P.F. Unstructured citizen science data fail to detect long-term population declines of common birds in Denmark. *Divers. Distrib.* **2016**, *22*, 1024–1035. [CrossRef]
20. Kosmala, M.; Wiggins, A.; Swanson, A.; Simmons, B. Assessing data quality in citizen science. *Front. Ecol. Environ.* **2016**, *14*, 551–560. [CrossRef]
21. Wiggins, A.; Newman, G.; Stevenson, R.D.; Crowston, K. Mechanisms for Data Quality and Validation in Citizen Science. In Proceedings of the 2011 IEEE Seventh International Conference on e-Science Workshops, Stockholm, Sweden, 5–8 December 2011; IEEE: Piscataway, NJ, USA, 2011; pp. 14–19. [CrossRef]
22. Bishr, M.; Mantelas, L. A trust and reputation model for filtering and classifying knowledge about urban growth. *GeoJournal* **2008**, *72*, 229–237. [CrossRef]
23. Grewling, Ł.; Frątczak, A.; Kostecki, Ł.; Nowak, M.; Szymańska, A.; Bogawski, P. Biological and Chemical Air Pollutants in an Urban Area of Central Europe: Co-exposure Assessment. *Aerosol Air Qual. Res.* **2019**, *19*, 1526–1537. [CrossRef]
24. Sheng, N.; Tang, U.W. The first official city ranking by air quality in China—A review and analysis. *Cities* **2016**, *51*, 139–149. [CrossRef]
25. WHO. *Air Quality Guidelines—Particulate Matter, Ozone, Nitrogen Dioxide and Sulphur Dioxide*; WHO Europe Publication: Geneva, Switzerland, 2005; pp. 67–105.
26. Enemark, S.; Rajabifard, A. Spatially Enabled Society. *Geoforum Perspekt.* **2011**, *10*, 1–8. [CrossRef]
27. Ionita, A.; Visan, M.; Niculescu, C.; Popa, A. Smart Collaborative Platform for eLearning with Application in Spatial Enabled Society. *Procedia Soc. Behav. Sci.* **2015**, *191*, 2097–2107. [CrossRef]
28. Liu, H.; Li, Q.; Yu, D.; Gu, Y. Air Quality Index and Air Pollutant Concentration Prediction Based on Machine Learning Algorithms. *Appl. Sci.* **2019**, *9*, 4069. [CrossRef]
29. Liang, J. *Chemical Modeling for Air Resources*; Academic Press, Elsevier: Oxford, UK, 2013; p. 298. ISBN 9780124081352. [CrossRef]
30. Kelly, F.J.; Fussell, J.C. Air pollution and public health: Emerging hazards and improved understanding of risk. *Environ. Geochem. Health* **2015**, *37*, 631–649. [CrossRef] [PubMed]
31. Zwozdziak, A.; Sówka, I.; Willak-Janc, E.; Zwozdziak, J.; Kwiecińska, K.; Balińska-Miśkiewicz, W. Influence of PM1 and PM2.5 on lung function parameters in healthy schoolchildren—A panel study. *Environ. Sci. Pollut. Res.* **2016**, *23*, 23892–23901. [CrossRef] [PubMed]

32. Bastl, K.; Berger, M.; Bergmann, K.-C.; Kmenta, M.; Berger, U. The medical and scientific responsibility of pollen information services. *Wien. Klin. Wochenschr.* **2017**, *129*, 70–74. [CrossRef] [PubMed]
33. Joseph, E.P.; Jackson, V.B.; Beckles, D.M.; Cox, L.; Edwards, S. A citizen science approach for monitoring volcanic emissions and promoting volcanic hazard awareness at Sulphur Springs, Saint Lucia in the Lesser Antilles arc. *J. Volcanol. Geotherm. Res.* **2019**, *369*, 50–63. [CrossRef]
34. Baldacci, S.; Maio, S.; Cerrai, S.; Sarno, G.; Baïz, N.; Simoni, M.; Annesi-Maesano, I.; Viegi, G. Allergy and asthma: Effects of the exposure to particulate matter and biological allergens. *Respir. Med.* **2015**, *109*, 1089–1104. [CrossRef]
35. Di Menno di Bucchianico, A.; Brighetti, M.A.; Cattani, G.; Costa, C.; Cusano, M.; De Gironimo, V.; Froio, F.; Gaddi, R.; Pelosi, S.; Sfika, I.; et al. Combined effects of air pollution and allergens in the city of Rome. *Urban For. Urban Green.* **2019**, *37*, 13–23. [CrossRef]
36. McInnes, R.N.; Hemming, D.; Burgess, P.; Lyndsay, D.; Osborne, N.J.; Skjøth, C.A.; Thomas, S.; Vardoulakis, S. Mapping allergenic pollen vegetation in UK to study environmental exposure and human health. *Sci. Total Environ.* **2017**, *599–600*, 483–499. [CrossRef]
37. Robichaud, A.; Comtois, P. Environmental factors and asthma hospitalization in Montreal, Canada, during spring 2006–2008: A synergy perspective. *Air Qual. Atmos. Health* **2019**, *12*, 1495–1509. [CrossRef]
38. Werchan, B.; Werchan, M.; Mücke, H.-G.; Gauger, U.; Simoleit, A.; Zuberbier, T.; Bergmann, K.-C. Spatial distribution of allergenic pollen through a large metropolitan area. *Environ. Monit. Assess.* **2017**, *189*, 169. [CrossRef]
39. Bédard, A.; Sofiev, M.; Arnavielhe, S.; Antó, J.M.; Garcia-Aymerich, J.; Thibaudon, M.; Bergmann, K.C.; Dubakiene, R.; Bedbrook, A.; Onorato, G.; et al. Interactions between air pollution and pollen season for rhinitis using mobile technology: A MASK-POLLAR study. *J. Allerg. Clin. Immun.* **2020**, *8*, 1063–1073.e4. [CrossRef]
40. Connors, J.P.; Lei, S.; Kelly, M. Citizen Science in the Age of Neogeography: Utilizing Volunteered Geographic Information for Environmental Monitoring. *Ann. Assoc. Am. Geogr.* **2012**, *102*, 1267–1289. [CrossRef]
41. Eitzel, M.V.; Cappadonna, J.L.; Santos-Lang, C.; Duerr, R.E.; Virapongse, A.; West, S.E.; Kyba, C.C.M.; Bowser, A.; Cooper, C.B.; Sforzi, A.; et al. Citizen Science Terminology Matters: Exploring Key Terms. *Citiz. Sci. Theory Pract.* **2017**, *2*, 1. [CrossRef]
42. Silvertown, J. A new dawn for citizen science. *Trends Ecol. Evol.* **2009**, *24*, 467–471. [CrossRef] [PubMed]
43. Dickinson, J.L.; Zuckerberg, B.; Bonter, D.N. Citizen Science as an Ecological Research Tool: Challenges and Benefits. *Annu. Rev. Ecol. Evol. Syst.* **2010**, *41*, 149–172. [CrossRef]
44. Kar, B.; Sieber, R.; Haklay, M.; Ghose, R. Public Participation GIS and Participatory GIS in the Era of GeoWeb. *Cartogr. J.* **2016**, *53*, 296–299. [CrossRef]
45. Bonney, R.; Balard, H.; Jordan, R.; McCallie, E.; Phillips, T.; Shirk, J.; Wilderman, C.C. *Public Participation in Scientific Research: Defining the Field and Assessing Its Potential for Informal Science Education*; A CAISE Inquiry Group Report; Center for Advancement of Informal Science Education (CAISE): Washington, DC, USA, 2009; p. 58. Available online: http://www.birds.cornell.edu/citscitoolkit/publications/CAISE-PPSR-report-2009.pdf (accessed on 14 August 2019).
46. Loreto, V.; Haklay, M.; Hotho, A.; Servedio, V.D.P.; Stumme, G.; Theunis, J.; Tria, F. *Participatory Sensing, Opinions and Collective Awareness*; Springer: Cham, Switzerland, 2017; p. 405. [CrossRef]
47. Grey, F. *The Age of Citizen Cyberscience*; CERN Courier, IOP Publishing: Bristol, UK, 2009; Available online: http://cerncourier.com/cws/article/cern/38718 (accessed on 31 May 2017).
48. Guo, B.; Wang, Z.; Yu, Z.; Wang, Y.; Yen, N.Y.; Huang, R.; Zhou, X. Mobile Crowd Sensing and Computing. *ACM Comput. Surv.* **2015**, *48*, 1–31. [CrossRef]
49. Capponi, A.; Fiandrino, C.; Kantarci, B.; Foschini, L.; Kliazovich, D.; Bouvry, P. A Survey on Mobile Crowdsensing Systems: Challenges, Solutions, and Opportunities. *IEEE Commun. Surv. Tutor.* **2019**, *21*, 2419–2465. [CrossRef]
50. Haklay, M.; Basiouka, S.; Antoniou, V.; Ather, A. How Many Volunteers Does it Take to Map an Area Well? The Validity of Linus' Law to Volunteered Geographic Information. *Cartogr. J.* **2010**, *47*, 315–322. [CrossRef]
51. English, P.B.; Richardson, M.J.; Garzón-Galvis, C. From Crowdsourcing to Extreme Citizen Science: Participatory Research for Environmental Health. *Annu. Rev. Public Health* **2018**, *39*, 335–350. [CrossRef]
52. Nimbalkar, P.M.; Tripathi, N.K. Space-time epidemiology and effect of meteorological parameters on influenza-like illness in Phitsanulok, a northern province in Thailand. *Geospat. Health* **2016**, *11*. [CrossRef]
53. Sheppard, S.A.; Terveen, L. Quality is a verb. In Proceedings of the 7th International Symposium on Wikis and Open Collaboration—WikiSym '11, Mountain View, CA, USA, 3–5 October 2011; ACM Press: New York, NY, USA, 2011; p. 29. [CrossRef]
54. Lin, Y.-P.; Deng, D.; Lin, W.-C.; Lemmens, R.; Crossman, N.D.; Henle, K.; Schmeller, D.S. Uncertainty analysis of crowd-sourced and professionally collected field data used in species distribution models of Taiwanese moths. *Biol. Conserv.* **2015**, *181*, 102–110. [CrossRef]
55. Parrish, J.K.; Burgess, H.; Weltzin, J.F.; Fortson, L.; Wiggins, A.; Simmons, B. Exposing the Science in Citizen Science: Fitness to Purpose and Intentional Design. *Integr. Comp. Biol.* **2018**, *58*, 150–160. [CrossRef] [PubMed]
56. Fritz, S.; Fonte, C.; See, L. The Role of Citizen Science in Earth Observation. *Remote Sens.* **2017**, *9*, 357. [CrossRef]
57. Maantay, J. Asthma and air pollution in the Bronx: Methodological and data considerations in using GIS for environmental justice and health research. *Health Place* **2007**, *13*, 32–56. [CrossRef]
58. Keddem, S.; Barg, F.K.; Glanz, K.; Jackson, T.; Green, S.; George, M. Mapping the urban asthma experience: Using qualitative GIS to understand contextual factors affecting asthma control. *Soc. Sci. Med.* **2015**, *140*, 9–17. [CrossRef]

59. Palmer, J.R.B.; Oltra, A.; Collantes, F.; Delgado, J.A.; Lucientes, J.; Delacour, S.; Bengoa, M.; Eritja, R.; Bartumeus, F. Citizen science provides a reliable and scalable tool to track disease-carrying mosquitoes. *Nat. Commun.* **2017**, *8*, 916. [CrossRef]
60. Penza, M.; Suriano, D.; Pfister, V.; Prato, M.; Cassano, G. Urban Air Quality Monitoring with Networked Low-Cost Sensor-Systems. *Proceedings* **2017**, *1*, 573. [CrossRef]
61. Kankanamge, N.; Yigitcanlar, T.; Goonetilleke, A.; Kamruzzaman, M. Can volunteer crowdsourcing reduce disaster risk? A systematic review of the literature. *Int. J. Disaster Risk Reduct.* **2019**, *35*, 101097. [CrossRef]
62. Choi, J.; Hwang, M.; Kim, G.; Seong, J.; Ahn, J. Supporting the measurement of the United Nations' sustainable development goal 11 through the use of national urban information systems and open geospatial technologies: A case study of south Korea. *Open Geospatial Data Softw. Stand.* **2016**, *1*, 1. [CrossRef]
63. Chmielewski, S.; Samulowska, M.; Lupa, M.; Lee, D.; Zagajewski, B. Citizen science and WebGIS for outdoor advertisement visual pollution assessment. *Comput. Environ. Urban Syst.* **2018**, *67*, 97–109. [CrossRef]
64. Flanagin, A.J.; Metzger, M.J. The credibility of volunteered geographic information. *GeoJournal* **2008**, *72*, 137–148. [CrossRef]
65. Antoniou, V.; Skopeliti, A. Measures and indicators of VGI quality: An overview. *ISPRS Ann. Photogramm. Remote Sens. Spat. Inf. Sci.* **2015**, *II-3/W5*, 345–351. [CrossRef]
66. Wu, P.; Ngai, E.W.T.; Wu, Y. Toward a real-time and budget-aware task package allocation in spatial crowdsourcing. *Decis. Support Syst.* **2018**, *110*, 107–117. [CrossRef]
67. Foody, G.; See, L.; Fritz, S.; Moorthy, I.; Perger, C.; Schill, C.; Boyd, D. Increasing the Accuracy of Crowdsourced Information on Land Cover via a Voting Procedure Weighted by Information Inferred from the Contributed Data. *ISPRS Int. J. Geo-Inf.* **2018**, *7*, 80. [CrossRef]
68. Gillooly, S.E.; Zhou, Y.; Vallarino, J.; Chu, M.T.; Michanowicz, D.R.; Levy, J.I.; Adamkiewicz, G. Development of an in-home, real-time air pollutant sensor platform and implications for community use. *Environ. Pollut.* **2019**, *244*, 440–450. [CrossRef] [PubMed]
69. Kosmidis, E.; Syropoulou, P.; Tekes, S.; Schneider, P.; Spyromitros-Xioufis, E.; Riga, M.; Charitidis, P.; Moumtzidou, A.; Papadopoulos, S.; Vrochidis, S.; et al. hackAIR: Towards Raising Awareness about Air Quality in Europe by Developing a Collective Online Platform. *ISPRS Int. J. Geo-Inf.* **2018**, *7*, 187. [CrossRef]
70. Commodore, A.; Wilson, S.; Muhammad, O.; Svendsen, E.; Pearce, J. Community-based participatory research for the study of air pollution: A review of motivations, approaches, and outcomes. *Environ. Monit. Assess.* **2017**, *189*, 378. [CrossRef]
71. International Organization for Standardization. *ISO 19157: Geographic Information—Data Quality*; International Organization for Standardization: Geneva, Switzerland, 2013.
72. Fonte, C.C.; Antoniou, V.; Bastin, L.; Estima, J.; Arsanjani, J.J.; Bayas, J.-C.L.; See, L.; Vatseva, R. Assessing VGI Data Quality. In *Mapping and the Citizen Sensor*; Foody, G., See, L., Fritz, S., Mooney, P., Olteanu-Raimond, A.-M., Fonte, C.C., Antoniou, V., Eds.; Ubiquity Press: London, UK, 2017; pp. 137–163. [CrossRef]
73. Chehregani, A.; Majde, A.; Moin, M.; Gholami, M.; Ali Shariatzadeh, M.; Nassiri, H. Increasing allergy potency of Zinnia pollen grains in polluted areas. *Ecotoxicol. Environ. Saf.* **2004**, *58*, 267–272. [CrossRef]
74. D'Amato, G.; Holgate, S.T.; Pawankar, R.; Ledford, D.K.; Cecchi, L.; Al-Ahmad, M.; Al-Enezi, F.; Al-Muhsen, S.; Ansotegui, I.; Baena-Cagnani, C.E.; et al. Meteorological conditions, climate change, new emerging factors, and asthma and related allergic disorders. A statement of the World Allergy Organization. *World Allergy Organ. J.* **2015**, *8*, 25. [CrossRef]
75. Karatzas, K.D. Informing the public about atmospheric quality: Air pollution and pollen. *Allergo J.* **2009**, *18*, 212–217. [CrossRef]
76. Sofiev, M.; Bergmann, K.C. *Allergenic Pollen*; Sofiev, M., Bergmann, K.-C., Eds.; Springer: Dordrecht, The Netherlands, 2013; ISBN 978-94-007-4880-4. [CrossRef]
77. Bonney, R.; Cooper, C.B.; Dickinson, J.; Kelling, S.; Phillips, T.; Rosenberg, K.V.; Shirk, J. Citizen Science: A Developing Tool for Expanding Science Knowledge and Scientific Literacy. *Bioscience* **2009**, *59*, 977–984. [CrossRef]
78. Ring, J.; Krämer, U.; Schäfer, T.; Behrendt, H. Why are allergies increasing? *Curr. Opin. Immunol.* **2001**, *13*, 701–708. [CrossRef]
79. Harlin, J.; Kloetzer, L.; Patton, D.; Leonhard, C. Turning students into citizen scientists. In *Citizen Science*; UCL Press: London, UK, 2018; pp. 410–428. [CrossRef]
80. Seymour, V.; Haklay, M. Exploring Engagement Characteristics and Behaviours of Environmental Volunteers. *Citiz. Sci. Theory Pract.* **2017**, *2*, 5. [CrossRef]
81. Kmenta, M.; Bastl, K.; Jäger, S.; Berger, U. Development of personal pollen information—the next generation of pollen information and a step forward for hay fever sufferers. *Int. J. Biometeorol.* **2014**, *58*, 1721–1726. [CrossRef] [PubMed]
82. Samoliński, B.; Raciborski, F.; Lipiec, A.; Tomaszewska, A.; Krzych-Fałta, E.; Samel-Kowalik, P.; Walkiewicz, A.; Lusawa, A.; Borowicz, J.; Komorowski, J.; et al. Epidemiologia Chorób Alergicznych w Polsce (ECAP). *Alergol. Pol. Polish J. Allergol.* **2014**, *1*, 10–18. [CrossRef]
83. Galesic, M.; Bosnjak, M. Effects of Questionnaire Length on Participation and Indicators of Response Quality in a Web Survey. *Public Opin. Q.* **2009**, *73*, 349–360. [CrossRef]
84. Malhotra, N.K. Questionnaire design and scale development. In *The Handbook of Marketing Research: Uses, Misuses, and Future Advances*; Grover, R., Vriens, M., Eds.; SAGE Publications Inc.: Thousand Oaks, CA, USA, 2006; p. 720. ISBN 1-4129-0997-X. [CrossRef]
85. Krosnick, J.A.; Presser, S. Question and Questionnaire Design. In *Handbook of Survey Research*, 2nd ed.; Wright, J.D., Marsden, P.V., Eds.; Elsevier: San Diego, CA, USA, 2009; p. 81.
86. Weijters, B.; Baumgartner, H.; Schillewaert, N. Reversed item bias: An integrative model. *Psychol. Methods* **2013**, *18*, 320–334. [CrossRef]

87. Bousquet, J.; Bewick, M.; Arnavielhe, S.; Mathieu-Dupas, E.; Murray, R.; Bedbrook, A.; Caimmi, D.P.; Vandenplas, O.; Hellings, P.W.; Bachert, C.; et al. Work productivity in rhinitis using cell phones: The MASK pilot study. *Allergy* **2017**, *72*, 1475–1484. [CrossRef]
88. Albuam, G.; Oppenheim, A.N. Questionnaire Design, Interviewing and Attitude Measurement. *J. Mark. Res.* **1993**, *30*, 393. [CrossRef]
89. Garbarski, D.; Schaeffer, N.C.; Dykema, J. The effects of response option order and question order on self-rated health. *Qual. Life Res.* **2015**, *24*, 1443–1453. [CrossRef]
90. Schaeffer, N.C.; Presser, S. The Science of Asking Questions. *Annu. Rev. Soc.* **2003**, *29*, 65–88. [CrossRef]
91. Boynton, P.M.; Greenhalgh, T. Selecting, designing, and developing your questionnaire. *BMJ* **2004**, *328*, 1312–1315. [CrossRef] [PubMed]
92. Arvidsson, M.B.; Lowhagen, O.; Rak, S. Allergen specific immunotherapy attenuates early and late phase reactions in lower airways of birch pollen asthmatic patients: A double blind placebo-controlled study. *Allergy* **2004**, *59*, 74–80. [CrossRef] [PubMed]
93. Galli, S.J.; Tsai, M.; Piliponsky, A.M. The development of allergic inflammation. *Nature* **2008**, *454*, 445–454. [CrossRef] [PubMed]
94. Gauvreau, G.M.; El-Gammal, A.I.; O'Byrne, P.M. Allergen-induced airway responses. *Eur. Respir. J.* **2015**, *46*, 819–831. [CrossRef]
95. Skoner, D.P. Allergic rhinitis: Definition, epidemiology, pathophysiology, detection, and diagnosis. *J. Allergy Clin. Immunol.* **2001**, *108*, S2–S8. [CrossRef]
96. Ferguson, B.J. Influences of Allergic Rhinitis on Sleep. *Otolaryngol. Neck Surg.* **2004**, *130*, 617–629. [CrossRef]
97. Kampa, M.; Castanas, E. Human health effects of air pollution. *Environ. Pollut.* **2008**, *151*, 362–367. [CrossRef]
98. D'Amato, G.; Liccardi, G.; D'Amato, M.; Cazzola, M. Outdoor air pollution, climatic changes and allergic bronchial asthma. *Eur. Respir. J.* **2002**, *20*, 763–776. [CrossRef]
99. Couper, M.P.; Traugott, M.W.; Lamias, M.J. Web Survey Design and Administration. *Public Opin. Q.* **2001**, *65*, 230–253. [CrossRef]
100. Ganassali, S. The influence of the design of web survey questionnaires on the quality of responses. *Surv. Res. Methods* **2008**, *2*, 21–32. [CrossRef]
101. Lupa, M.; Samulowska, M.; Chmielewski, S.; Myszkowska, D.; Czarnobilska, E. A concept of webgis pollen allergy mapping. In Proceedings of the 17th International Multidisciplinary Scientific GeoConference Surveying Geology and Mining Ecology Management, Albena, Bulgaria, 29 June–5 July 2017; SGEM: Sofia, Bulgaria, 2017; pp. 1141–1148. [CrossRef]
102. Kubik-Komar, A.; Piotrowska-Weryszko, K.; Weryszko-Chmielewska, E.; Kuna-Broniowska, I.; Chłopek, K.; Myszkowska, D.; Puc, M.; Rapiejko, P.; Ziemianin, M.; Dąbrowska-Zapart, K.; et al. A study on the spatial and temporal variability in airborne Betula pollen concentration in five cities in Poland using multivariate analyses. *Sci. Total Environ.* **2019**, *660*, 1070–1078. [CrossRef] [PubMed]
103. Caspari, G.; Donato, S.; Jendryke, M. Remote sensing and citizen science for assessing land use change in the Musandam (Oman). *J. Arid Environ.* **2019**, *171*, 104003. [CrossRef]
104. Nov, O.; Arazy, O.; Anderson, D. Dusting for science. In Proceedings of the 2011 iConference on iConference'11, Seattle, WA, USA, 8–11 February 2011; ACM Press: New York, NY, USA, 2011; pp. 68–74. [CrossRef]
105. McCrory, G.; Veeckman, C.; Claeys, L. Citizen Science Is in the Air—Engagement Mechanisms from Technology-Mediated Citizen Science Projects Addressing Air Pollution. In *Lecture Notes in Computer Science, 10673*; Springer: Cham, Switzerland, 2017; pp. 28–38. [CrossRef]
106. Farman, J. Infrastructures of Mobile Social Media. *Soc. Media Soc.* **2015**, *1*. [CrossRef]
107. Hube, C.; Fetahu, B.; Gadiraju, U. Understanding and Mitigating Worker Biases in the Crowdsourced Collection of Subjective Judgments. In Proceedings of the 2019 CHI Conference on Human Factors in Computing Systems—CHI '19, Glasgow, UK, 4–9 May 2019; ACM Press: New York, NY, USA, 2019; pp. 1–12. [CrossRef]
108. Eickhoff, C. Cognitive Biases in Crowdsourcing. In Proceedings of the Eleventh ACM International Conference on Web Search and Data Mining—WSDM '18, Los Angeles, CA, USA, 5–9 February 2018; ACM Press: New York, NY, USA, 2018; pp. 162–170. [CrossRef]
109. Alabri, A.; Hunter, J. Enhancing the Quality and Trust of Citizen Science Data. In Proceedings of the 2010 IEEE Sixth International Conference on e-Science, Brisbane, Australia, 7–10 December 2010; IEEE: Piscataway, NJ, USA, 2010; pp. 81–88. [CrossRef]
110. Langley, S.A.; Messina, J.P.; Moore, N. Using meta-quality to assess the utility of volunteered geographic information for science. *Int. J. Health Geogr.* **2017**, *16*, 40. [CrossRef]
111. Nowak, M.M.; Dziób, K.; Ludwisiak, Ł.; Chmiel, J. Mobile GIS applications for environmental field surveys: A state of the art. *Glob. Ecol. Conserv.* **2020**, *23*, e01089. [CrossRef]
112. Geoghegan, H.; Dyke, A.; Pateman, R.; West, S.; Everett, G. *Understanding Motivations for Citizen Science*; Final Report on Behalf of the UK Environmental Observation Framework; University of Reading, Stockholm Environment Institute (University of York) and University of the West of England; UK Centre for Ecology & Hydrology, Lancaster Environment Centre: Lancaster, UK, May 2016; pp. 1–4. Available online: http://www.ukeof.org.uk/resources/citizen-science-resources/citizenscienceSUMMARYReportFINAL19052.pdf (accessed on 12 November 2018).

MDPI
St. Alban-Anlage 66
4052 Basel
Switzerland
Tel. +41 61 683 77 34
Fax +41 61 302 89 18
www.mdpi.com

ISPRS International Journal of Geo-Information Editorial Office
E-mail: ijgi@mdpi.com
www.mdpi.com/journal/ijgi

www.ingramcontent.com/pod-product-compliance
Lightning Source LLC
Chambersburg PA
CBHW061511210926
51850CB00008B/1880
9783036537139